Secrets of Creation

Volume 1

The Mystery of the Prime Numbers

Secrets of Creation

Volume 1

The Mystery of the Prime Numbers

Matthew Watkins

Winchester, UK
Washington, USA

First published by Liberalis Books, 2015
Originally published by The Inamorata Press, 2010
Liberalis Books is an imprint of John Hunt Publishing Ltd., Laurel House, Station Approach,
Alresford, Hants, SO24 9JH, UK
office1@jhpbooks.net
www.liberalisbooks.com

For distributor details and how to order please visit the 'Ordering' section on our website.

ISBN: 978 1 78279 781 4
Library of Congress Control Number: 2014958368

A CIP catalogue record for this book is available from the British Library.

The frontispiece is based on the "Flammarion woodcut" (anonymous).

www.secretsofcreation.com

Printed and bound by CPI Group (UK) Ltd, Croydon, CR0 4YY, UK

table of contents

an introduction 2

1. numbers and counting 5
and how they've taken over the world

2. how to build the number system 35
five simple rules that take you out to infinity

3. prime numbers 51
something anyone could have noticed

4. prime factors 81
and the most important thing we know about numbers

5. a philosophical interlude 105
and a journey into space

6. addition versus multiplication 117
they're surprisingly different things

7. an infinity of primes 135
how we can be sure there isn't a biggest one

8. patterns and formulas 149
the questions everyone seems to ask

9. spirals 167
the central image in this story

10. the distribution 187
the curiously random-looking arrangement of primes

11. staircases 203
a useful way to picture the distribution

12. the deviation 231
isolating the deviant behaviour of the primes

13. harmonic decomposition 249
breaking everything down into waves

14. spiral waves 267
which no one's bothered to name properly

15. mysterious frequencies 287
the inevitable cliffhanger ending

notes 296

appendices 1–9 312

...upon looking at these numbers one has the feeling of being in the presence of one of the inexplicable secrets of creation.

Don Zagier
Bonn University, 1975

For Alan, who kept encouraging me to write this book
and for Stef, who refused to accept that it couldn't be done.

The Mystery of the Prime Numbers is the first volume of the *Secrets of Creation* trilogy. Volume 2: *The Enigma of the Spiral Waves* and Volume 3: *Prime Numbers, Quantum Physics and a Journey to the Centre of Your Mind* will eventually follow.

Although reading Chapter 1 is not necessary in order to follow the ideas in the rest of the book, it sets the scene and presents certain issues which will be revisited at the end of the third volume so I'd encourage you to read it.

I've avoided using any mathematical formulas or equations in the main text although some appear in a few of the notes and appendices. The appendices are aimed at readers who want to explore certain ideas in more depth. Each appendix has a level of difficulty stated below its title.

The notes contain a number of website addresses. If you find any of these to have become inaccessible, the Internet Archive's "Wayback Machine" at www.archive.org is a useful tool for recovering old versions of webpages.

More information on the *Secrets of Creation* trilogy, further web links, additional resources and an ever-expanding list of acknowledgments can be found at **www.secretsofcreation.com**.

Matthew Watkins
Canterbury, 2009

an introduction

Is there *anything* we can all agree about?

For just about any idea, ideology, theory or proposal you can think of, there's going to be *someone* who seriously disagrees with it. Even the most seemingly "commonsense" suggestions will be challenged by some obscure philosophy or other, whether academic, mystical or political. The details of historical events are continually being called into question. There's no shortage of conflicting attitudes about the "meaning of life", the best ways to live, the causes of suffering and how they might be alleviated. Rival philosophies and religious beliefs seem to be multiplying endlessly rather than moving towards greater integration and unity.

Science presents itself as a uniquely valid approach to universal truth, but of course many religious believers reject key scientific theories such as Darwinian evolution and the Big Bang. Even *within* science, many widely accepted views are countered by a small but serious-minded minority of scientists who are prepared to challenge them, regardless of how unpopular that might be.

So where, if anywhere, is the common ground in this vast, confusing patchwork of clashing views? Wouldn't it be somehow comforting, in this time of widespread conflict, cultural fragmentation and general confusion, if there were something that *everyone*, from every possible background, however contrary, argumentative or ideologically rigid, could agree about?

Well there *is* something.

I'm going to tell you about it.

Happily, the "something" I'm going to tell you about appears to be a kind of gateway into *a world of profound mystery and wonder*. Yet, unlike most things which get described in such terms, it has the quality of being (as far as anything can be) *indisputable*. So it's a real pleasure to bring into being a book (a trilogy, in fact) which not only deals with just about the only thing *everyone* can agree about but also spreads awareness of something which can awaken feelings of awe and delight in almost anyone willing to make the small effort necessary to follow.

In this, the first volume, I'll carefully explain the fundamental ideas that are involved, interspersing a few (perhaps less indisputable) thoughts about "what it all might mean". If you reach the end and wish to explore further, the second and third volumes will delve deeper into the mystery and then plunge into some *very* strange territory indeed.

1
2
3
4
5
6
7
8
9
10
11
12
13
14
15
16
17
18
19
20
21
22

chapter 1

numbers and counting

From the title of this volume you'll no doubt have guessed that *numbers* are somehow involved in the indisputable "something" which the introduction referred to. Indeed, the basic ideas of *numbers* and *counting* will act as our starting point. In the chapters that follow this one we'll be looking at them in a way which was inspired by my academic background in mathematics – this will lead us into the "indisputable" territory. Although presented in a gentle and accessible way, this approach to numbers may be unlike anything you've come across before. First, though, I think it's important that we take some time to have an informal look at numbers and counting from the perspective of ordinary human experience rather than from a strictly mathematical point of view. If you find yourself disputing some of what I have to say about this, don't worry, we haven't really started yet!

So, what exactly *are* numbers?

Because we all learnt about numbers and counting when we were very young, these ideas have come to be strongly associated with early childhood – they might even seem an unnecessarily and almost embarrassingly "childish" topic to be considering[1]. But this brings us to the first thing that we should pay attention to: the fact that young children routinely and easily grasp the basic ideas of numbers and counting. Counting is one of the very first practical things a child learns to do and adults take it for granted that it's a sensible and appropriate thing to teach them. As children we all gradually learnt how to recite numbers in sequence, count things with them, recognise and draw the symbols that our culture uses to represent them, combine them by adding and multiplying, and so on. Some of us picked it up faster than others, but with very few exceptions (due to, for example, certain neurological conditions), children's minds absorb the *basic ideas* very easily. Some people end up very comfortable and capable working with numbers in adulthood. Some struggle. Most just get by. But it's almost unheard of for someone to remain completely baffled by the *very idea* of numbers and counting – everyone *gets it*.

And yet, if you think hard about numbers and *what they really are*, you'll probably get quite confused. You no doubt know how to work with them (at least to some extent), but if you spend enough time contemplating what they *really are*, you may well end up concluding that *you don't know*. The more you think about it, the more confusing it seems to become. Perhaps it all seems perfectly clear to you. But philosophers have been debating this matter for centuries and they're still far from providing us with a clear answer. You might not be able to imagine why, but it would be fair to say that there's still no straightforward consensus surrounding this issue at the deepest levels of philosophical discussion.

There are philosophical factions such as *Platonists* and *social constructivists* who continue to debate whether numbers and related concepts exist somehow independently of us and we "perceive" them with our minds or whether they are merely mental, social or cultural "construct". Much has been written about this question over the years. But these rigorous attempts to pin down exactly *what numbers are* would almost certainly confuse matters rather than clarify them, if presented to the "ordinary person in the street" who unproblematically deals with bus fares, football results, temperatures and shoe sizes.

Despite this puzzling situation, everyone should be able to agree that numbers are *the common property of all*. No one can be excluded from access to them. No one can take ownership of them. They're there for everyone equally. Wherever you find yourself in space or time, you'd expect the numbers to be there, accessible to you. But where is this "there"? They have this peculiar status of sort-of-existing (we're continually dealing with numbers of objects) but sort-of-not-existing (numbers don't

exist in the way that actual objects do).

The fact that young children have no problem accepting numbers suggests to me that number concepts may be in some sense *built into our minds*. That is, a child learning about numbers is in fact *recognising* something which is already present within her or his mind. But even if I'm right in my vague suggestion that "it's in there" somewhere, there's still no agreed understanding of what "it" is, in what sense it's "in there" or even where "there" is.

Anthropologists have reported examples of cultures with extremely limited counting abilities which, once in contact with Western traders and the use of money, have suddenly switched into a highly competent number usage (without the introduction of any Western-style education). In *The Emergence of Number* [2], John Crossley suggests that "*the idea of counting lies dormant until evoked*". Having considered accounts of various indigenous peoples of Latin America, Polynesia, Australia and Malaya he concludes that "*non-verbalized ideas of particular numbers appear to be present long before they may be needed in a verbal form and once counting is established there seems little difficulty in advancing rapidly*".

The significant word here is "present". Present *where*?

I suggested that number concepts may be "built into our minds" but to some extent it now seems that they may be built into our *brains*. The relationship between the mind and the brain is another important matter which philosophers are unable to agree on. Certainly, the brain is the "physical part" with physically describable regions and components, while the mind is the "non-physical part" which is somehow related to the brain but in a way that no one is entirely sure about.

In recent years, neuropsychologists [3] such as Stanislas Dehaene have been carrying out experimental work in the area of "numerical cognition" to explore the possibility that physical structures exist in the brain which relate to counting and basic operations with numbers, these having possibly evolved for survival-related reasons.

Other researchers have carried out experimental work involving non-human animals, demonstrating the abilities of some to distinguish between various small numbers [4].

Despite the *extremely* widespread use of numbers in Western culture, the sense in which they "exist" and their relationships with the mind and the brain are rarely discussed – these are surprisingly marginal subjects. I find this situation strange, especially if we consider the incredible range of subjects which humans *have* explored in the most minute detail.

QUALITY- AND QUANTITY-BASED VIEWS OF NUMBER

Children first learning about numbers often describe *feelings* they have about each of the first few: 1, 2, 3, 4, 5, 6, 7, 8, 9, 10, 11, 12,… Perhaps you have faint memories of something like this. I can still clearly remember sitting next to my friend Paul at school, aged six or seven, casually discussing our feelings about various numbers while we were working on our simple arithmetic problems. There were likes and dislikes, favourite numbers and numbers which seemed to have some sort of *personality* which we couldn't express clearly but we could somehow sense or feel. It felt as normal as discussing our feelings about various colours, songs or storybook characters.

In many cases, this kind of feeling might be linked to the shape of the numeral or the sound of the word associated with the number. Or it might be due to some association with an age, a birthdate, a house number or the shirt number of a favourite athlete. But I suspect that there may be something deeper going on with the *overall phenomenon* of these feelings, as suggested by the accounts of people with severe autism and related conditions, some of whom can perform baffling, almost superhuman feats of mental arithmetic and, at the same time, describe having a direct inner experience of numbers as having textures, colours and/or personalities [5]. The combination of these people's extraordinary abilities with numbers and their claimed "inner perceptions" of them suggests that they might know something about

number which the rest of us don't.

However, a "sensible grown-up" outlook dictates that there is no value in dwelling on these "childish" number-related feelings. In state-sanctioned systems of Western education, numbers are presented to children in a systematic, unemotional way. They are treated solely as *quantities* to be added, multiplied and so on. *Their properties and interrelations are entirely unaffected by our feelings about them.*

This brings into focus the distinction between two very different approaches to number. If I say "seven is between six and eight" or "seven is an odd number", those statements concern seven's properties as a *quantity*. But if I say, "seven is a lucky number", or "seven feels smooth, like a pebble", those are claims regarding supposed *qualities* of seven.

Prior to the emergence and expansion of science-based Western civilisation, many cultures had a kind of *reverence* for certain numbers or a belief that numbers have a *qualitative* aspect (a quality, personality or "meaning" of some kind) as well as the more obvious and mundane *quantitative* aspect (a quantity, an amount of something)[6].

This distinction between the "qualitative" and "quantitative" approaches to number has hardly been discussed by academics outside a small fringe of thinkers. It seems that the unspoken, almost unconscious, belief among Western intellectuals is that because arithmetic and all higher mathematics involve the quantitative (and most definitely *not* the qualitative) approach, the qualitative *obviously* lacks any serious value and the quantitative is the "correct" view, so there's nothing to discuss. In this way, the quantitative view – the view used by mathematicians, scientists, stockbrokers, bookmakers and pocket calculators – triumphs.

Many Western children seem instinctively drawn to a qualitative approach to number but they are systematically directed away from this by their formal education. Wherever pre-Westernised cultures have gravitated to the qualitative approach[7], this tendency has been similarly countered by the nearly universal introduction of

Western-style educational practices, part of that questionable ongoing global project sometimes called "progress".

Perhaps you're thinking "well, yes, this *is* progress, this *is* the correct way" – you may have no problem at all with completely dismissing the qualitative approach to number. After all, a dozen different people could "feel" a dozen different "qualities" associated with a number, so there's not much point trying to study this sort of thing, is there? Or perhaps you feel that there *is* something behind the qualitative approach to number worthy of more serious attention. In any case, as we proceed, try to keep in mind this distinction between the "qualitative" and "quantitative" approaches to number, ideally *without judgement*. Just remember that these two very different perspectives exist and try to avoid thinking about them in terms of true/false, right/wrong or valid/invalid.

> You might find that I'm using the word "number" in an unfamiliar way (for example, when I write "approaches to number"). "Numbers" is easy – we all understand how that word is meant to be used, and there's no lack of examples: 1, 2, 3, 5074, 95 million, 0, 191, *etc*. But when I use the word "number" in this way, I mean the *overall concept* of numbers, rather than isolated, individual numbers. It's very much like the way an artist might talk about the use of *colour* in general, rather than about individual colours.

THE QUALITATIVE VIEW

The qualitative view has no serious role in organised society. Still, remnants can still be seen at the level of individuals and their idiosyncrasies.

Any Western-style mathematics education is entirely based on the quantitative approach to number. In order to be considered "successful" on its own terms, it would

have to involve any number-related feelings being "educated out" of children. But despite educators' best efforts, feelings of this type can persist into adulthood, and do, far more widely than some people would like to think [8]. There are many curious remnants of "number mysticism" in our modern, scientific culture. Many people have lucky numbers, seven being the most notable for some reason. Fear of the number thirteen is still widespread in the Western world. Some major hotel operators routinely number their floors $\ldots, 11, 12, 14, 15, \ldots$ for practical economic reasons – an economically significant proportion of their customers don't feel comfortable staying in a thirteenth floor room. Numerology books continue to proliferate. Telephone numerology consultations are commonly advertised in the back pages of popular newspapers. Websites and lucrative workshops abound. I've heard of a variety of eccentrically ritualistic and quasi-mystical ways in which people choose their lottery numbers – numbers which they see as the keys to a kind of salvation. And a significant number of people now struggle with variations of obsessive-compulsive disorder which involve an urgent need to repeat certain actions certain numbers of times.

There's a huge gulf between the dominant "scientific" (that is, quantitative) approach to number and the qualitative "folk beliefs" regarding numbers which can still be found throughout Western populations. This is similar to the gulf between the culturally dominant "scientific" view of numbers which now prevails and the views which were held throughout most of human history.

Although the social phenomena I've described could be worth examining for various reasons, they're still very marginal in the overall workings of the Western world. The powers-that-be (bankers, corporate leaders, politicians, economists, scientists, *etc.*), if they were to give the matter any thought, would certainly be of the opinion that such beliefs are nonsensical remnants of a pre-rational, pre-scientific age. Western science simply denies the validity of anything "numerological" and assumes the thinking behind it to be fundamentally misguided. Although relatively new in historical terms, this perspective is now firmly established as the dominant one.

I've used the terms "Western world" and "Western science", and I'll continue to use this terminology, so it would help if I explained exactly what I mean. I'm the first to admit that it's quite ridiculous language since we live on an approximately spherical planet and a sphere has no "west". But the already-familiar terms "the West" and "the Western world" will be useful shorthand to mean those parts of the human world which have been heavily influenced by Western European culture. So that includes all of Europe, as well as the places which remain colonised by Europeans such as Canada, Australia, the USA and New Zealand. Also, many urbanised areas in the rest of the world are becoming increasingly "Westernised", so there's no clearly defined edge where "the West" begins and ends. I intend these terms to be understood in a cultural sense: "the Western world" describes more of a mindset than a geographical region and "Western science" refers to a set of practices (and beliefs[9]) which can be adopted anywhere on the planet.

THE QUANTITATIVE VIEW

Having mastered the basics of arithmetic as children, most people give very little thought to numbers beyond their immediate use in financial transactions and other such practical matters. There is a strong tendency to take them for granted. But if they were to stop and consider the extent to which numbers have become woven into their lives, many people would be quite surprised.

Suppose you were to switch on a radio and catch the end of the hourly news. You, together with possibly millions of other people, are listening to a publicly sanctioned source of information. You may well hear some new government statistics on crime, education or unemployment, the stock exchange index and the number of points it's gone up or down, some sports results (in the form of numbers), the time, a few temperatures, the identifying numbers of some major roads and junctions, the

speeds of traffic in their vicinities and, finally, the frequency of the station you're listening to.

And it's not just actual numbers which you begin to notice everywhere once you've started looking, it's also the tendency for Westernised humans to *measure* and *quantify* the things they encounter. In almost every area of our lives, attempts are being made to reduce everything to measurements, which take the form of numerical data. We'll look at the main examples of this after a quick explanation of how I'm going to be using certain words.

By *quantification*, I mean the process of assigning a number to something. So quantification includes simple counting and all familiar forms of measurement (using a ruler, a stopwatch, a thermometer, *etc*.). But the word is more commonly applied to all of the other ways in which numbers get assigned to things which aren't obviously measurable – things like human intelligence, the "value" of a painting or the "performance" of a school or hospital. These things can be *quantified* when someone finds a way to measure them – an IQ test, an art auction, a governmental evaluation procedure.

By *counting*, I mean the application of number to the physical world, by means of *agreed-upon categories of things-to-be-counted*.

Eh?

OK, try this: look around you and *count everything you can see*.

It's not so obvious is it? No, in order for counting to be meaningful, there must be an agreement as to a "category of thing" which you're going to count (person, grain of sand, item of furniture, hexagon, occurrence of the letter "j", *etc*.). This might seem like an obscure philosophical observation, but the implication is that *counting things* and *breaking the contents of the world down into categories of things* are very closely related activities, an important point which we will return to in Volume 3.

Measurement is really just a more abstract form of counting. When measuring something, you're counting a "unit of measurement" (an inch, a degree Fahrenheit, a volt, a kilogram, a megahertz, *etc.*). With a tape measure, you can count the number of inches or centimetres between two locations. If the distance ends up being, say, 182.57 centimetres, then the measurement has involved counting centimetres – your "unit" – as far as you could get with them (182), then switching to tenths of your unit (which would be millimetres), counting as far as you could with *them* (5), finally switching to count tenths-of-millimetres and finding there to be exactly seven of *them*. The distance is 182 centimetres plus 5 tenths of a centimetre plus 7 "tenths of tenths of a centimetre" (182.57). Don't worry if you found that last bit confusing. The main point is that when you *measure* something, you're actually *counting* something else: units of measurement (and subdivisions of those units).

A great diversity of things can be measured – temperature, the passage of time, electrical resistance, volume, weight, voltage, the intensity of light, radioactivity, pressure, *etc*. In each case, you (with the help of your instruments) are *counting* a clearly defined unit of measurement.

A shepherd in ancient Greece counting his sheep and a physicist in 21st century Switzerland using an ultra-complex bank of instruments to measure the mass of an elusive subatomic particle are ultimately both engaged in the same thing – they're *counting an agreed-upon category-of-thing* (sheep and "units of mass", respectively).

The essence of quantification, in the commonly used sense of the word, is to reduce something complex (psychological, sociological, ecological or whatever) to a quantity or quantities – to numbers, numerical data. Governments, economists, psychologists, sociologists and market researchers do a lot of quantifying. But whether they're assessing value, cost, risk, "quality of life", "performance" or anything else, there must always be an agreed unit or scale according to which this is done – so we're back to measurement and counting again.

Counting, measurement and quantification, then, are just different forms of the same thing – a kind of "mapping out" in terms of numbers of the world we experience. This is what characterises the quantitative approach to number, which is unquestionably the dominant one in the Westernised world. We'll now look at three of the main areas where it is evident.

THE DOMINANCE OF QUANTITATIVE NUMBER: MONEY AND ECONOMICS

Although "the economy" is a system of human activity – production, consumption, distribution, exchange – as it becomes ever more complex, it's being dealt with more and more as a huge system of numbers. We're encouraged to think that these numbers refer to something "real" but if you start to look into what that is, you'll find that it's not at all obvious.

An amount of money always involves a number, but that number changes if you convert your money from one currency to another. Currencies act as different units of measurement in this context (converting US dollars to Euros is the same sort of process as converting inches to centimetres or ounces to grams, although the exchange rate may change from day to day). This system works quite effectively – the problem is that *it's not clear what the units are measuring*. Prior to the existence of money, people bartered, directly exchanging commodities. Money emerged as helpful tokens of exchange and has since gradually transformed from…

☆ units of "hard" currency (coins of various metals and sizes, often with numbers stamped on them)

…to…

☆ units of "soft" currency (pieces of paper with numbers printed on them, supposedly "representing" quantities of precious metals which are kept far from view in fortresses or bank vaults, or which, in many cases, don't even exist)

…to…

☆ "plastic" (nothing physical is exchanged, only numbers, and this being done via plastic cards which are themselves emblazoned with many digits)

…and now…

☆ more sophisticated "virtual" and "smart" mechanisms for carrying out financial transactions via the electronic transmission of numbers (no tokens whatsoever).

Apart from a relatively small number of stateless people, everyone on the planet lives in a state with some kind of currency. For the purposes of comparison these currencies can all be converted into Euros (or Yen, Rupees, Canadian dollars or whatever you like) – this is the same process of "converting to common units" as you would use if you wanted to compare various lengths which had been given in

centimetres, yards and nautical miles. Almost everyone alive, then, theoretically has (or owes) a number of Euros, a number which locates them on a scale measuring wealth. "Top 10" and "Top 100" lists of wealthy people often feature in glossy magazines these days based on this kind of comparison. People's numbers go up and down as they buy, sell, inherit, work and spend. Almost everyone is striving to increase their number (usually causing other people's numbers to decrease) and this has arguably become *the* central feature of human life in recent times.

The situation is getting ever more abstract, with people buying and selling not only shares in corporations and amounts of various currencies but also less familiar things like *futures*, *derivatives*, *options*, *swaptions*, *volatility swaps*, *quantos*, *lookbacks* and other such "exotic financial instruments". Glancing through recent issues of prominent economics journals, I find that most of the articles have titles like "*Idiosyncratic shocks and the role of nonconvexities in plant and aggregate investment dynamics*", "*Tests for asymmetric threshold cointegration with an application to the term structure*", "*Backward integrated information gatekeepers and independent divisions in the product market*" and "*Rigidity in bilateral trade with holdup*". As these things become further and further removed from traditional human activities, and as it becomes increasingly difficult to explain to a non-specialist *what it is that the numbers involved are actually "counting"*, all that we can be entirely sure about in any financial matter are *the numbers themselves*.

Let's now consider one of the most noticeable forms of economic activity – mass production. A rapidly growing proportion of the material objects humans come in contact with these days are mass produced. All over the world, large numbers of near-identical objects are being produced mechanistically and transported elsewhere, almost exclusively for economic motives: movements of large numbers of near-identical objects – "units of production" – away from the manufacturer, ideally resulting in movements of large numbers of "currency units" back towards the manufacturer. The flippant expression "shifting units", used in the "busyness" world to describe sales of products, suggests that "economic activity" is no longer about creating and distributing necessary objects and commodities but rather *just moving numbers around* (which is exactly what most people working in the financial sector appear to be doing).

THE DOMINANCE OF QUANTITATIVE NUMBER: SCIENCE AND TECHNOLOGY

Another place where quantitative number-based thinking is rife is Western science and all that it has spawned. That, of course, includes everything involving computers and other digital technology.

Science seeks to quantify and measure all phenomena in order to gain a complete mathematical description of reality. If you take away the ability to *measure*, Western science is reduced to almost nothing. Beyond basic measurement, science depends heavily on mathematics, and at the root of all mathematics we have the basic quantitative ideas of numbers and counting.

Computers (in their various forms) are becoming involved in almost all aspects of our lives, whether we choose to notice it or not. And whether they're being used by transnational banking networks, surgeons in the midst of delicate operations, nuclear power station safety inspectors, musicians mixing tracks, graphic designers designing cereal boxes or bored teenagers playing computer games, they're all essentially doing the same thing – they're *manipulating large amounts of numbers*. At the most basic level of their electronic circuitry, everything is reduced to 0's and 1's, as you may know.

All Internet content, anything a computer can handle, in fact all digital technology is ultimately *numerical* (digital = digits = numbers – yes?). In recent times, much long-standing technology has been replaced by digital versions (digital cameras, radio, TV and music recording systems have all appeared during my lifetime). It's getting to the point that if something can't be handled in this way – can't be captured in alphanumeric text, digitally imaged or in some other way represented as data – then it doesn't really exist, or at least it isn't worth considering. Subtle intuitions and feelings, dreams, "vibes", empathy, beauty, joy and love are still hugely important to the vast majority of the population but with the ever-expanding dominance of quantitative science they're being marginalised because they can't be measured or "captured" as numerical or digital information – that is, *they can't be quantified* [10].

THE DOMINANCE OF QUANTITATIVE NUMBER: GOVERNMENT DECISION MAKING

The third and final instance of "quantitative thinking" which we shall consider is government decision making, something which is extensively, and to an ever greater extent, based on numbers. Many aspects of society and the environment (commerce, employment, health, education, crime, agriculture, pollution, voting trends, *etc.*) are continually being measured or quantified. The resulting numerical data is analysed by statisticians in order to calculate an "index" measuring some social, economic or environmental tendency, or to reveal *correlations* between various factors involved in the issue at hand. Governments who employ these statisticians use their conclusions to establish targets and "league tables" which assist in the making of decisions to increase or decrease the amounts of money being spent in various areas, raise or lower taxes, adjust interest rates, restructure school curriculums or introduce new laws.

The decisions made by our leaders are dictated to a very great extent by the numerical data which is presented to them. As more and more data is gathered, more and more "number crunching" must occur before they're able to do anything with it, and so the further the decision making becomes removed from the realities of individual citizens' lives [11].

In past ages, important decisions made on behalf of organised groupings of people would have been based to a large extent on things which *could not be quantified*: religious scriptures and teachings, the interpretations of omens and dreams, priestly deliberations about the wishes of gods, spirits or ancestors. Despite any talk of "values" or "ethics", the meanings of which are continually being manipulated by people interested in power, it's really economics and science that now act as the bedrock of social infrastructure and government decision making. Number, understood quantitatively, is the common thread.

Quantitative economics is central in government decision making and is becoming progressively more dominant to the exclusion of all other considerations. *Free trade* is held up as the highest of ideals by our leaders. The need for *economic growth* – the continual increase of a very particular, and particularly dubious, measurement called a *gross domestic product* – is now taken as a fundamental axiom and is the most pressing concern for the leaders of every major nation-state.

Quantitative science is actively promoted by some governments as the ultimately reliable source of truth. I've even heard it *evangelised* about in speeches by political leaders – during his time as the UK's Prime Minister, Tony Blair spoke of the country's path to the future being "lit by the brilliant light of science"[12]. To some extent, this phenomenon reflects the influence of busyness interests on government (the biotechnology industry, in particular, in the early 21st century), but I sense that it also has a deeper significance...

SCIENCE AND RELIGION

Centuries ago, leaders of Western populations using that kind of language ("lit by the brilliant light of...") would have been talking not about science but about Christianity or God. Interestingly, various commentators have pointed out that both economics *and* science can be seen to parallel religion in some ways. Considering that both are underpinned by number, we should pay attention to this.

We'll consider science first – "the brilliant light of science". In certain ways, "quantitative science" is emerging as the unacknowledged religion of the modern Western world. And I'm not the first person to draw this parallel – the extremely lucid moral philosopher Mary Midgley, for example, has written extensively on this matter [13].

There will undoubtedly be defenders of science and opponents of religion who would vigorously dispute this, arguing that science is the polar opposite of religion. However, I think that even such people would have to concede certain points. For example, it would now be considered appropriate to turn to astrophysicists, cosmologists, microbiologists and geneticists, not sages, priests or theologians, when asking the "big" questions about the origin and destiny of life, the planet and the universe. These are questions which religion, in its various forms, has addressed for thousands of years. Science only took over this role relatively recently, historically speaking.

The young scientists of today undergo a rigorous period of study and initiation into the ways of science comparable to young men entering a priesthood, learning special symbols to be manipulated and actions to be carried out in order to arrive at cosmic truths. The labcoat-clad scientist performing experiments in a laboratory has replaced the robed high priest carrying out rites in a temple.

So is science just a new form of religion? The pro-science, anti-religion crowd would argue no, it's an entirely different kind of thing since it relies not on faith but on the *scientific method*. But if we examine this, the very foundation of science, we find that there are philosophical problems relating to the idea of repeatable experiments (since no set of conditions can ever be restored *exactly*), the assumption of some sort of "uniformity through time" [14] and something called the *Law of Large Numbers* [15]. These problems are currently swept aside by the adoption of certain beliefs about "the way things are", beliefs which seem perfectly reasonable to almost all Westernised humans and which have allowed science to develop in the way that it has. But they're still *beliefs*. The scientific method, resting as it does on these beliefs, can never be used to prove *their* validity.

Even if you really can't accept the science-religion analogy as I've presented it, I'm sure that, at the very least, you'll agree that science is filling some parts of the void which is being left wherever the influence of traditional religion is declining.

ECONOMICS AND RELIGION

Economists have come to resemble a priesthood, discussing among themselves and quantifying such mysterious things as *confidence*, *growth*, *rigidity*, *liquidity* and *volatility*. Through the sheer power of capital, such quantification can lead to market activity which has profound, tangible effects on the world. Their forecasts and public statements are reminiscent of the pronouncements made by ancient priests who consulted oracles. There's real power invested in these people, with their highly specialised, mysterious knowledge, and none of us can entirely escape being affected or somehow involved.

As the University of Maryland economist Robert Nelson puts it in his book *Economics as Religion* [16], economists

> "*think of themselves as scientists, but…they are more like theologians.* [One] *basic role of economists is to serve as the priesthood of a modern secular religion of economic progress that serves many of the same functions in contemporary society as earlier Christian and other religions did in their time.* [17] *Economic efficiency has been the greatest source of social legitimacy in the United States for the past century, and economists have been the priesthood defending this core social value of our era.*"

In an earlier book [18], Nelson argued that

> "[*b*]*eneath the surface of their formal economic theorizing, economists are engaged in an act of delivering religious messages. Correctly understood, these messages are seen to be promises of the true path to a salvation in this world – to a new heaven on earth. Because this path follows along a route of economic progress, and because economists are the ones – or so it is believed by many people – with the technical understanding to show the way, it falls to the members of the economics profession assisted by other social scientists) to assume the traditional role of the priesthood.*"

In a 1997 paper, the moral philosopher John McMurtry goes further, putting forward a convincing list of ten criteria which can be used to characterise a "fundamentalist theology" and then showing how global free market theory and practice satisfies all of them [19].

Western (and Westernised) people's highest aspirations are continuing to drift away from divine salvation and towards the acquisition of vast material wealth. Their devotion has shifted from saints and deities to specific products and brands, offered in the form of desire, shopping and consumption, as opposed to prayer, offerings and sacrifice. I'm not the first person to suggest that shopping malls have become the new cathedrals [20].

Just about everything we do these days seems to be somehow tied into "the economy". But to the extent that this "economy" is built out of numbers, it feels to me that it's largely held in place by our willingness to continue believing in it. "The economy", it could be argued, has replaced "God" in the modern, secular West. Meanwhile, both Christians and nonreligious anticapitalists talk about the worship of "Mammon", a deity embodying the pursuit of wealth above all else.

NUMBER AND RELIGION

Science corresponds to the "explanatory" content of religion (*How did the world come into existence, what steers its course and what will be its fate?*). Economics – sometimes described as a branch of science but really something very different, according to thinkers like Nelson and McMurtry [21] – corresponds to the "devotional" content (*What are our deepest desires and aspirations, what should we be striving for, what is the route to our salvation, individually and collectively?*).

This convergence of science and economics as two faces of an unacknowledged religion is also evident in the twin usage of the word *materialism*. On one hand, it refers to a serious philosophy, associated with thinkers such as Thomas Hobbes, which argues that the only thing whose existence we can be sure of is matter – that's the "science" side. On the other hand, the same word is used to describe the rampant "shop 'til you drop" consumer culture which has emerged in the most affluent parts of the modern world – that's the "economics" side. Philosophical materialism's denial of non-physical or "spiritual" realities is wholly compatible with the increasingly widespread approach to life currently found throughout the "materialistic" Westernised world, which is to *get as much stuff as you possibly can*. This is an understandable approach to life if material "stuff" is all that's really going on but it stands in stark contrast to the approach of all those cultures who have believed that, and behaved as if, much of what's going on around them is occurring in nonphysical realms.

Now, what's the connection between science and economics in their present forms? It is, as I have mentioned, that they're both built on the foundation of number- or quantity-based thinking.

So, if we accept the science-religion and economics-religion analogies, then it could be argued that number is woven into not only Westernised humanity's social, political, scientific and economic life but into a sort of (unconscious) religious life as well. The "rites" performed by the scientific and economic "priesthoods" involve the manipulation of numbers. These "priests" don't say prayers, cast spells, sacrifice animals or make invocations – they analyse data, create mathematical models, solve equations, *etc.* The subject at the heart of all this, the central "mystery" of this scientific-economic religion is *number*.

You may be struggling to accept my use of the word "mystery" (especially in its religious sense!) in connection with something as familiar and seemingly innocuous as numbers. But, as discussed earlier, even the nature of the *existence* of numbers remains problematic to philosophers. And, far beyond that, I'm confident that what will gradually be revealed to you about number in the chapters which follow will strike you as both indisputable and *deeply* mysterious.

My aim is to show that the system of counting numbers is *not what you thought it was*. It seems to me that it's high time for a reconsideration of number and its central role in our current way of experiencing reality. By the end of this volume, it will start to become clear that *there's something truly weird going on with the system of counting numbers*. Readers who reach the end of the second volume will have acquired a fairly detailed perspective on this weirdness but by then an even more profound (yet indisputable) layer of mystery will have begun to be revealed. The final volume will look into this and consider what it all might mean (without offering any definitive answers).

So have number and quantification acquired a deep but unacknowledged "religious significance" in the modern world? Even if we avoid talking in terms of religious significance, it seems that something quite noteworthy and poorly understood is going on involving humans and numbers.

To bring this further into focus, I would argue that one of the features best characterising modern Western culture is its relationship with number. The scientific West distinguishes itself from other cultures by its complete rejection of the qualitative in favour of the quantitative approach to number. And this quantitative approach has been embraced with such fervour that it has nearly *consumed* the West. Earlier cultures related to number in both a quantitative *and* a qualitative way, at a time when the distinction between religious and secular life was much less clear (or non-existent) [22]. Western culture has emphasised this distinction and has effectively tried to *secularise* number by reducing it to its quantitative aspect. And while any qualitative experience of number must now seek refuge in some kind of marginalised mystical/religious framework, the *quantitative* approach seems to have itself (ironically) acquired some religion-like features.

HISTORICAL PRECEDENTS

A few historical and cultural observations may be helpful here.

Abraham Seidenberg's *diffusion theory*, put forward in a 1962 paper [23] presents evidence suggesting that, contrary to the widely held view, the use of counting did *not* arise spontaneously among many different peoples but rather arose *just once* and then "diffused" around the ancient world. In his paper, Seidenberg proposes that the unique instance of humans introducing this supremely important innovation – the use of numbers – was central to a *ritual* "*and that participants in ritual were numbered*". He goes on to focus more specifically on rituals which involve the enacting of creation myths (such activities being central to the religious life of the people in question).

Seidenberg also considers certain passages in religious scriptures (the Bible and Qur'an as well as ancient Greek, Babylonian and Buddhist texts). He unpicks a complicated tangle of interrelated analogies involving shepherds counting sheep, the moon counting stars and "the Lord's people" being "sand upon the seashore" to deduce that the Creator was once characterised as an entity which *counts*.

Many examples of religious art and architecture around the world involve numerical themes encoded into their geometry [24]. The geometry of temples, churches and other sacred sites worldwide is a fascinating topic, one deserving of far wider attention (the lack of which has allowed the subject to be overrun by careless people writing nonsense).

In 1916, the highly influential psychologist Carl Jung proposed the existence of what he called the *collective unconscious*, a deep level of mind which is somehow "shared", accessible to every human consciousness [25]. Jung arrived at this idea, which came to underlie his theory of *archetypes*, through having studied folktales, myths and religious symbolism from around the world and observing common themes and images. In particular, he noticed how certain *numbers* recur in similar contexts in religious scriptures and symbols, fairy tales, superstitions, oracles, *etc.*,

leading him to the idea that there are *number archetypes* which inhabit the collective unconscious. This particular line of research was continued after Jung's death by his student Marie-Louise von Franz, who emphasised the importance of recognising both the quantitative and qualitative aspects of number [26].

The numerological superstitions still common in the West and these instances of number mysticism in scripture and "folk religion" studied by von Franz and Jung have their roots in similar places. Although it's too fragmented and disorganised to be called a religion, modern numerology is surely closer to being a religious activity than a scientific or mathematical one. In my experience, outspoken atheists and sceptics will generally denounce both religion and numerology on very similar grounds.

Examining the historical roots of Western mathematics and quantitative science, we are inevitably confronted with Pythagoras, the 6th century BCE Greek thinker who had one foot in the pre-rational world of shamanic/animistic religion and the other in the modern world of rational scientific thought. Although sometimes described as the first real mathematician, Pythagoras was primarily a number mystic. He and his followers were at least as concerned with the qualitative as the quantitative aspects of number. His name is now inextricably linked with possibly the most famous of all mathematical theorems [27], but he was also the central figure in what was effectively a religious movement dedicated to the contemplation and "worship" of numbers. He attained mythical status among his followers who actively continued his cult for at least a century after his death. Pythagoras is supposed to have promoted the belief that "all is number" (as well as the belief that eating beans is immoral).

The "Pythagoreans" were very much an underground phenomenon, subjected to suspicion and persecution. But they appear to have won in some sense, for here we are some twenty-five centuries later, and who is collectively entrusted by society with the role of safeguarding the dominant notion of "absolute truth"? It's their descendants, the quantitative scientists who use numbers and number-related concepts as their main tools for probing the workings of the physical universe.

Although today's "priestly" attempts to decipher the laws of nature with number-based thinking appear to have little to do with the human-centred, mundane realities of economics and government statistics, we can't just dismiss the "deciphering" work of scientists as irrelevant or esoteric. This "priesthood" has had much practical success, as can be seen in the accelerating proliferation of "amazing" technology which has occurred in the early 21st century. The circuitry in all the new electronic gadgets which now create so much excitement in the Westernised world only exists as the result of a deep theoretical (that is, mathematical) understanding of quantum physics which has developed over decades. Such inarguable, material successes have led to a belief in some quarters that quantitative science is the "correct" way for humans to understand and relate to the world. Scientists are quietly entrusted with a kind of ultimate authority and in this way the quantitative number-based approach to reality has become almost unchallengeable in its dominance.

WHERE'S THIS ALL GOING?

Some scientists have commented on the surprising extent to which the physical universe appears to be *inherently mathematical* – that is, compatible with our number-based approach to reality. Because quantitative science is able to describe and predict features of the universe which appear to have been here long before we humans (along with our concepts and calculations), it has strengthened its claim to be the "correct" way in which to understand the universe. In this context, mathematics has been described as a kind of "language of Creation".

But this overlooks something important. It's not the universe as a whole which appears to be "inherently mathematical" but rather just *those aspects of the universe which Western science has chosen to focus on*. And those are the *quantifiable* aspects – the ones which can be measured and converted into data. The *unquantifiable* aspects, those which *can't* be reduced to numerical data, are much less accessible to science and, as a result, very little attention has been paid

to them. In fact, they've been getting marginalised at an alarming rate.

It's as if we're rapidly replacing the world around us with a crude replica, a world built out of numbers. Seemingly unstoppable economic and technological forces are transforming the world into a place where nearly every aspect of our experience is mediated by quantification, measurement and counting. This is largely taken for granted, or seen as a perfectly ordinary way for things to be progressing.

How and why has this happened? Who, if anyone, was responsible? And how do we feel about it? Compared to the enormous (and rapidly expanding) body of mathematically-based science and economics literature, very little is ever said about this matter.

One book I would highly recommend, *Descartes' Dream: The World According to Mathematics*, by Philip Davis and Reuben Hersh, *does* deal with this issue in some depth:

> "*There is occurring today a mathematization of our intellectual and emotional lives. Mathematics is not only applied to the physical sciences where successes have been thrashed out over the centuries but also to economics, sociology, politics, language, law, medicine. These applications are based on the questionable assumption that problems in these areas can be solved by quantification and computation. There is hardly any limit to the kind of things to which we can attach numbers or to the kinds of operations which are said to permit us to interpret these numbers.*" [28]

Although there has not yet been any organised resistance to this phenomenon, there is, I believe, a widespread and growing disillusionment with the extent to which things are being quantified, mathematicised, digitised, computerised and technologised. In other words, there's a disillusionment with the extent to which the world is being replaced by a world of numbers, even if people haven't quite realised that that's what it is. Davis and Hersh conclude the preface to their book with what they describe as a "moral" and it reflects this feeling:

"*The social and physical worlds are being mathematized at an increasing rate… We'd better watch it, because too much of it may not be good for us.*" [29]

This sentiment would be increasingly endorsed by members of Western societies, at least as far as they recognise that the phenomenon (this "mathematization" of the social and physical worlds) is occurring. But the authors also offer some consolation and reassurance:

"*To find things that cannot be mathematized, then, we must look away from the physical world. What other world is there? If you are a sufficiently fanatical mechanical materialist you may say none. Period. Discussion concluded.*

If you are more of a human being, you will be aware that there are such things as emotions, beliefs, attitudes, dreams, intentions, jealousy, envy, yearning, regret, longing, anger, compassion, and many others. These things – the inner world of human life – can never be mathematized." [30]

Some people (the "fanatical mechanical materialists") might not accept this last assertion, certain that even these things could ultimately be quantified and captured by mathematical formulas. Personally, I hope that Davis and Hersh are right. But even if they are, *because* these subtle, unquantifiable aspects of experience cannot be "mathematized", the "scientific" Westernised world has, to a worrying extent, gone from forgetting to look at them

…to…

forgetting how to see them

…to…

forgetting that they ever existed

…to…

actively denying that such things could have any sort of "real" existence.

Finally, let's compare the extent of this "mathematization (quantification) of everything" with the most common psychological reactions to mathematics. Beyond any feelings of alienation or disillusionment of the type just discussed, *many* people in the West, simply due to their experiences at school, see mathematics as something frightening, even repulsive. It's like a *poison* for some people. This is really very common, I've found. And unlike any other major subject of study, it is acceptable, even *desirable*, at all levels of society, to express a dislike for, or ignorance of, maths. In Volume 3, we'll return to this issue, for surely something is seriously out of balance when *we build our world around something and then collectively cultivate a deep distaste for it.*

IN SUMMARY

We've seen a few different reasons why it might be worth reexamining number and our relationship with it:

☆ It's common property, it's somehow *universal* and it may be "built in" to our minds and/or brains in a way we don't fully understand.

☆ Philosophers still can't agree on the sense in which numbers *exist*.

☆ The world's dominant civilisation distinguishes itself from all other civilisations by its total emphasis on the "quantitative" approach to number and rejection of the "qualitative" approach.

☆ Via science and economics, number seems to have taken on (unnoticed) an almost religious significance. This has interesting historical precedents.

- ☆ The overemphasis on quantification may be causing us, collectively, to lose touch with the many *unquantifiable* aspects of human experience.
- ☆ Despite so many aspects of our lives now involving the system of counting numbers, *many* people feel strong negative emotions in connection with the study of it (that is, with mathematics).

Going back to ancient Greece at the time of Pythagoras, we find that modern mathematics evolved out of what we now call "numerology" or "number mysticism", much as astronomy evolved out of astrology and chemistry out of alchemy – in each case, we have a science emerging from something which is now widely seen as a supernatural belief system. Ideas about numbers have changed considerably over the millennia, so it's not at all certain that the currently held view is going to last very long in historical terms. In the book *The Emergence of Number* cited earlier, author John Crossley presents a convincing argument that "*our present age is no more in possession of the ultimate characterization of number and numbers than any previous age.*"[31]

We shall now (at least for the time being) put aside all of our feelings, opinions and speculation about number and numbers, put our quantitative heads on and have a look at how, according to the experts, our "number system" is put together.

chapter 2

how to build the number system

Back in 1880s Italy, a gentle, optimistic[1] man called Giuseppe put together a list of five rules. Although my account of this is entirely true and perfectly accurate, I shall translate his rules from the original Italian in my own particular way. The rules are about some things, which may or may not exist, which we'll call *squiffles*.

Rule A. There's a squiffle called "uug",

Rule B. If you've got a squiffle, then I can find a unique squiffle which is called the "ob-squiffle" of your squiffle.

Rule C. Uug is not the ob-squiffle of any squiffle.

Rule D. Different squiffles can't have the same ob-squiffle.

Rule E. Suppose you have a bucket of squiffles. If your bucket contains uug and it also contains the ob-squiffle of *every squiffle in the bucket*, then your bucket contains *all* possible squiffles.

So, what you could you deduce from these rules?

Rule A speaks for itself – there are some squiffles and one of them's called "uug". Uug will have to be our starting point, then.

Rule B tells us that uug, being a squiffle, must have a "unique squiffle" which is called its "ob-squiffle". This means there's just one "ob-squiffle of uug", not more.

Could uug be the ob-squiffle of itself? No, because Rule C says that uug cannot be the ob-squiffle of *any* squiffle, and that includes itself.

So, there's another squiffle, different from uug, which is uug's ob-squiffle. We'll call it "duug". So, we have uug and duug, with duug being the ob-squiffle of uug.

Rule B tells us that duug must also have a unique ob-squiffle. Itself? No, because it's the ob-squiffle of uug, so: if it were also the ob-squiffle of itself, Rule D would be violated. Think about it.

Could uug be duug's ob-squiffle? No, by rule C, it's not the ob-squiffle of any squiffle. So there must be *yet another* squiffle acting as duug's ob-squiffle. We'll call it "gluug" (we can think of duug as "ob-uug" and gluug as "ob-duug").

Continuing on like this, even if we run out of silly names, we can keep producing new squiffles, different from those we've already found, each one being the ob-squiffle of the last squiffle produced, and this can go on forever.

Translated slightly differently from the Italian, the list of rules looks like this:

Rule A. There's a counting number called "1".

Rule B. For every counting number, there's a unique counting number known as its *successor*.

Rule C. No counting number has 1 as its successor.

Rule D. Different counting numbers can't have the same successor.

Rule E. Suppose you have some selection of counting numbers that includes 1 and also has the property that *if a number is in the selection, then its successor is also in the selection*. Then your selection must contain *all* counting numbers.

In fact, the amazing infinite-squiffle-generating rules put together by Giuseppe Peano (for that was his name) are better known as the *Peano axioms* and they lie at the very foundation of mathematics, being the standard way in which mathematicians define the system of counting numbers!

We'll be looking at "axioms" in a more general context in Chapter 7. Until then, Peano's five rules will be the only ones we'll need to concern ourselves with. Although his list of axioms could be described as the "official" starting point for working with numbers, it's probably quite far removed from your early childhood number experiences. We'll return to these axioms shortly, but for now let's consider the process by which we came to accept the concept of numbers at an early age.

For almost all of us, our first experience of consciously dealing with numbers involved gradually learning to chant the sequence

"one, two, three, four, five, six, seven,..."

or the equivalent in some other language.

This chanting would sometimes be linked to our fingers or other collections of objects and sometimes not.

Next, we learnt to count to bigger numbers – first to 10, then perhaps to 20, 50 or 100. Children at this stage will sometimes wonder out loud or ask an adult "what's the biggest number?" in the same way that they might ask what the biggest animal or the highest mountain is. Growing older, they come to terms with the idea that the numbers go on forever and so realise that this question has no answer.

Some well-meaning but misguided adults confuse children with the answer that there *is* a biggest number and that it's called "infinity". "Infinity" is a useful *concept* in some areas of mathematics, but *it's not a number* in the sense that 7, 13, 2089 and 1 471 788 003 are numbers. You can (eventually, at least in theory) *count* to any of these numbers, but you can't ever (even in theory) count to infinity.

Numbers which can eventually be counted to – 1, 2, 3, 4, 5, 6, 7, ... – are called *natural numbers* by mathematicians. Notice that 0 is *not* included. You can't count to zero. Children don't learn to chant "zero, one, two, three, ..." as this would lead to confusion when they tried to count things. As with my second translation of Peano's rules, we're going to call 1, 2, 3, 4, ... *counting numbers*, which seems a bit clearer. Mathematicians also talk about *integers* – these include the counting numbers, but also 0, and *negative* numbers too. These will come up later, but don't panic if you're not comfortable with such ideas – all will be made clear! For now, we'll restrict our attention to the counting numbers (also known as the "positive integers" as well as the "natural numbers").

So, almost every adult in the modern Westernised world has accepted the fact that *the counting numbers go on forever*. Although we may realise that we'll eventually run out of word-based names for them, we're all quite sure that the numbers themselves will *just keep going*.

But this isn't something we can know by experience. You can walk the length of a river from its mouth to its source and, through this experience, come to know that it *doesn't* go on forever. But that something *does* go on forever – you can't ever come to know that by means of direct experience. It might seem like a silly question, but how *do* we know that the counting numbers go on forever?

It's obvious, but quite possibly, you've never given it much conscious thought.

Think about it for a minute.

The answer is most easily summarised like this:

There's always another one.

Give me any counting number, and I can add 1 to it. So, if you think you've got the biggest counting number, you're wrong. I can produce a bigger number by adding 1. And this means that you can't ever have a biggest counting number. There's always another one that's bigger. Easy.

This might seem so obvious to you that you feel it's stupid to have to read it in a book. But, in fact, the property...

given any counting number, you can add 1 to it to get another counting number

...turns out to be crucially important if you're going to try to describe the sequence of counting numbers 1, 2, 3,... as a whole – as a *number system* – in a completely thorough and logical way. Obviously *you can never actually list them all*, so if you do want to be completely thorough and logical about describing the whole lot of them in one go, then you'll have to come up with some clever way to deal with the fact that they go on forever.

The first person to do this was Giuseppe Peano, in the 1880s. Humans had been busily counting and using numbers in connection with agriculture, commerce, building and governance for thousands of years without any need for this, but by Peano's day, there was a widespread tendency well under way in the sciences (including mathematics) to try to pin down *everything* in the most logical and systematic way possible.

Peano came up with a foolproof "recipe" for the entire collection of counting numbers in the form of those five simple rules which I presented earlier and called the "Peano axioms" (they're also sometimes called the "Peano postulates").

So, it's time to forget everything you know about numbers, apart from the fact that:

There might be some things called "counting numbers"

and then to consider these rules again:

A. There's a counting number called "1".

B. For every counting number, there's a unique counting number known as its *successor*.

C. No counting number has 1 as its successor.

D. Different counting numbers can't have the same successor.

Let's try to imagine that we had never encountered the counting numbers before, yet were endowed with extremely logical minds. If we were presented with these four rules, what could we deduce?

Well, by Rule A, we know there's something called a "counting number" and it's called "1", so we ought to start there.

Next, we observe that by Rule B, every counting number has a unique counting number known as its "successor". Since we know that 1 is a counting number, there must be a unique counting number which we'll call "the successor of 1". We can be sure that this is not the same as 1 itself, because 1 cannot be the successor of any number (including itself) by Rule C. So we know that our counting numbers must include "1" and "the successor of 1".

As we know that this thing we're calling "the successor of 1" is a counting number, we can again use Rule B to deduce that it must have a unique counting number as its successor. This can't be 1, by Rule C. And it can't be itself, by Rule D ("the successor of 1" can't be the successor of any other counting number than 1). So we'll need a new counting number to act as successor of "the successor of 1".

We now have the counting numbers "1", "the successor of 1" and "the successor of the successor of 1". These names are quickly going to get quite cumbersome so let's introduce some shorthand. In place of "the successor of 1", we'll write "ob-1". Similarly, we'll represent "the successor of the successor of 1" (also known as "the successor of ob-1") as "ob-ob-1". The successor of that, of course, would be written "ob-ob-ob-1".

By continuing to apply Peano's rules in this way, we keep generating new counting numbers. In practice, we may run out of time, patience, brain-capacity or paper, but *in theory* there's no reason why this should ever stop. So we have a system for generating an endless sequence of counting numbers starting from 1.

This all suddenly seems a lot more familiar if we just think of the successor of a counting number as *that number plus 1*. Without yet concerning ourselves with the broader notion of *addition*, we can agree that a counting number with "+1" written after it is shorthand for the successor of that number, exactly the same as writing "ob-" before it. So "1 + 1" means "the successor of 1" (which we can call "2"), "2 + 1" means "the successor of 2" (which we can call "3"), and so on.

Rule C guarantees that 1 really is the beginning of the sequence, so we can't have this situation:

→ 1 → 2 → 3 → 4 → 5 → 6 → 7

Rule D means that every number except 1 must have a *unique* number immediately preceding it, which means that we can rule out building a system of counting numbers that looks like this...

...where we have a succession of "alien" numbers joining the usual sequence like the tributary of a river. Notice how Rules A, B and C still work, but Rule D is violated, because "8" does not have a unique counting number immediately before it – it has "7", and another number represented by what looks suspiciously like a "smiley face" symbol. "8" acts as the successor of both of these. Without Rule D, you could have countless "tributaries" joining the usual sequence of counting numbers in this way.

Notice that Rule B requires that successors are "unique", so we can similarly rule out situations like this...

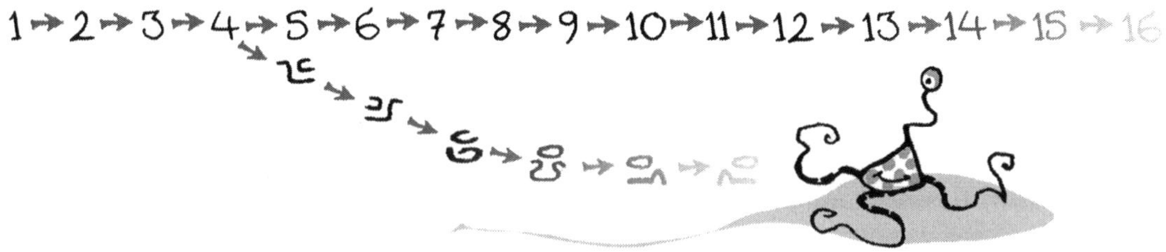

...where "4" has two different successors ("5", and an "alien" number) but all of the other rules are obeyed. Without this requirement of uniqueness in Rule B, you could have countless branches forking off from the usual sequence of counting numbers.

Rule A is necessary to avoid this…

…that is, there *not being any* counting numbers, which, if you think about it carefully, doesn't actually violate Rules B, C or D.

But wait – wasn't there also a Rule E – a longer, more complicated one?

There was. We needn't worry too much about it, but it's there to avoid situations like this:

It said…

E. Suppose you have some selection of counting numbers that includes 1 and also has the property that *if a number is in the selection, then its successor is also in the selection*. Then your selection must contain *all* counting numbers.

Notice how the situation in the last picture doesn't violate any of Rules A, B, C or D – every number has a unique successor (which isn't the successor of any other number) and there's a number called 1 which isn't the successor of any number – but the illustrated situation *does* violate Rule E. Why? Well, suppose that we took our "selection" to be just those counting numbers in the second line (the familiar-looking counting numbers). This selection *does* include 1 and *does* have the property that "if a number is in it, then so is its successor". But it *doesn't* contain *all* counting numbers, because there are those other supposed "counting numbers" on the other lines and loops which aren't included in it. Rule E is there to prevent the existence of hidden streams and loops of counting numbers which are missed by the familiar sequence 1, 2, 3, 4, 5, 6, 7,....

With his five simple rules then, Giuseppe Peano captured the entire collection of counting numbers[2].

If you didn't get that last bit about the fifth rule, try re-reading it a couple of times. If you still don't get it, don't worry, we're not actually going to *use* these rules. Although Peano's axioms effectively back up all the work they ever do, quite a few mathematicians would struggle to give an accurate account of them! They're usually discussed briefly during undergraduate studies and then put aside because, with one important exception[3], they're almost never really "used" in mathematical research. Only in the seriously deep but marginal study of *mathematical philosophy* (sometimes called the *foundations of mathematics*) are the axioms ever a direct concern. Here's a glimpse of what that area of study typically looks like:

Given a proof $\mathfrak{P}$ in $(\mathfrak{F})$ of $A(b)$, a proof in $(\mathfrak{F})$ of $A(0^{(n)})$ consists of $\mathfrak{P}$ with the formula $A(0^{(n)})$ attached; suppose this has the number $\lambda(n)$. Then

$$Prov\ [\lambda(n),\ \alpha(n)],$$

and hence from (1)

$$\rightharpoondown Prov\ \{\pi(n),\ e[\alpha(n)]\}.$$

By (iii)

$$A(n).$$

Note in passing that Ackermann[1] uses only transfinite induction *up* to ω_p $(\omega_1 = \omega,\ \omega_{n+1} = \omega^{\omega_n})$ to establish

$$P(m)\ \&\ P(p)\ . \rightarrow .\ \rightharpoondown Prov\ (m,\ n)\ \mathrm{v}\ \rightharpoondown Prov\ [p,\ e(n)],$$

where $P(m)$ is a (suitable) arithmetisation of the proposition: the proof with number m contains $< P$ quantifiers. Thus, given a proof in $\mathfrak{F}$ of $A(b)$ with $< P$ quantifiers, both $\lambda(n)$ and $\pi(n)$ are numbers of sequences of formulae with $< P$ quantifiers, and hence $A(n)$ can be proved by recursive arithmetic with transfinite induction up to ω_p only.

This is taken from Georg Kreisel's article "A variant to Hilbert's theory of the foundations of arithmetic", which appeared in a 1953 edition of *The British Journal for the Philosophy of Science*.

In that setting, Peano axioms would be presented something like this...

...but we won't have to concern ourselves with such (potentially terrifying) notation.

THE SIGNIFICANCE OF PEANO'S RULES

The significant point here is that everything in mathematics can eventually be traced back to the counting numbers and the counting numbers can be related back to the Peano axioms.

The Peano approach to the counting numbers is undeniably very thorough but it might also seem a bit *dull*. Apart from the different (but arbitrary) symbols we choose to represent them, every counting number seems to be just like all of the others (with the exception of 1, which is distinguished by that fact that there is no counting number preceding it).

Peano's system is rather like a "mass production" approach to the counting numbers. Apart from 1, which acts as a kind of starting point for the system, the same "manufacturing process" is used to produce all of the counting numbers: *just add 1 to the previous number*.

Surely, we can't expect anything mysterious or surprising, any "secrets of creation" to come out of this!

But an important conceptual leap has been made here. As mortal creatures, we humans can have no direct material experience of anything infinite or endless. Yet, in this abstract way, our minds are able to deal very precisely with the endlessness of the counting numbers. Despite the seeming dullness, Peano's apparently innocuous list of rules somehow has *the infinite* contained within it. All of the counting numbers produced are finite but *infinitely many of them can be produced*. Peano's list of axioms is like a small, innocent-looking box with an infinitely long spring somehow coiled up inside it – quite a surprise to anyone who opens it.

I've found that there's a lot of misunderstanding and confusion surrounding the concept of *infinity*. Certainly, only a tiny fraction of the population has a clear understanding of the theory behind this concept (which is usually passed over quickly in undergraduate maths study). At the same time though, the idea of there being an infinite sequence of counting numbers seems to sit quite comfortably in the Western mind, having achieved near universal acceptance and never seriously being called into question. As our practical, physical experience only ever involves finite numbers of things, it's perhaps worth paying attention to this.

PEANO'S RULES AND THE "NUMBER SYSTEM"

Besides the somewhat dull, repetitive Peano approach, there's a much more interesting way to construct the entire collection of counting numbers. We'll explore this in Chapter 4. You might be quite surprised at what strangeness has been lurking there, hidden within this number system which you've lived with all your life!

I've used this term *number system* a few times already. It will be helpful for you to start thinking in terms of a single unified "system" rather than just a collection of separate counting numbers. By "number system" I just mean the entire sequence of counting numbers *together with the usual rules for adding and multiplying them*[4]. We're going to treat this as a single entity, almost like an organism, and look at its properties – its *anatomy*, if you like.

I've claimed that all of mathematics is ultimately founded on the Peano axioms. As the number system we're going to be looking at only involves adding and multiplying, we'll now quickly consider how these can be related back to Peano's rules.

"$3+5$" can be interpreted in the following way. "3" is the successor of "2", so we can write it as "$2+1$". And since 2 is the successor of 1, or "$1+1$", we can think of "3", the "successor of the successor of 1" as "$1+1+1$". In the same way, we can think of "5" as "the successor of the successor of the successor of the successor of 1", or "$1+1+1+1+1$". If we now rewrite "$3+5$" with "$1+1+1$" in place of 3 and "$1+1+1+1+1$" in place of 5, it becomes

$$1+1+1+1+1+1+1+1,$$

that is, "the successor of the successor of the successor of the successor of the successor of the successor of the successor of 1". This is the counting number more commonly known as "8".

So, clearly, any addition of counting numbers can be broken down in terms of the Peano approach (which only deals with the adding of 1's).

Multiplication can be thought of in terms of addition: "5×7" just means "add five sevens together". It's shorthand for $7+7+7+7+7$, and each of those additions can be broken down into a series of acts of adding 1. So multiplication, indirectly[5], can also be related back to the Peano "just add 1" approach to the number system.

Both addition and multiplication are processes that take any two counting numbers and combine them to produce a counting number (the "answer" or "result"). Notice that subtraction and division aren't like that: although "$4-7$" and "$5\div 11$" *do* mean something mathematically, they *don't* produce counting numbers.

So, we have an infinite sequence of counting numbers built using Peano's rules. We can choose any pair of them and add or multiply them together to produce a counting number. That's it, basically – the "number system" which we'll be looking at in this book. We're all using it all the time and our potential to do so is one of very few things that we have in common with everyone else. Its "reality" is debated and its "location" is unclear, but when looking into its inner workings (quantitatively speaking) *everyone finds the same thing*.

Let's have a look then.

chapter 3

prime numbers

The number system which I've described can be imagined as something like this:

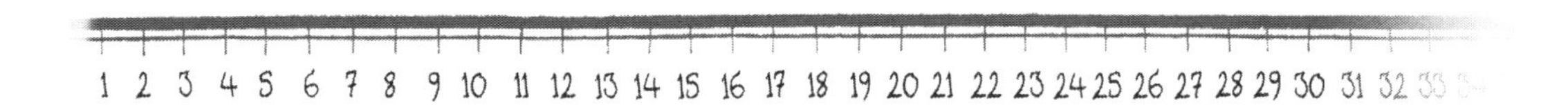

The "mass production"-style construction and apparent uniformity of the number system which we just saw bring to mind such mundane images as these:

The homogenous, uniform nature of the thing we've constructed might lead us to think that nothing particularly interesting or mysterious could possibly lie hidden within it. People have been known to wonder what it is exactly that mathematicians *do* (I've been asked this a few times) since surely there's nothing left to find out about *numbers*. Nothing could be further from the truth. Quite possibly the *most* interesting and mysterious thing you'll ever encounter with the rational part of your mind awaits your discovery. So, without further ado, let's take this number system apart and see what makes it tick.

We've got all these counting numbers and we know that we can take any pair and add them to produce another counting number. This can ultimately be related back to just adding 1's together. We can also take any pair of counting numbers and multiply them to produce a counting number. As we've seen, this is less directly connected to the "just add 1" approach but it can be related back to it by understanding multiplication as the adding of numbers together a certain number of times.

The leap from adding to multiplying (or, more technically speaking, from "addition-based thinking" to "multiplication-based thinking") is one of the most important things we'll consider. I'll not say too much about it now – I'll just make a couple of observations.

Before we introduce multiplication, we can describe amounts of "things" and add amounts of them together (by "combining piles").

By introducing multiplication, we make the sudden leap from describing amounts of "things" (three acorns, eight books) to describing amounts of *numbers* (3×7 is "three sevens", 8×12 is "eight twelves").

This transition adds a whole new layer of structure to the number system.

We're so used to multiplication that it seems perfectly "normal" or "natural", but it's quite a major (and far from obvious) innovation. It's very useful, of course: the busyness world would be completely lost without the ability to multiply – how do you find the total profit from 18 containers full of 2400 boxes, each box containing 12 packs of a product which sell at a profit of 18p each?

Usefulness is not something we're going to dwell on, though. We'll be more concerned with the fact that before you have multiplication, the number system is largely featureless [1], but once you introduce multiplication, *something truly weird starts to happen*. You may doubt this claim, or find it extravagant, but if you make it to the end of this volume, I can assure you that you no longer will.

DISSECTING SOME COUNTING NUMBERS

Let's pick a counting number of a manageable size, say 14.

If we were going to start from fundamentals and use Peano's approach to the number system, we would think of 14 as

$$1+1+1+1+1+1+1+1+1+1+1+1+1+1$$

or "the result of starting with 1 and adding 1 to it thirteen times" (thirteen, in turn, is defined in terms of twelve, and so on, all the way back to $2 = 1 + 1$ or "2 is the successor of 1").

If you think about it, Peano's recipe for 14 actually has fourteen steps and goes something like this:

- ☆ start with the number 1,
- ☆ add 1 to it,
- ☆ add 1 again,
- ☆ add another 1,
- ☆ and another 1,
- ☆ one more,
- ☆ another one,
- ☆ once again,
- ☆ once more,
- ☆ add 1 to that,
- ☆ add 1 again,
- ☆ one more 1,
- ☆ and another 1,
- ☆ now add a final 1.

That's about the most interesting way you could describe the rather boring, "mass-production", Peano-inspired way of "building" the number 14.

Let's put this into practice. Suppose we now sit down at a table with a bag of dried beans and place a single bean on the table to represent "1". By taking a single bean at a time and adding it to the pile, the size of the pile will eventually reach 14 (or any other counting number anyone cares to name, within the obvious bounds of patience, mortality, beans available, the size of our table, the size of the universe, and so on).

So we can make a pile of fourteen beans. You may well be unimpressed.

14 beans

Look at the pile. That's a pretty straightforward representation of the counting number we call "14", isn't it? It's about as clear as you can get. Anyone who deals with numbers should be able to understand it.

We're going to try to pull the idea of "14" apart, and see if it's got any internal structure or "anatomy" more interesting than $1+1+1+1+1+1+1+1+1+1+1+1+1+1$.

Look at the pile again. Apart from the trivial facts that it's a pile and that it's made of beans, what can we say about it? How does the "14-ness" come into it?

What could we say about the pile that's purely about the "14-ness" and unrelated to the "bean-ness"? That is, what would still be true if the beans were all suddenly turned into feathers, matchsticks, meteorites, dragonflies or transistors?

We could consider the "size" of 14 in relation to the other counting numbers. Obviously, if we add more beans, we'll have a bigger pile, and if we take beans away we'll have a smaller pile. So 14 is a "smaller" counting number than 15, 16, 17,… but a "larger" one than 13, 12, 11,…

That's all very well, but it's not exactly a revelation. Every counting number (except 1) is bigger than some counting numbers and smaller than some others. How about something a bit more particular to "14"?

One obvious thing (at least to some people) which we can try is splitting our pile into a number of smaller, equal-sized piles.

The pile splits into two piles of seven beans...

...or seven piles of two:

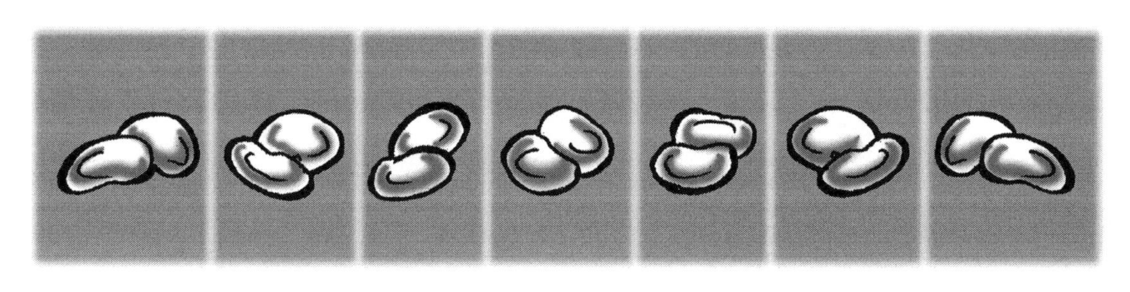

"Well, yes," you may be thinking, "seven is half of fourteen – but that's hardly a profound insight."

Nevertheless, we can note down the fact that $14 = 2 \times 7$. And that $14 = 7 \times 2$. Notice how we need to use *multiplication* to express this "equal piles" stuff.

Let's try this with some other pile sizes. Add another bean to our pile of fourteen.

We call this pile "15". How can we split it?

"Easy," you say, "Three piles of five, or five piles of three."

We note down that $15 = 3 \times 5$ and that $15 = 5 \times 3$.

Add another bean. Sixteen. Again, it's quite obvious. Four piles of four.

We note down the fact that $16 = 4 \times 4$. But we can also see that the pile could be split differently, to get $16 = 2 \times 8$ or $16 = 8 \times 2$.

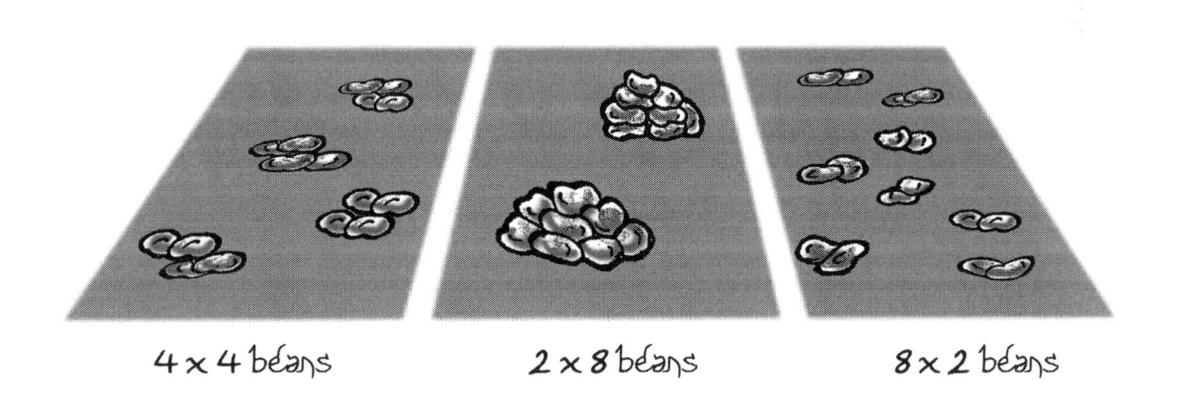

At this point, it would be useful for you, the reader, to go and get a handful of beans, pebbles, acorns, buttons, pennies or whatever small numerous objects you happen to have around you and actually *try this*. In my experience, the process of discovery seems to sink in deeper if you are physically experimenting with objects rather than just reading from a page.

So we add another bean to get a pile of seventeen. Now try. I guarantee that however long and hard you try, you won't manage to separate a pile of seventeen beans into several equal piles. (Cutting beans in half isn't allowed!)

"Seventeen piles of one!" you may cleverly argue, but *any* pile can be divided into "piles of one". We can't allow that or we'd really be wasting our time – we'll certainly not discover anything of interest. In this case, we would just be saying that $17 = 1+1+1+1+1+1+1+1+1+1+1+1+1+1+1+1+1$ and we're back to that dreary, repetitive "just add 1" approach to the number system.

To disallow this, we'll have to agree to the quite sensible view that a single bean cannot be considered a "pile of beans". To qualify as a "pile" there must be more than one bean involved.

So 17 stands out from all the numbers that we've looked at until now.

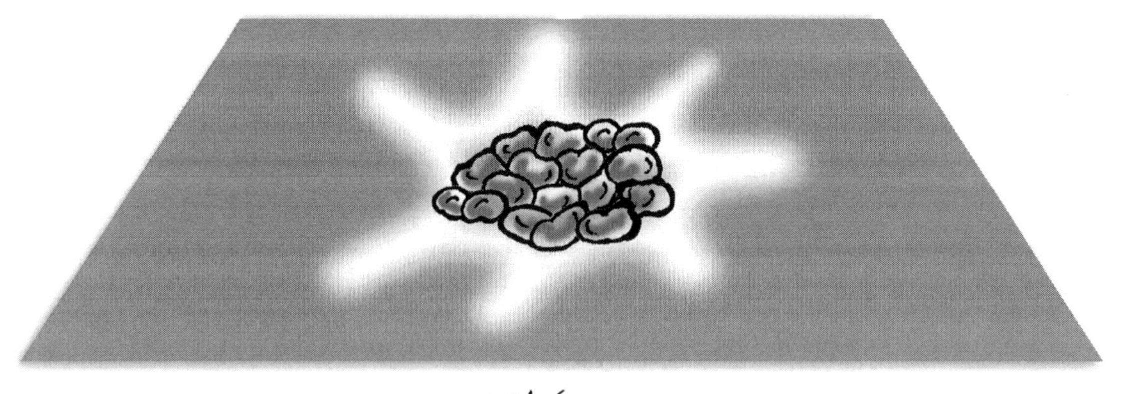

17 beans

Continuing on in this way...

Looking at 18, we find that it can split as three piles of six (3 × 6) or two piles of nine (2 × 9).

19, on the other hand, refuses to be split.

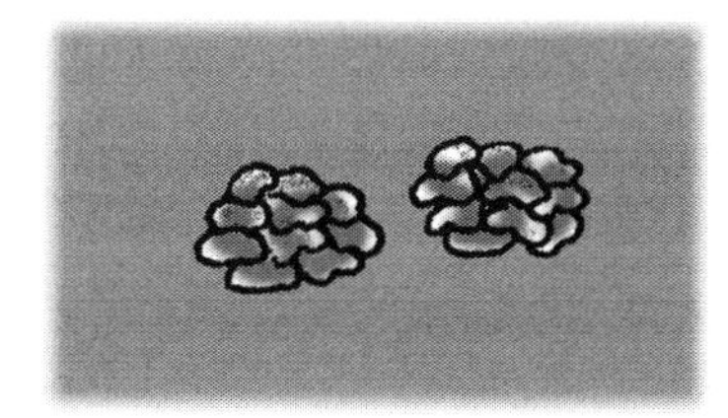

20 can be split as 2 × 10 or 4 × 5,

21 can be split as 3 × 7,

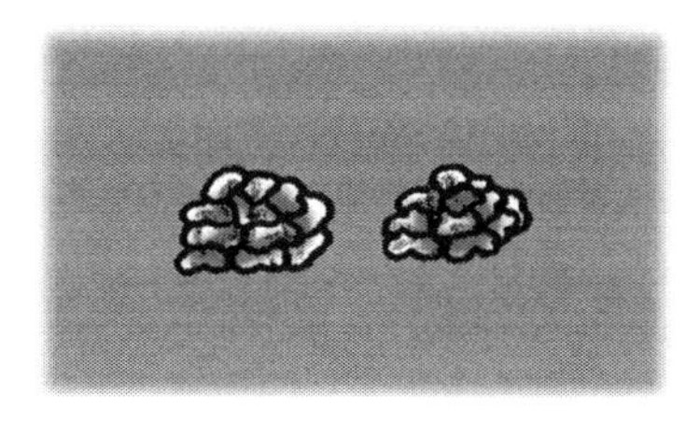

22 as 2 × 11.

23 will not split.

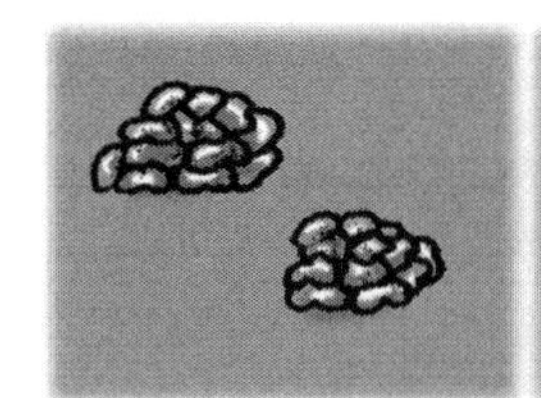
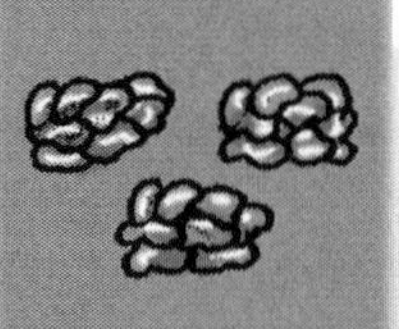
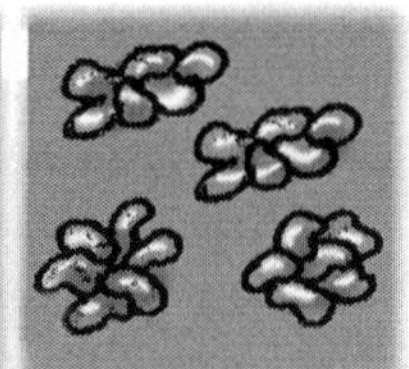

24 splits in a few ways:
2×12, 3×8, 4×6.

25 splits as 5×5,

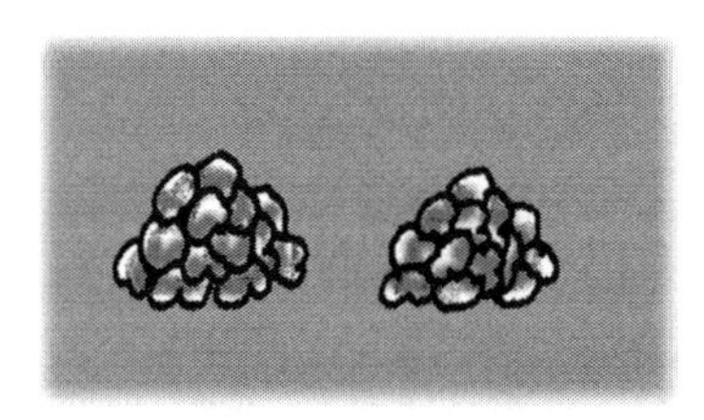

26 as 2×13,

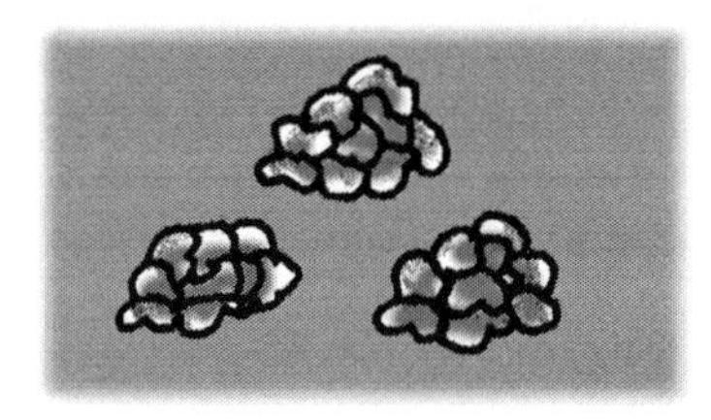

27 as 3×9,

28 as 2×14 or 4×7.

29 does not split.

It looks like we've found the beginnings of something interesting, or at least *not entirely obvious*, going on in the number system.

Clearly, we could continue on down the "number line" taking note of which counting numbers can be split into equal piles and which can't.

At this point, we can see that there's something *irregular* going on. And if we didn't have multiplication (the idea of describing amounts of numbers, along with all the other things we can describe amounts of), then, in some sense, this irregular phenomenon *wouldn't be there for us* – keep this in mind.

"PARTITIONING" COUNTING NUMBERS

It's a separate matter, but there is an irregular pattern which is noticeable even *before* multiplication gets involved. To see this, divide your piles of beans into piles of *any size you like* (including 1). In how many ways can you do this?

Here's an example: 6 splits as 1 + 1 + 1 + 1 + 1 + 1, 1 + 1 + 1 + 1 + 2, 1 + 1 + 2 + 2, 2 + 2 + 2, 1 + 2 + 3, 1 + 1 + 1 + 3, 3 + 3, 1 + 1 + 4, 2 + 4, 1 + 5 and 6 (just leaving the pile alone also counts as a "splitting" in this scheme). That's a total of eleven ways. Notice how we could reorder, but 1 + 1 + 4, 1 + 4 + 1 and 4 + 1 + 1 are all considered to be the same "splitting". Looking at the first few counting numbers, we find this...

size of pile	1	2	3	4	5	6	7	8	9	10	11	12	...
number of splittings	1	2	3	5	7	11	15	22	30	42	56	77	...

...and, a bit farther down the line,

size of pile	...	31	32	33	34	35	36
number of splittings	...	6842	8349	10143	12310	14883	17977

size of pile	37	38	39	40	41	42	...
number of splittings	21637	26015	31185	37338	44583	53174	...

Finding a formula for this turns out to be a ridiculously difficult task[2]. The irregularity we see here, like the irregularity which appears when you count equal-sized piles, results from counting, but not counting *things* – we're not applying the number system to the world around us, we're *turning it on itself*. "5 × 7" is "five sevens" so we're describing an amount of sevens – we're describing amounts of *numbers*. The "11" below the "6" in the table above is the result of counting not numbers but *ways of combining numbers* to get a certain total (6), so we're even further removed from the simple counting of objects.

We'll now have to put aside this intriguing issue of "additive partitions", as number theorists call them, and return to the question of splittable and unsplittable numbers in the sequence of counting numbers.

HISTORICAL AWARENESS

There's no way we can know how long ago people first paid attention to this issue of "splittability". It could well have been long before anyone started writing things down. The *Ishango bone*, found in central Africa in 1960 and currently believed to be at least 20 000 years old, has groups of notches carved into it in three rows. The numbers in one of these rows are 11, 13, 17 and 19 – all the unsplittable numbers between 10 and 20. Of course, it's impossible to know whether or not this was intended, but the arrangements of notches in the other two rows do suggest some kind of wider, but unclear, number-related scheme, providing evidence in favour of that possibility. The bone is an interesting discovery but unfortunately doesn't lead us to any definite conclusions.

It's something any culture on Earth *could* have noticed [3]. As soon as anyone starts counting and then dividing things into equal piles (which, if you think about it, is the reverse of multiplication), they could potentially notice the puzzlingly irregular sequence of unsplittable numbers. But written records displaying an awareness of this are surprisingly limited. There's indirect evidence of a limited awareness in ancient Egypt [4], China and India from writings which have survived, but it's all rather patchy. The earliest systematic study of this phenomenon of which we have any record was carried out in the ancient Greek world around 300 BCE.

Beyond this lack of clear evidence from early literate civilisations, it's hard to know who noticed these numbers, as the vast majority of the history of human thought is unrecorded. We can't know, but it does seem plausible that no one systematically examined this phenomenon before the ancient Greeks. It's plausible (there being no evidence to the contrary), but it would also be *surprising* since it seems such an obvious thing to notice. Wouldn't someone, in all those millennia, have picked up on it?

Perhaps earlier cultures did notice this irregularity in the counting numbers (as far back as 20 000 years ago). We'll probably never know. But if they didn't, as the almost total lack of evidence suggests, then this would be very interesting. We know that people worldwide, for millennia, have noticed all kinds of intricate patterns in the night sky, the motions of sun and moon, the weather, the development and behaviour of animals and plants, *etc*. So why not in this (universally accessible) number system? It might suggest that what we see as "obvious" may not be as obvious as we think. In Western culture, the way we currently think about the number system, like so much of the framework within which we think and through which we relate to the world, has its roots firmly in ancient Greece. So, if looking at which numbers can be divided into equal amounts seems like an "obvious", "natural" or "normal" thing to do, then we should remember that our thought patterns have been shaped by that curious little culture which stood out from the rest of humanity by its tendency to explore these kinds of things in systematic detail.

Anyway, that's enough vague, pseudo-academic speculation – let's get back to the indisputable stuff.

PRIME NUMBERS DEFINED (AT LAST)

As you may have guessed, these *indivisible* or "unsplittable" numbers – 17, 19, 23, 29, *etc*. – are what are known as *prime numbers*. Working backwards, we can use the same pile-splitting test to find that 13, 11, 7, 5, 3 and 2 are all prime numbers. Working forwards, we find 31, 37, 41, 43, 47, 53, 59, 61, 67, 71, 73, 79, 83, 89, 97, 101, 103, 107, 109, 113, 127, 131, 137, 139, 149, 151, 157, 163, 167, 173, 179,...

There's a good chance that you learnt about prime numbers at school (the UK National Curriculum currently introduces the idea to 10- or 11-year-olds, although much younger children can often grasp the idea). The definition you would have been given probably didn't involve piles of beans. Many people I talk to can remember (and repeat somewhat mechanically) that a prime number is

a number which can only be divided by 1 and itself

but then they usually have to stop and think what that actually means.

Seventeen is prime because a pile of seventeen beans can only be "divided" into one pile of seventeen or seventeen "piles of one". A number which isn't prime isn't prime because it can be divided by something other than 1 and itself.

It's a perfectly acceptable definition then, but obviously not one which easily fires the imagination of children or of anyone who isn't already inclined towards thinking about mathematical ideas.

There are a few pleasantly visual ways of getting the idea across aside from the "dividing into piles" approach which we've seen.

The first one we'll look at is incredibly simple: a prime number is a number of beans which you can't arrange in a rectangular grid (where the words "rectangular grid" are being used in such a way that *a single row doesn't qualify*).

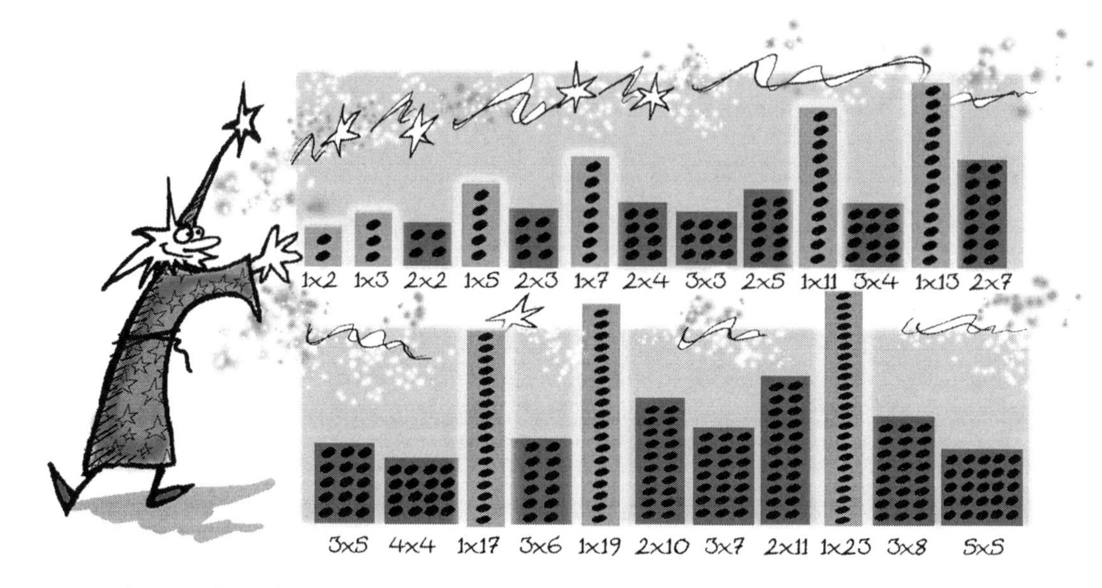

Here, the numbers from 2 to 25 are separated into those which can be arranged into a rectangular grid and those which cannot – prime numbers correspond to the lighter blocks.

This way of thinking about prime numbers features in *Uncle Petros and Goldbach's Conjecture,* a novel about an elderly disillusioned Greek mathematician and his nephew, by Apostolos Doxiadis [5].

The second approach is a visualisation which might allow you to get a better "spatial" sense of the prime numbers as they fall along the number line. Imagine that you are standing on the first plank of an infinitely long rope bridge with wooden planks across it, labelled with the counting numbers. You're standing on plank 1 and your task is to keep jumping to the next available plank. But each time you do so, it will knock out some of the planks which lie ahead of you.

It works like this. You jump onto plank 2, and that sets up a sort of wave which knocks out every second plank after it (4, 6, 8, 10, 12,...) as shown:

But plank 3 is still there, so you jump onto that. By doing so, you create another wave which passes through the position of every third plank after it (6, 9, 12, 15,…), knocking out any that are left (9, 15, 21,…), as shown:

You're now standing on plank 3, and plank 4 is gone. So you have to jump a bit farther this time, to plank 5. The wave that's created then passes through the position of every fifth plank (10, 15, 20, 25,...), knocking out any that are left (25, 35, 55, 65,...).

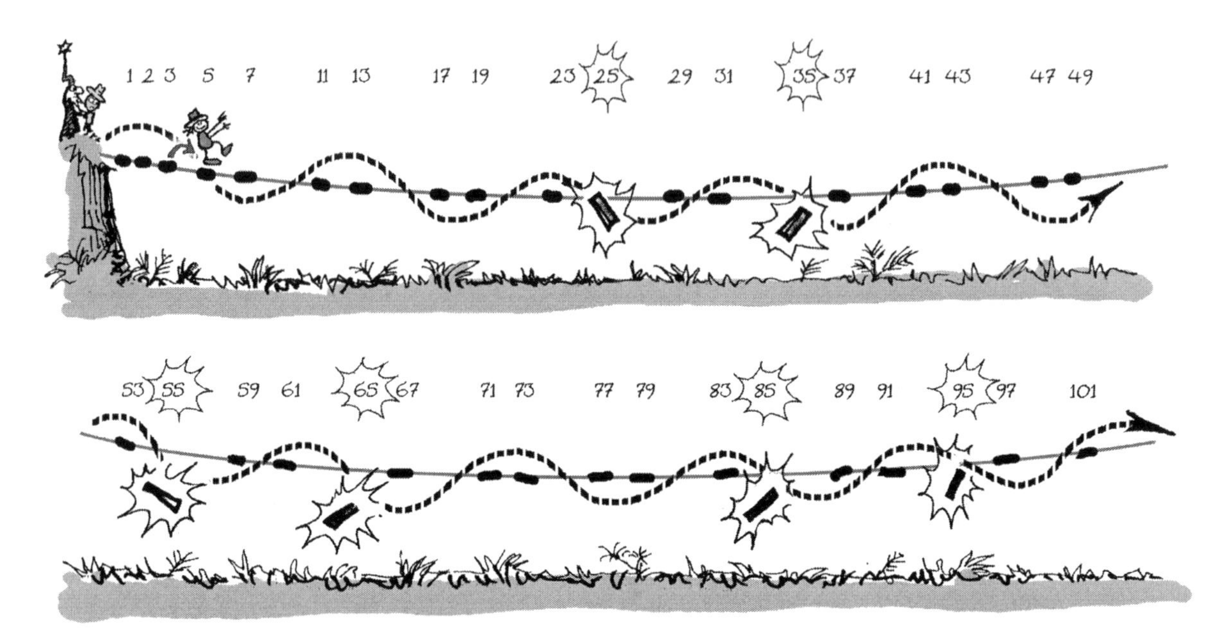

Now you're on plank 5, and plank 6 is gone. Another short hop takes you to plank 7, sending a wave through the position of every seventh plank (many of them now gone as the result of earlier waves – 14, 21, 28, 35, 42,...), taking out 49, 77, 91,...

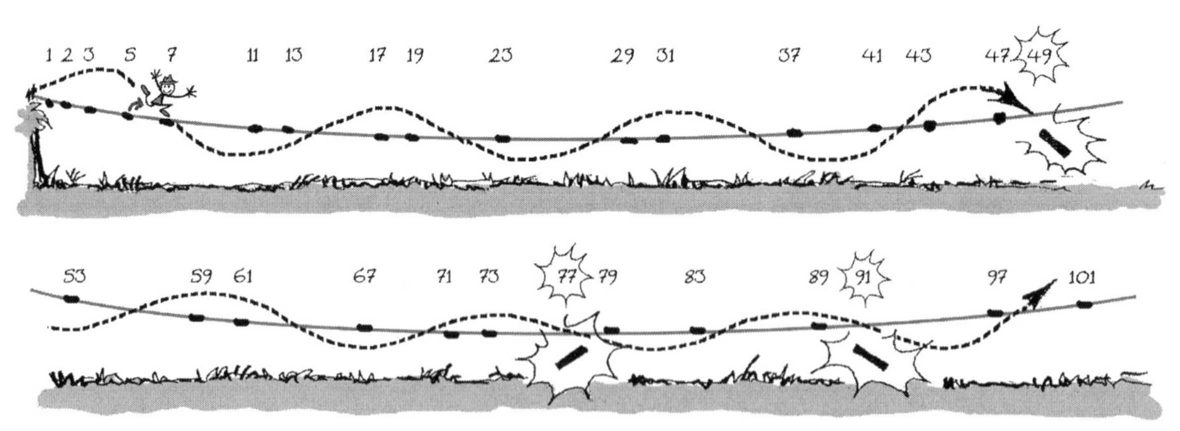

Planks 14, 28, 42, 56, 70, 84 and 98 were all taken out by the wave caused by jumping on plank 2, while planks 21, 35 and 63 were similarly taken out by the wave caused by jumping on plank 3.

Planks 8, 9 and 10 are all gone, so this time you have to take a bigger leap to plank 11, which will set up a wave through 22, 33, 44, 55, 66, *etc.* – those first few are already gone, but the wave will eventually take out 121, 143, 187.

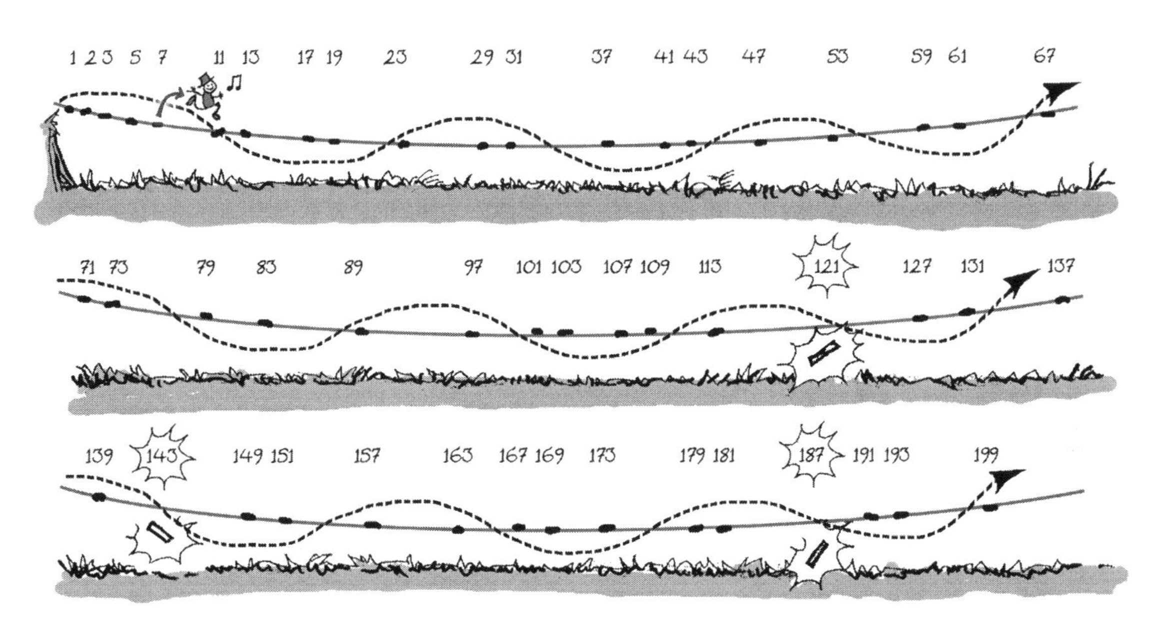

Planks 22, 44, 66 and 88 were all taken out by the "2" wave, while planks 33 and 99 were taken out by the "3" wave, planks 55 and 110 by the "5" wave and plank 77 by the "11" wave.

Plank 12 is gone, but 13 is still there, so there's an easy jump from 11 to 13. From 13, you must jump to 17 (14, 15 and 16 all now being gone). From there you jump to 19, then on to 23 from which you must make your biggest jump yet, to 29,...

It turns out that every jump you make will knock out *infinitely many planks* farther down the line. So the farther we go, the more "thinned out", the fewer and farther between, we can expect the planks to become. Clearly, we'll have to make some considerably bigger jumps as we carry on along the bridge. Some *very* big jumps, as it happens – as big as you care to name.

Notice that all of the planks you jump onto are numbered with prime numbers. The complete collection of prime numbers would perfectly match the numbers of those planks that would be left on the bridge "at the end of time" (that's only meant in a vague and playful sense, this being a fanciful visualisation). The variation in the gaps between these planks – that is, the distances you have to jump – directly illustrates the "irregularity" of the prime numbers to which I referred earlier.

The visualisation we've just seen is based on something called the *Sieve of Eratosthenes*. Eratosthenes was from what is now Libya and worked in the 3rd century BCE as librarian of the Library of Alexandria, then the world's greatest centre of learning. His "sieve" is the standard way of systematically producing a list of prime numbers, still taught to schoolchildren (in the somewhat less picturesque form of crossing out every second, every third, every fourth, every fifth,... number in a list). A computer can quickly produce an extensive list of prime numbers with this method, although even the fastest computer is soon stretched to its computational limits once the numbers reach a certain size. More specialised and efficient forms of "sieving" have been developed, allowing computers to find ever larger prime numbers. In fact, there's a whole sub-branch of number theory called *sieve theory*.

The final method also involves visualising a series of actions. We imagine an "elastic line" attached to a "rigid line". The rigid line is marked with numbers and joined to the elastic line at its "0" point, as shown opposite. On the elastic line, 1 is marked with a white dot, whereas 2, 3, 4, 5,... and all the other counting numbers are marked with black dots. We're going to stretch this systematically so that the white dot ends up at 2, then at 3, then at 4, *etc*. At each stage, we'll print our "stretched" elastic number line onto the rigid one (the black dots leave imprints, whereas the white "1" dot doesn't). So we'll end up with an overlay of all possible "counting number stretches" or "magnifications" of the number line (where 1 is excluded). If our white dot was instead black, then we'd get the whole sequence of counting numbers, except 1. By excluding it, you produce a "reversed image" of the sequence of prime numbers: they're the gaps that are left after all of the dots are printed.

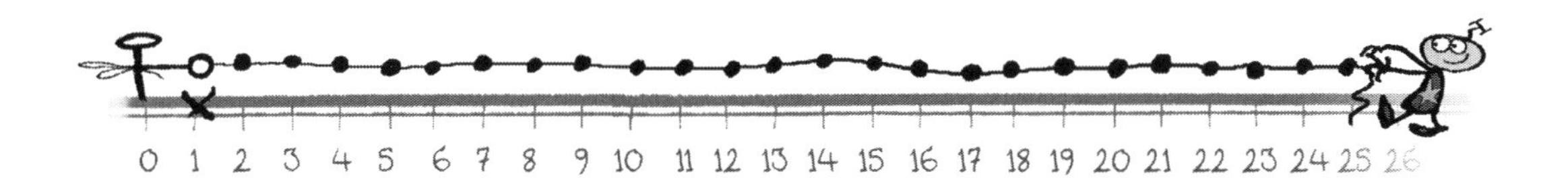

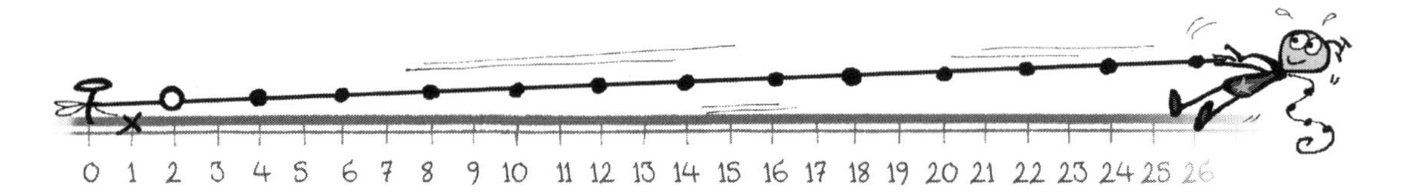

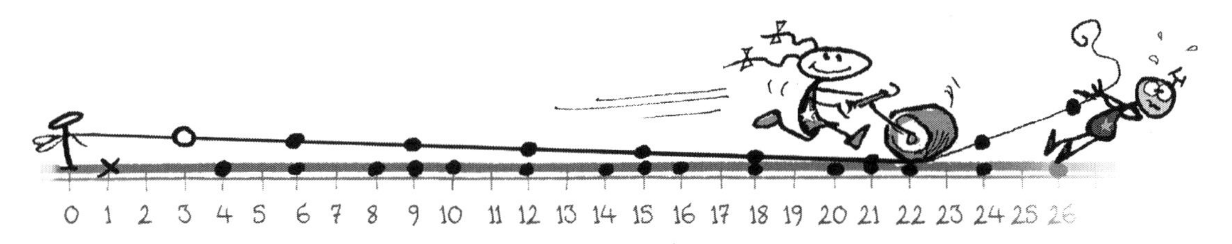

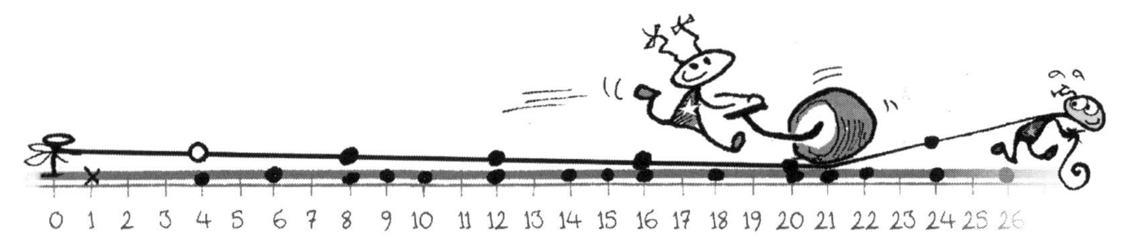

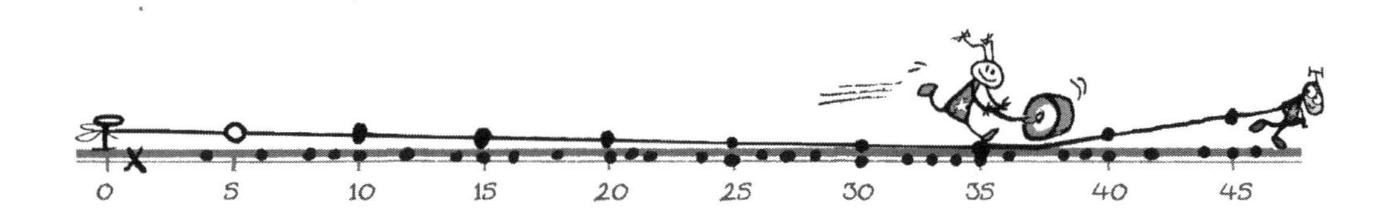

Notice that in the "4" stretch (second from the bottom), no new black dots are printed on the line as the "2" stretch has already taken care of all even numbers. As a rule, no new dots will be printed when the stretch takes the white dot to a number which has already been printed.

The prime numbers are what are left when you superimpose all possible "counting number stretches" of the counting numbers (2×, 3×, 4×, 5×,...). As with the discussion of multiplication earlier, we are here looking at what happens when you *turn the number system on itself*.

You can easily communicate the idea of prime numbers to young children or to people with whom you share no common language.

Imagine sitting at a long table next to someone who has no previous experience of prime numbers and no language in common with you. You have a sack of beans by your side, a knife and a pen. You proceed to place one bean on the table, and then carve a notch into the edge of the table, writing "I" above it. You pause, then place a second bean and carve a second notch to the right of the first. Momentarily contemplating the pair of beans, you mark an "X" above the new notch. You place a third bean, carve a third notch and mark an "X" above it. The person beside you looks puzzled, not understanding what you're doing.

Now, having placed a fourth bean and carved a fourth notch, you manually demonstrate that your pile of four divides into two piles of two and proceed to mark an "O" above the notch. The fifth bean produces an X, whereas the sixth produces a pile which subdivides into a pair of threes, so that notch gets an O. The seventh bean produces an X, whereas the eighth, ninth and tenth all produce O's.

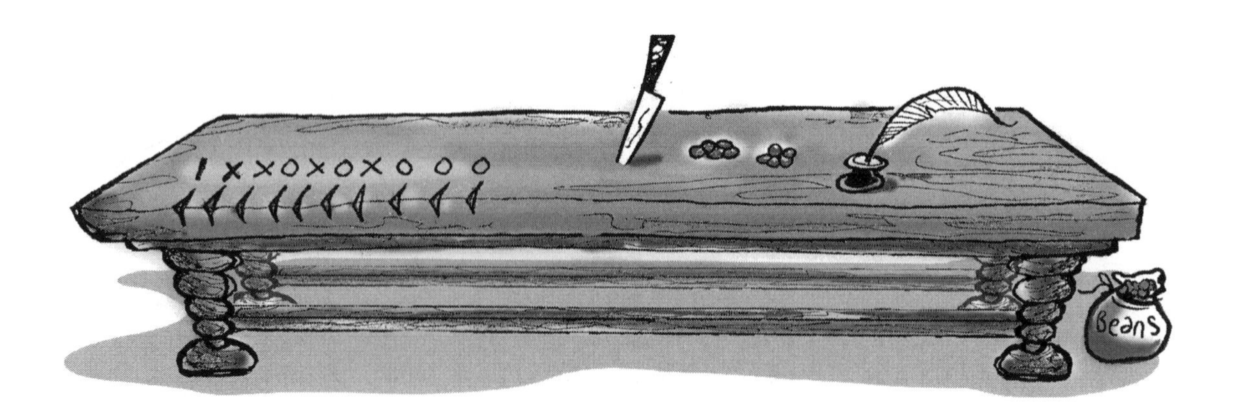

Soon enough, the person beside you recognises that an X means the pile of beans in question can be subdivided into equal piles. Although they need not have any further grasp of the "prime number" concept, they could eventually take over from you, continuing from where you left off. If they were patient and thorough enough, they could correctly mark the prime numbers out along a number line for as far as they were prepared to go. The irregular pattern made by the prime numbers would become evident and, though they may have no idea what exactly they were doing, or why, they'd instinctively know that someone else sitting at another table doing the same thing with a sack of beans (or anything else) should produce exactly the same pattern of O's and X's. When confronted with bean piles of the appropriate sizes, the facts that 17 is a prime number and 18 is not become truly indisputable!

At this point, the person has, at some level (if not fully consciously) realised the curious but indisputable fact that, etched into our number system, by the very nature of reality, *is something highly irregular*.

THE STATUS OF 1 AND 2

Prime numbers are often just called "primes", so from now on I'll generally call them by this abbreviated name.

All primes bigger than 2 are odd which means that 2 is the only even prime. People sometimes dwell on this fact, but there's really nothing remarkable about it. We just happen to have a word in our language ("even") which, when applied to a counting number, means *can be divided into two equal amounts to produce another counting number*. All even numbers larger than 2 (that is, 4, 6, 8, 10, 12,...) can be divided by 2 – *that's what makes them even!* And because they're divisible by 2 – you can divide them into two equal piles – they can't be prime numbers. 2 is even, but it's divisible into two equal "piles of one" and you'll remember that this doesn't count, so 2 *is* prime.

If we invented the word *treven* to mean "can be divided into 3 equal amounts to produce another counting number", then 3, 6, 9, 12, 15,... would all be "treven numbers" and 3 would be the only "treven prime". For some reason, though, we have no such word. But there's nothing to stop us inventing similar words for 5, 7, 11 or any of the other primes.

So there's really nothing particularly noteworthy about the fact that 2 is the only even prime. More relevant are the facts that it's the only prime whose successor is also a prime and that it's the smallest prime.

You may have noticed that I haven't listed 1 as a prime number and have avoided the issue of whether it is or not. This continues to be a minor source of confusion for people learning about the primes. Certainly, it satisfies the definition quoted earlier ("can only be divided by 1 and itself"). Also, you can't make a "rectangular grid" from a single bean. And in our "elastic number line" visualisation, it seems like 1 belongs with the primes (remaining "unprinted" on the rigid line). But the worldwide community of mathematicians is now of the opinion that 1 should *not* be considered a prime number, stating the definition as

a counting number larger than 1 which can only be divided by 1 or itself,

which is just a bit more precise than the oft-remembered-from-schooldays definition "a counting number which can only be divided by 1 or itself". The former definition explicitly excludes 1 from the list of primes. Another definition you sometimes see is

a counting number with exactly two divisors.

A *divisor* is a counting number which the original counting number can be divided equally into piles of. So 5 is a divisor of 15. So is 3. And 1 and 15 are, too. So 15 has four divisors. 4 has three: 1, 2, and 4. All of the primes have just themselves and 1 as divisors. *But 1 has only one divisor* (itself), so this definition cleverly excludes 1.

For centuries, 1 *was* considered by mathematicians to be prime. Then, towards the end of the 19th century, things started to change. Having hugely expanded their understanding of the number system, many good reasons emerged why we should not consider 1 to be in the same category as 2, 3, 5, 7, 11, 13,... We'll be exploring one of these later. Despite this, for some reason, some people feel driven to dispute the assertion that "1 is not a prime number". These tend to be mathematical enthusiasts presenting elaborate arguments on webpages rather than professional mathematicians contributing to the academic literature. I've even seen one argument invoking the Old Testament to "prove" that 1 is a prime! [6] But, ultimately, like arguing about whether or not something is "art", *it comes down to the definition of a word*. And whether 1 is considered prime or not won't make that much difference to us in much of this book anyway.

TRANSCENDING CULTURAL DIFFERENCES

The explanation involving piles of beans should clear up another source of confusion. That is, *whether or not a number is prime has nothing to do with the name we give it*. Nineteen is a prime number whether we write it as 19 (Arabic-derived Western digits), XIX (Roman), 𝋳 (Mayan), 十九 (Chinese), ||||||||| ∩ (ancient Egyptian), ԺԹ (Armenian) or 00010011 (binary). Changing your symbols is clearly not going to help you to divide a pile of nineteen beans into equal-sized piles.

Prime numbers transcend all cultural differences. A pile of beans is a pile of beans and *it consists of a fixed number* of beans, no matter where you come from or how you write this number down (if indeed you do). Whether or not the pile splits into equal amounts is unaffected by such things. *The prime numbers are the same for everyone.*

We're starting to get a glimpse at the reasons why Don Zagier, a senior mathematician at the Max Planck Institute in Bonn, would declare in a high-profile public lecture:

> *"...there is no apparent reason why one number is prime and another not. To the contrary, upon looking at these numbers one has the feeling of being in the presence of one of the inexplicable secrets of creation."* [7]

Perhaps to you, "inexplicable secrets of creation" sounds a little bit grandiose for something so easily defined and seemingly inconsequential as prime numbers. But Zagier's remarks are backed up by his extremely deep knowledge of the number system. As we proceed, I hope that you'll come to see what he's getting at.

There are two immediate observations I'd like to make, even if they seem obvious.

- ☆ Because the facts of whether or not a number is prime leave no space for speculation or opinion, this is one thing which *no one can seriously argue about*.
- ☆ The fact that a particular number is prime (or not) cannot change, so the prime numbers have always been the same and will always be the same.

In 1940, the Cambridge University number theorist Godfrey Hardy wrote

> *"317 is a prime, not because we think so, or because our minds are shaped in one way rather than another, but because it is so, because mathematical reality is built that way."* [8]

Or, in the blunter language of late 20th century North America [9],

It's like THAT, and that's the WAY IT IS.

Whether a particular number is prime or not is also independent of our feelings towards the number, of course. Many people around the world consider 7 to be a favourable or lucky number and in Western culture 13 is widely felt to be unfavourable or unlucky. This is an interesting phenomenon, but it's not something "absolute". Someone may genuinely feel the number 7 to be unfavourable or unlucky and 13 to be favourable or lucky, and these feelings would be entirely valid for them. On the other hand, 7 and 13 are prime numbers *for everyone*, regardless of what anyone might feel about them. This requires no faith or specialised knowledge and *it is as certain as anything is ever going to be*.

QUESTIONS OF "BASE TEN" AND NEGATIVE NUMBERS

You may be familiar with the fact that the way we usually write numbers is known as *base ten* and that it's equally possible to write numbers in other "bases". This relates to the fact that we use a system of ten digits (0, 1, 2, 3, 4, 5, 6, 7, 8, 9) , but systems involving more or fewer digits can also be used. People sometimes ask if you'll get different prime numbers when using a different base. If you have no idea what that question means, and don't want to know, then just keep reading – you won't miss anything important. If you *are* interested, an explanation of what the question means, and why the answer is an emphatic "no", can be found in Appendix 1.

A similar question that is often asked concerns *negative* prime numbers: are there primes among the negative numbers and do they differ in their "spatial arrangement" from the usual "positive" primes? You may not be familiar or comfortable with the idea of negative numbers, in which case that's not a question you're likely to ask, so you can rest assured that the answer is "no, there's nothing interesting to be found here". But we'll need negative numbers later on and they're really not that difficult to understand, so we'll cover the issue now.

Although they're *not* counting numbers, we can think of negative numbers as living on the "number line" too, as a sort of mirror image of the (positive) counting numbers, like this (0, which is also not a counting number, being the "mirror"):

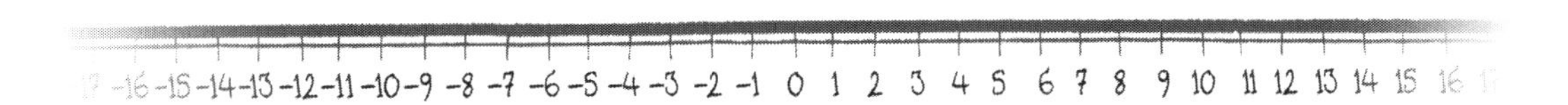

If you want to understand a bit more about negative numbers, then (unfortunately) it's probably easiest to think about *debt*. If a group of people all marked points on a number line to show how much money they had (in some agreed currency), then, clearly, the richer people would mark their points farther to the right. The farther to the left you go, the poorer you are. If you're completely broke, you mark your point at zero. Now, what's poorer than being broke? Being in debt, of course. If you're in debt, it makes sense that your point should be marked to the left of zero. The more in debt you are, the farther to the left of zero your point will be. People gaining money will move their points to the right by the appropriate amount (in some cases going from being in debt to being in credit, or at least being less in debt). People losing money will move their points to left by the appropriate amount (in some cases going into debt, becoming increasingly indebted or at least becoming less wealthy). In this context, the idea of negative numbers and their placement on the number line should become clearer.

So we have a number line, with zero acting as a kind of balance point, positive numbers stretching off in one direction and negative numbers stretching off into the other.

It's hard to imagine making a pile from a negative number of beans and dividing it up into equal parts. But we can multiply negative numbers easily enough. Having a sixty Euro debt can be thought of as "having −60 Euros". Three 60 Euro debts = 3 × (−60) = −180 = one 180 Euro debt. So it's straightforward to think in terms of dividing negative numbers into equal amounts (imagine paying off a debt in monthly instalments).

The definitions quoted on page 74 (the ones accepted across the whole mathematical community) are such that a prime number must be positive, so from this point of view, negative primes are not an issue – they're ruled out by definition.

We could break with convention, remove this constraint and see what happens when we look at negative numbers, but we find nothing interesting. Like 21, the negative number -21 would not be prime, because it can be divided into three, that is, $-21 = -7 \times 3$. Similarly, like 37, -37 *would* be prime because it can't be divided, except by itself, 37, 1 and -1[10].

This all means that if we were to allow negative primes, they would just be a mirror image of the positive ones.

No new insight is gained. There's nothing of interest here. It's really just a matter of definition and, for our purposes, it's more appropriate to limit the idea of "prime" to the positive numbers, so we'll stick with that. After all, it's the system of counting numbers which, through the introduction of multiplication, gave birth to the primes.

PROCEEDING WITH CAUTION

> "*Prime numbers are the most basic objects in mathematics. They also are among the most mysterious, for after centuries of study, the structure of the set of prime numbers is still not well understood.*" (A. Granville and A. Jackson[11])

It's hard to imagine how earlier modes of human consciousness would have related (or perhaps *did* relate) to the concept of "primeness". Spanning the millennia from at least ancient Greek times, possibly much earlier, the concept has now come to dwell within the framework of an evolving body of ideas related to logic, "set theory" and higher mathematics. Prior to anything remotely like this, the experience of "divisibility" and primeness was, it would seem, just as accessible to human beings. Although the concept of primeness could have fit quite easily into a worldview which understood numbers qualitatively (each number having its own personality and characteristic qualities), there is, as I've explained, almost no evidence of pre-Greek cultures paying any attention to it. We can only speculate, but any speculation about other modes of consciousness is unavoidably going to be rooted in our current mode of consciousness and so should be regarded with appropriate caution.

The idea of prime numbers was around long before the emergence of our current formulation of that idea. And, unless lost entirely due to cataclysmic future events, the idea of prime numbers will persist in some form as human consciousness evolves and the way we think about numbers and "primeness" changes. *Humility*, then, is probably the most appropriate stance here. As in all previous times, we humans tend to be limited by the currently dominant views of, and assumptions about, reality.

chapter 4

prime factors

You'll recall how, in the previous chapter, we noticed the existence of prime numbers through our attempts to divide piles into equal parts. We didn't pay as much attention to the counting numbers which *weren't* prime so we'll have a look at these now.

We'd got as far as 29. Let's continue splitting piles of beans. Add one bean, bringing the pile to 30. That divides as three piles of 10. But we don't have to stop there. We can continue splitting piles because each of those piles of 10 can be split into two piles of 5. This can be summarised by the statement that "$30 = 3 \times 2 \times 5$". As 3, 2 and 5 are all prime numbers, no more splitting can occur so we must stop at this point.

The pile with 31 beans we find to be "unsplittable" so 31 is a prime number.

32 splits as two 16's. 16 isn't prime as it splits into two 8's. So $32 = 2 \times 16 = 2 \times 2 \times 8$. But we can continue further. $8 = 2 \times 4 = 2 \times 2 \times 2$. So, putting all of this together, $32 = 2 \times 2 \times 2 \times 2 \times 2$. Since 2 is prime, that's as far as we can go.

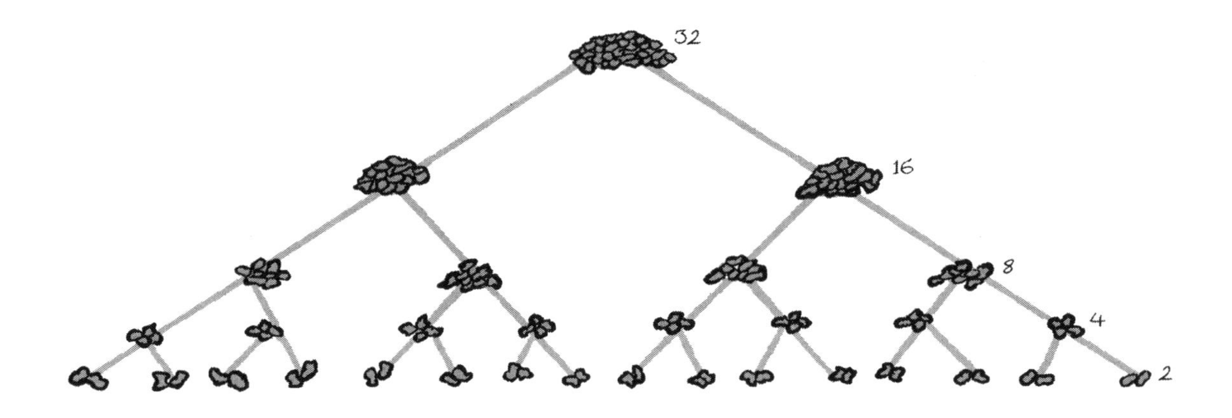

33 splits into 3 piles of 11. We can't split either of these numbers, they're both prime, so we must stop there.

34 splits into 2 piles of 17. Again, these are both prime numbers.

35 splits into 5 piles of 7. Both primes, once more.

36 splits into 2 piles of 18, the piles of 18 split into 2 piles of 9, and the piles of 9 into 3 piles of 3 each. That can be written out as:

$$36 = 2 \times 18,$$
$$36 = 2 \times 2 \times 9,$$
$$36 = 2 \times 2 \times 3 \times 3.$$

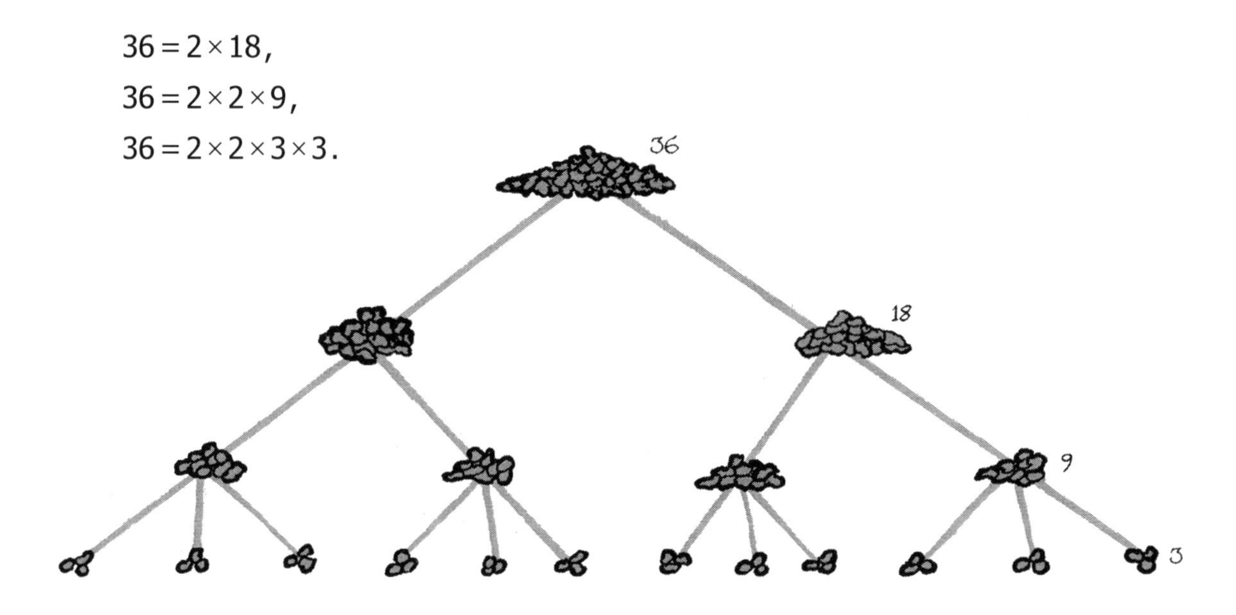

2 and 3 are both prime numbers, so that's as far as we can go.

37 is prime.

38 is 2 piles of 19, both primes.

39 is 3 piles of 13, also both primes.

40 is 2 piles of 20, and the piles of 20 spilt into 2 piles of 10, themselves each 2 piles of 5. So, we have:

$$40 = 2 \times 20,$$
$$40 = 2 \times 2 \times 10,$$
$$40 = 2 \times 2 \times 2 \times 5.$$

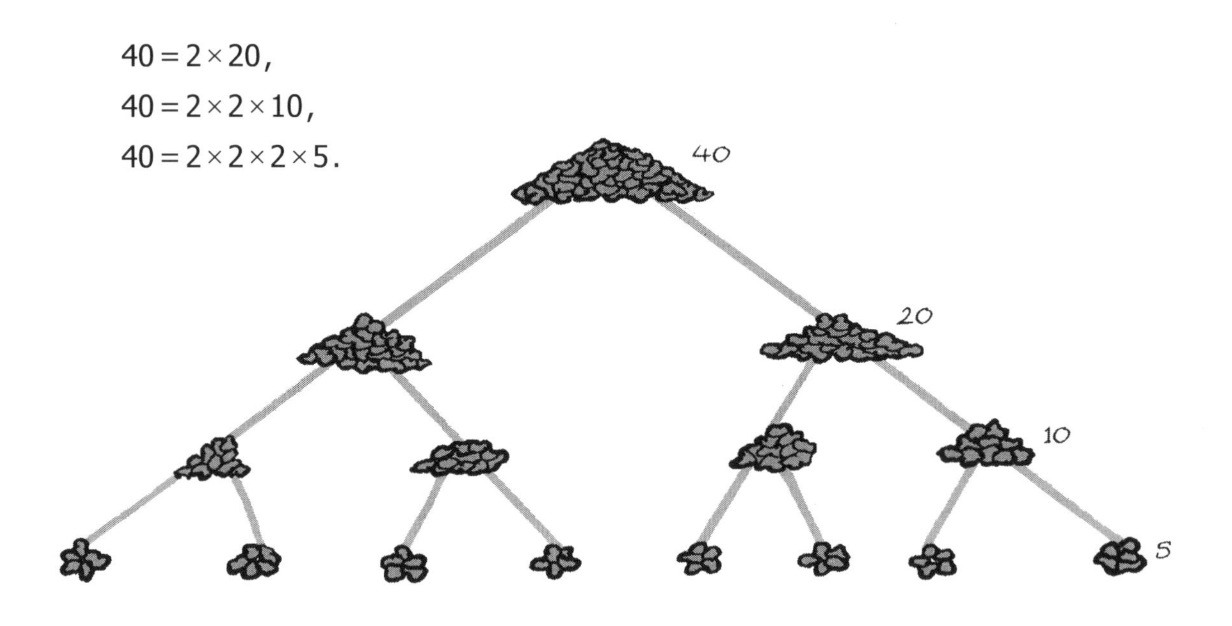

2 and 5 are both prime, so we must stop here.

41 is prime.

42 is 2 piles of 21, and the piles of 21 are each 3 piles of 7, so

$$42 = 2 \times 21,$$
$$42 = 2 \times 3 \times 7.$$

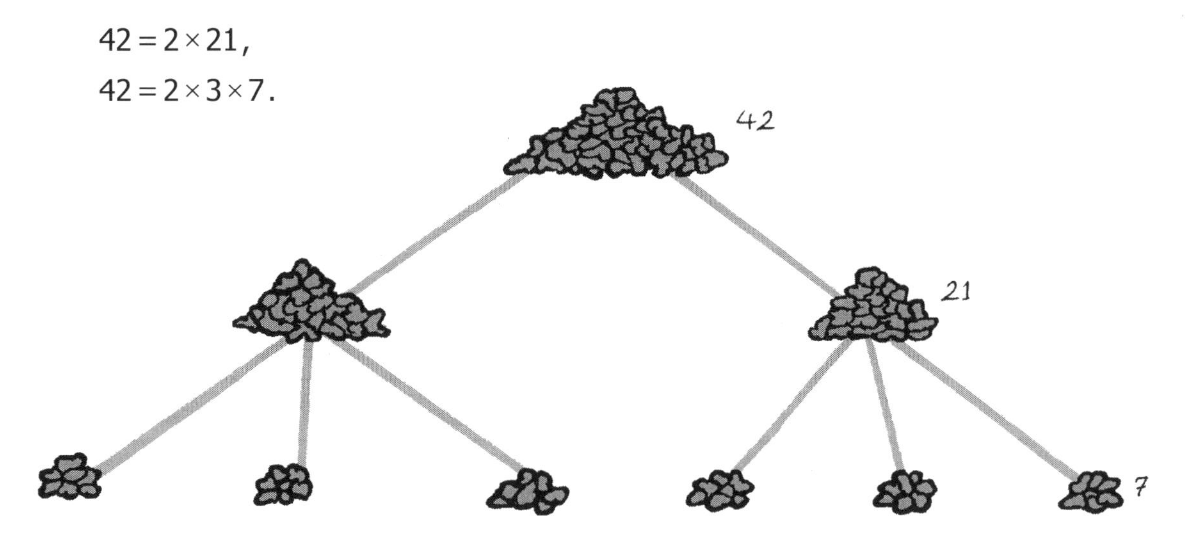

2, 3 and 7 are all primes, so we stop at this point.

In each of these cases, if the number wasn't prime to begin with, then we kept splitting piles into smaller piles until all of the numbers involved were prime numbers, at which point it became impossible to do any more splitting. We were left with one or more primes (sometimes with multiple occurrences of the same prime) which, when multiplied together, produce the number we started with.

So, if a number you encounter at any stage in your pile-splitting isn't prime, then you can split it (if you couldn't then it *would* be prime). The numbers keep getting smaller so eventually they have to stop (they're not allowed to get any smaller than 2).

Once the process has stopped, the prime numbers you're left with are called the *prime factors* of the number you started with. 30 isn't prime, and can be written as $2 \times 3 \times 5$. This means that 2, 3 and 5 are the prime factors of 30.

A counting number which isn't prime is called *composite*. This word was chosen to convey the fact that such numbers are "composed of" some smaller "component parts". These component parts are the prime factors and the composite number in question is composed of them through the act of multiplying them together.

We could also think of the adding of 1's together as an act of composition. 13 is equal to $1+1+1+1+1+1+1+1+1+1+1+1+1$, so 13 is "composite" in the sense that it's composed of some 1's added together. But *every* counting number has that property. We could call it the property of being "additively composite", but as it's an entirely unremarkable property (possessed by all counting numbers), we'll immediately discard it. The "composite" we're going to concern ourselves with (in the sense of being "not prime") means "multiplicatively composite".

Recall that 1 is considered not to be prime. Perhaps confusingly, *it's not considered to be composite either* (you certainly can't obtain it by multiplying primes together). 1 is considered as inhabiting a class of its own – this will be explained shortly. The counting numbers, then, can be separated into the primes, the composites and 1.

Now, back to the rest of those examples.

31 is prime. We can think of 31 as having just one prime factor – itself.

$32 = 2 \times 2 \times 2 \times 2 \times 2$, so 32 has just one prime factor, that is, 2 – but it's a *repeated factor*, in this case repeated five times.

$33 = 3 \times 11$, so 3 and 11 are the prime factors of 33.

$34 = 2 \times 17$, so 2 and 17 are the prime factors of 34.

$35 = 5 \times 7$, so 5 and 7 are the prime factors of 35.

$36 = 2 \times 2 \times 3 \times 3$, so 2 and 3 are the prime factors of 36, both repeated twice.

37 is prime – its only prime factor is 37.

38 has prime factors 2 and 19.

39 has prime factors 3 and 13.

$40 = 2 \times 2 \times 2 \times 5$, so its prime factors are 2 (repeated three times) and 5.

Let's now look at a larger number, say 21294. With a little bit of trial-and-error experimentation, we can work out that

$$21294 = 2 \times 3 \times 3 \times 7 \times 13 \times 13.$$

The prime factors of 21294 are therefore 2, 3 (repeated), 7 and 13 (repeated). 21294 is said to be *divisible* by each of these factors. This is because 21294 can be (with some patience) separated into

2 piles of 10647 (or 10647 piles of 2)
3 piles of 7098 (or 7098 piles of 3)
7 piles of 3042 (or 3042 piles of 7)
13 piles of 1638 (or 1638 piles of 13)

Ultimately, this means you could divide 21294 into

2 lots

of 3 sets

of 3 piles

of 7 groups

of 13 heaps

of 13 beans each...

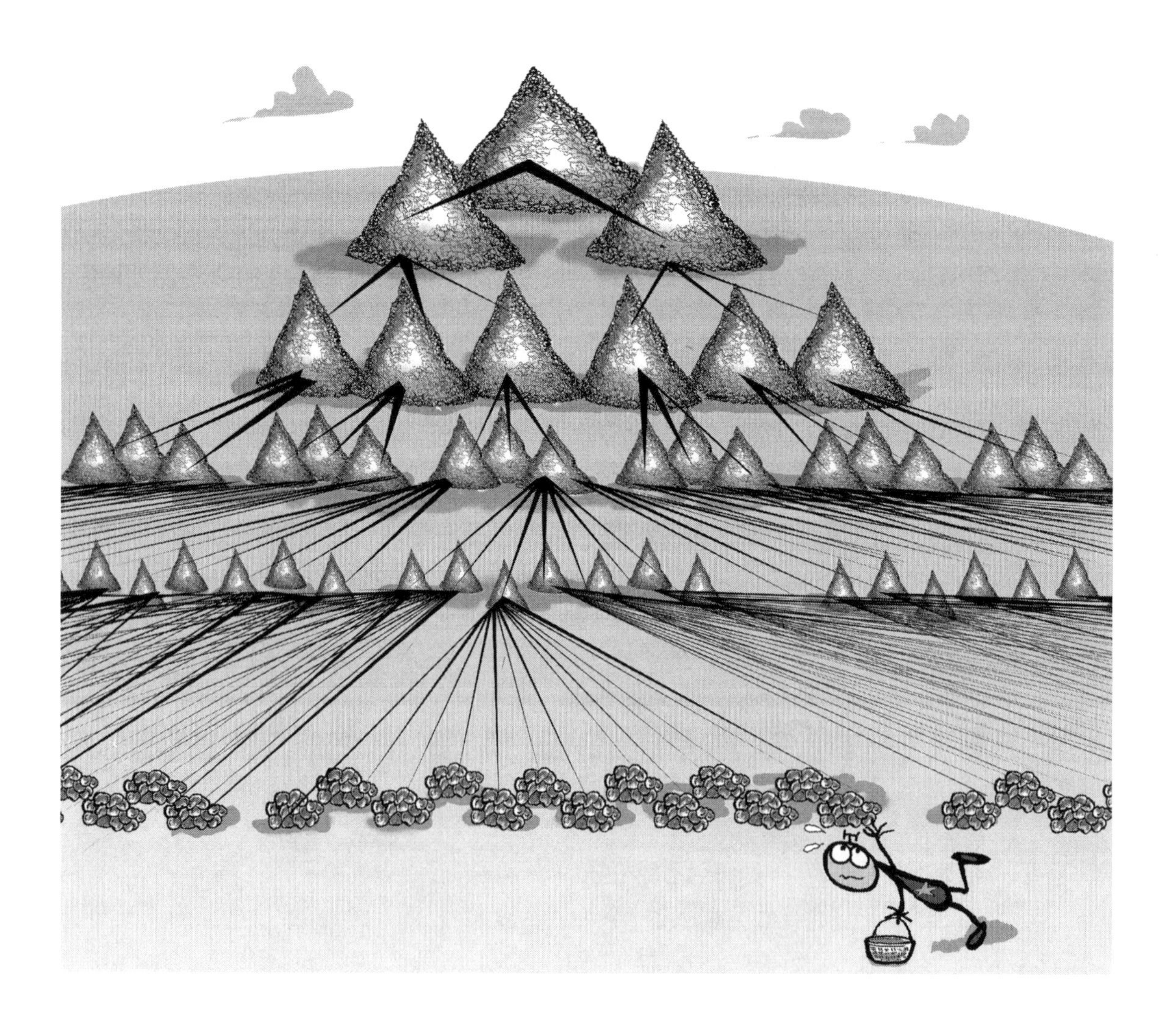

...or into 3 lots of 13 groups of 2 collections of 13 sets of 3 piles of 7 beans each...

...or in numerous other ways along the same lines.

If you could be bothered to check, you'd find that apart from 2, 3, 7 and 13, there are no other prime numbers by which 21294 can be divided.

FACTORISATION

This process, the breaking down of a counting number into a definitive collection of prime factors, is known as *factorisation*. It turns out that it's possible to factorise any counting number *uniquely*. I'm using the words "definitive" and "uniquely" here to make an important point. That is,

every counting number can be factorised in only one way.

Of course, we can re-order the factors, so $30 = 2 \times 3 \times 5$, but it also equals $3 \times 2 \times 5$ or $5 \times 2 \times 3$ or $5 \times 3 \times 2$. But the ingredients in the recipe for "multiplicatively composing" 30 from prime numbers are a two, a three and a five (and nothing else).

Think in terms of describing the contents of a bowl of fruit: "four apples, a pear, two bananas and three kiwis" – the order you describe it in doesn't matter, you're always describing the same collection of fruit. Similarly, if you had a "bowl of primes" and I asked you what it contained, you might say "four 2's, a 3, a 5, a couple of 11's and a 19". If you listed them in a different order ("I've got a 3, a 5, an 11, a 2, a 19, another 2, another 11 and another couple of 2's"), you're still describing the same collection. Your description contains the same prime numbers and the same amounts of each. If we take these to be prime factors, we end up with the counting number 551760, which is $2 \times 2 \times 2 \times 2 \times 3 \times 5 \times 11 \times 11 \times 19$.

The point here is that this is the *only* way we can build 551760 out of prime factors.

When I say "every counting number" I mean *all of them* – *not* just the ones which people have managed to check up to the time of my writing this. This "unique factorisation" thing has been mathematically proved. Through the careful application of logic it's possible to reach a state of total certainty – a kind of certainty not possible in any of the sciences or any other aspect of life – about certain features of the number system which continue on forever, quickly leaving behind the limited range of our direct perception and awareness. That's the nature of mathematical proof. I'll have more to say about this issue later.

Primes can be thought of as being "already factorised". The unique factorisation of 23 is just "23". A prime number always has exactly one prime factor (itself).

1, you'll recall, is an exception. It's a counting number, but it's considered to be *neither a prime nor a composite*. We can think of its factorisation as consisting of NO prime factors. Think of an empty fruit bowl. 1 is *the absence* of any prime factors. This is a factorisation of sorts, but 1 isn't "composed of" any prime factors – it has *zero* component parts – so it's not considered to be "composite". You may still be wondering why 1 isn't considered to be a prime number, but we'll be coming to that soon.

The first known observation of the fact that every counting number has a unique prime factorisation was made by Euclid of Alexandria in the 3rd century BCE[1]. As with the earlier discussion about the general phenomenon of prime numbers, we'll probably never know whether anyone earlier had noticed this, but Euclid's is the first written record we have of anyone doing so[2]. Carl Gauss, often described as the greatest mathematician ever, *proved* it at the very end of the 18th century. Unlike most things which were being written about in the ancient Greek world (and even in 1790s Europe), this is *not* something that becomes less relevant as time rolls ever onward. Not even slightly. It's as valid today as it was on the day Euclid first noticed it or on the day which Gauss proved it. Having been *proved*, it is a mathematical *certainty* whose truth has been logically deduced. Unlike a point of philosophy or theology this is not going to be debated, questioned seriously or reconsidered.

This particular property of counting numbers, the fact that you can split each one into prime factors in a unique way, is known to mathematicians as the *Fundamental Theorem of Arithmetic*.

THE FUNDAMENTAL THEOREM OF ARITHMETIC

The word *fundamental* has the same roots as the word *foundational*. A *theorem* is a mathematical truth, that is, something we know with complete certainty about mathematics [3]. The word *arithmetic* is here being used to mean the system of counting numbers together with addition and multiplication. These days, "arithmetic" is generally used to mean "simple problems with numbers which schoolchildren are taught to solve". In the context of "Fundamental Theorem of Arithmetic", it's being used in its original sense, that is, "the study of the system of counting numbers and its properties" (these days, that's usually called *number theory*).

So, "Fundamental Theorem of Arithmetic" translates roughly into everyday language as "the most important thing we know with certainty about the number system".

To reiterate, the Fundamental Theorem of Arithmetic (I'll call it the "FTA" from now on), *the basic truth on which the theory of the number system is founded*, says:

> *Every counting number bigger than 1 is either prime or else it can be "built" by multiplying prime numbers together. Also, a counting number can never be built in two different ways (involving different collections of prime numbers).*

The second sentence is saying that a number will always split into exactly the same group of factors, that is, there's only one way of splitting it – different collections of factors can't produce the same counting number. If this isn't already clear to you, here's another example: if you somehow found yourself in a situation where you were convinced that...

$$3 \times 11 \times 59 \times 2418671 = 3 \times 17 \times 4079 \times 22637$$

...then you'd have to go back and check your calculations, because the FTA guarantees that this just *isn't possible*.

If you look through a selection of introductory number theory books and webpages, you'll find the FTA stated in a variety of ways. They all say the same thing, just stating it in a slightly different way. Here are some examples (the word "product" here means "the result of multiplying numbers together", in the sense that "12 is the product 3×4"):

> *Any counting number greater than 1 can be decomposed in a unique way, up to the order of the factors, as a product of prime numbers.*
>
> *Except for the arrangement of factors, every counting number greater than 1 can be expressed uniquely as a product of primes.*
>
> *Each counting number greater than 1 can be uniquely factorised into primes. Here the order of the factors is considered irrelevant.*
>
> *Every counting number greater than 1 is a product of prime numbers, and moreover, in a unique way.*
>
> *There is exactly one way of expressing a counting number as a product of prime factors of increasing size.*
>
> *Every counting number greater than one can be written uniquely as a product of primes, with the prime factors in the product written in order of nondecreasing size (each factor larger than or equal to the factor written before it).*

The first three of these make direct reference to the fact that the "uniqueness" being invoked requires us to disregard the *order* of the prime factors. The last couple use the fact that if we require the factors to be listed in order of size (as in $30 = 2 \times 3 \times 5$), *then* the factorisation is unique.

You might wonder why I'm dwelling on this fact at such great length.

One reason is that it strikes me as rather odd that what is basically *the most important thing we know about the number system* is probably known to much less than 1% of the population of the number-obsessed Western world. We've built a civilisation around this number system and yet the most fundamental fact about it is unknown to almost everyone. How did such a situation come to be? Many of us will have learnt this as children but not in a way which made it seem very interesting or relevant, so it was passed over quickly and forgotten. It requires no specialised knowledge to understand. And *it applies to us all*, it's *universal*, unlike almost anything else we ever encounter.

It also produces something surprisingly irregular, something with (literally) endless unpredictable content. If we look at the prime factorisations of a succession of counting numbers, however long, we will see no clear pattern in the way the quantity of factors increases and decreases (this is illustrated a bit later in this chapter). What you're starting to see here is the "interesting" content of the number system which Peano's Axioms in no way prepare you to expect.

A thorough proof of the FTA is given in Appendix 2, for readers who already have a certain amount of experience with mathematical concepts and notation. Here, we'll just sketch out some basic reasoning as to why every counting number after 1 has a prime factorisation (demonstrating why the factorisation is *unique* turns out to be quite a bit trickier).

Given a counting number, ask "Is it prime?". If it is, you can stop. If it isn't, you must try to convince yourself that it can be split into primes. The fact it isn't prime means, by definition, that it must be divisible by something other than 1 or itself. So you can split it into two smaller counting numbers multiplied together (a certain number of equal-sized piles, each containing a certain number of beans). We now apply the same process to each of these numbers – if it's prime, then leave it alone, if it's not, split it into two smaller numbers multiplied together. As you continue to split the non-prime numbers, the numbers you split them into will get smaller and smaller. But these numbers aren't allowed to be smaller than 2, so the process will eventually have to stop, leaving you with a collection of prime numbers (and nothing else). Try a few examples and you'll get the idea.

WHY 1 ISN'T CONSIDERED TO BE A PRIME NUMBER

You'll recall that 1 is considered to be neither prime nor composite, so that we can think of the counting numbers as splitting into three classes: the primes, the composites and 1 (which is in a class of its own).

Related to this unique status among the counting numbers, 1 has the following unique properties:

☆ You can multiply anything by 1 (or *vice versa*) without changing it. Notice that in the context of addition, 0 plays this role: you can *add* anything to 0 (or *vice versa*) without changing it. 0 could be thought of as the "additive nothing" and, similarly, 1 could be thought of as a kind of "multiplicative nothing" [4].

☆ 1 is the only counting number which is mentioned in Peano's axioms.

☆ 1 is the only counting number which the FTA makes an exception for ("*Every counting number greater than 1…*")

☆ You can build *any* counting number by adding 1's together.

☆ 1 is the "edge" of the counting numbers – they continue forever in one direction, but if you start counting backwards (moving to the left along the number line), you'll eventually reach an "edge" which we call "1".

Perhaps now you're getting a feeling as to why 1 isn't considered to be a prime.

An often-stated reason for this is that if 1 *were* considered prime, then the FTA wouldn't work. For if 1 were prime, then 10 could be factorised not only as 2×5 (or 5×2) but also as $1 \times 2 \times 5$, $1 \times 1 \times 2 \times 5$, $1 \times 1 \times 1 \times 2 \times 5, \ldots$ You might argue that these are all really the same. Well, they are *representations of the same number* (10), but they're *different factorisations*, just as 3×7, $3 \times 3 \times 7$, $3 \times 3 \times 3 \times 7, \ldots$ are all different. A factorisation is described not just in terms of *which* prime numbers are involved but also *how many times* each one shows up. $7 \times 7 \times 11$ is not the same thing as $7 \times 11 \times 11$, just as a bowl containing two apples and an orange is not the same as a bowl containing an apple and two oranges.

As mentioned above, if we wanted to describe a "factorisation of 1", it would contain *no* prime factors. It would be the *absence* of prime factors – you might think that this would be 0 rather than 1, but we're working in the system of counting numbers, where 0 isn't included. I'd encourage you to think of 1 in terms of an absence or emptiness, a sort of background against which factorisations take place, an empty fruit bowl into which the fruit is placed. The 1 is always there, invisibly, in any factorisation. 231 has the factorisation $3 \times 7 \times 11$, but that can be understood as $1 \times 3 \times 7 \times 11$, where the 1 acts as a "starting point" rather than a prime factor. The same thing could be done with any factorisation, of course.

In that last example, we can think of the 7 "multiplying into" the 3 and the 11 as "multiplying into" the 3×7. But what about the 3? We can understand that as being "multiplied into" the 1 – that is, $3 = 1 \times 3$. Think of it like this: when we're sticking prime factors together to build counting numbers, we need something for the first factor to "stick to" – that's the role of 1.

The special status of 1 will come up again shortly, as well as in Volumes 2 and 3.

In an attempt to clarify some of the concepts we've met in this chapter, I'm going to present three images of prime factorisation. You can mentally hang on to whichever one you most like, but all three should add something to your understanding.

IMAGE 1: BUILDING BLOCKS

We've already seen that prime numbers are, in a sense, the "building blocks" of the counting numbers.

Starting with the prime "blocks", you can "build" any counting number. The FTA tells us this. In fact, it tells us more – that there's a unique "block recipe" for any counting number. A prime number's recipe will consist of a single block, whereas the recipe for a composite number will involve two or more blocks. Different collections of blocks (in the "fruit bowl" sense) will produce different counting numbers.

In this portrayal of prime factorisation, 1 is the absence of any building blocks, so we could perhaps think of it as *the ground on which the blocks are resting*. It's always there, implied in any factorisation (any assemblage of blocks representing a number). If asked "when there are no blocks, what number is represented?", the answer could only be "1". "Why not 0?" you might still be wondering. It's because the act of introducing a block involves *multiplying* that prime by what's already there. So if we start with 0, however many blocks we introduce (that is, primes we multiply), we can never represent anything other than 0. Anything multiplied by 0 is 0.

> "*To some extent the beauty of number theory seems to be related to the contradiction between the simplicity of the* [*counting numbers*] *and the complicated structure of the primes, their building blocks. This has always attracted people.*" (Andreas Knauf [5])

IMAGE 2: AN ALPHABET

Here's another useful image. Think of the primes as making up an "alphabet" from which we build each counting number as a "word". For example, if we were on a distant planet where the number 2 were represented by the symbol "ɣ", 3 by the symbol "ƕ", and 7 by the symbol "ϴ", then the number 126 (which equals $2 \times 3 \times 3 \times 7$) could be written as the word "ɣƕƕϴ". In this approach to representing numbers,

6 would be the "word" spelled "ɣƕ", 4 would be spelled "ɣɣ" and 81 would be spelled "ƕƕƕƕ". 2, 3 and 7 would all correspond to one-letter words: "ɣ", "ƕ" and "θ".

Like familiar alphabets, the members of this "alphabet" have a definite order: 2, 3, 5, 7, 11, 13, 17,...

Perhaps you're a bit confused by the fact that all of the primes after 7 are made up of more than one digit (11, 13, 17, 19, 23,...). We must be clear that it's the *primes*, not the digits 0, 1, 2, 3, 4, 5, 6, 7, 8, 9 which are the "letters" of this "alphabet". So 13 is *not* a two-letter word – it's a *one*-letter word (because it's a prime number). 6, on the other hand, would be a *two-letter* word, since it is 2×3.

If we had unlimited time, we could devise a unique symbol for *every* prime number and could then represent any counting number by "spelling" it with our "prime alphabet".

Since we can rearrange prime factors (for example, $126 = 2 \times 3 \times 3 \times 7 = 2 \times 3 \times 7 \times 3 = 2 \times 7 \times 3 \times 3 = 3 \times 7 \times 2 \times 3 = \cdots$), counting numbers can have many possible "spellings". To get around this, we can introduce the rule that each word must be spelled with its letters in "alphabetical order" (that is, with the prime factors in order of increasing size). Alternatively, we could just accept that any words which are "anagrams" (like *post, opts, spot, tops* and *stop* in English) are actually the same word – they're just alternative "spellings" for the same counting number.

In this portrayal of prime factorisation, 1 is the "0-letter word", the word with no letters, the *absence* of any letters. So we can perhaps best think of it as the blank page onto which a word is written. Any word made of letters from our alphabet represents a unique counting number. 1 is there, invisibly implied in any factorisation. If asked "what number is represented by a blank page?", the answer could only be "1". Remember, introducing a letter to the page involves *multiplying* a prime number into what is already there. This is why blankness corresponds to 1 rather than 0.

IMAGE 3: ATOMS

As well as "the building blocks of the counting numbers", the prime numbers are sometimes described as the "atoms of arithmetic".

This analogy involves atoms and molecules, but don't worry, *it's just an analogy*. We're not going to get into the details of chemistry or physics here. And I'm not about to suggest that prime numbers are directly linked to the physical realities of atoms and molecules. If any of this makes you uncomfortable or confused, it's probably best to skip ahead to the next subsection (page 99). If you followed the "building blocks" and "alphabet" descriptions, then you'll have got the general idea by now.

You're probably aware that water is known to chemists as "H_2O". This means that water is made up of tiny little things which they call *water molecules.* There are about 150 000 000 000 000 000 000 000 molecules in a teaspoon of water. Molecules are made up of even tinier things called *atoms*. Each water molecule is made up of three atoms: two hydrogen (H) atoms and one oxygen (O) atom – that's what "H_2O" means.

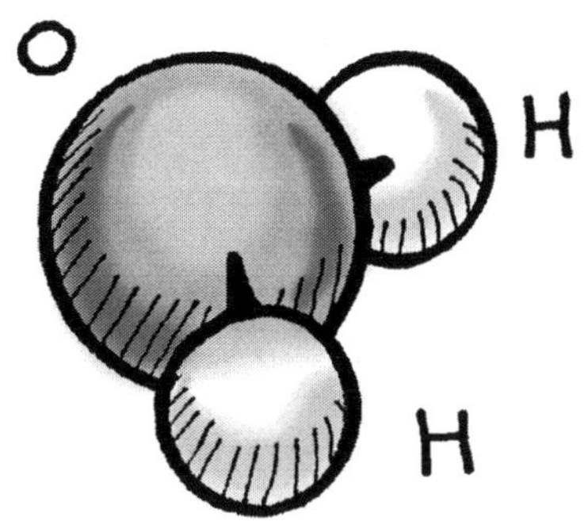

The number 20 can be compared to a water molecule. We can think of 20 as $2 \times 2 \times 5$. That's its prime factorisation. The prime factors of 20 are 2 and 5, but 2 is a repeated factor. The primes 2 and 5 play the role of the "atoms" in this analogy, with multiplication (*not* addition) acting as the "glue" bonding them together. Similarly, sugar is described by chemists as "$C_{12}H_{22}O_{11}$". Sugar molecules are much bigger and more complicated than water molecules, being made out of 39 atoms:

there are 12 of carbon (C), 22 of hydrogen (H) and 11 of oxygen (O). There are about 10 000 000 000 000 000 000 000 sugar molecules in a teaspoon full of sugar.

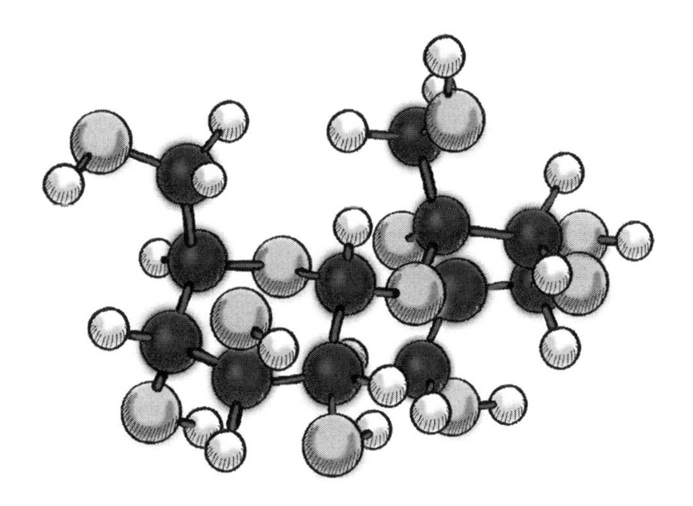

We could build a counting number according to the structure of a sugar molecule:

$2 \times 2 \times$
$5 \times 5 \times 5 \times 5 \times 5 \times 5 \times 5 \times 5 \times 5 \times 5 \times 5 \times 7 \times 7 \times 7 \times 7 \times 7 \times 7 \times 7 \times 7 \times 7 \times 7 \times 7 \times 7$

(here, “H” has been replaced by 2, “O” by 5 and “C” by 7). Mathematicians would write this more compactly as $2^{22}5^{11}7^{12}$ – you should be able to see how this notation works.

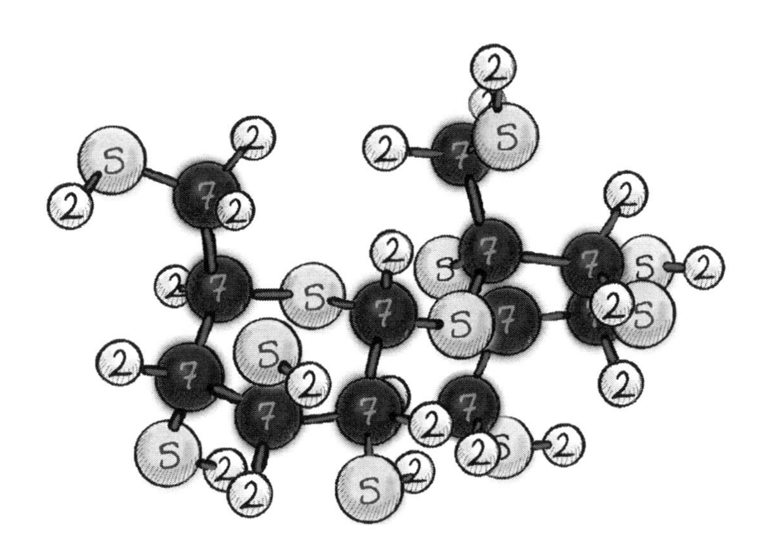

H_2SO_4 is sulphuric acid, which suggests a factorisation like $2 \times 2 \times 5 \times 5 \times 5 \times 5 \times 11 = 27\,500$ (if we use 2 for H, 5 for O and 11 for S).

An oxygen molecule (what your body needs to breathe in) is of the form O_2. That would correspond to $5 \times 5 = 25$.

A carbon dioxide molecule (what your body breathes out) is of the form CO_2. This would correspond to $5 \times 5 \times 7 = 175$.

A hydrogen molecule, made from a pair of hydrogen atoms (the ill-fated *Hindenburg* was filled with these) is represented as "H_2". That would correspond to $2 \times 2 = 4$.

A single atom of sulphur (match heads have traditionally contained these) would be represented simply as "S", corresponding to our 11.

A carbon atom is represented "C" (enough of these put together could be a lump of graphite or a diamond, depending on how they're packed), which would also correspond to a single prime, here 7.

So, from four different kinds of atoms – carbon (C), hydrogen (H), oxygen (O) and sulphur (S) – we've managed to get water, a highly corrosive acid, two invisible gases we breathe in and out, graphite, diamonds, another invisible gas that makes balloons rise into the air and the yellow crystalline stuff which can be used to make matches ignite. There are about ninety different kinds of atoms which make up all of the different molecules which exist.

Similarly, all of the counting numbers that exist can be built, in the same sort of way, from the prime numbers. A small collection of primes – 2, 3, 5 and 7 – can already produce an infinite variety of different counting numbers as "numerical molecules". Given *all* the primes, we can build ANY counting number and, in each case, there's only one way of doing it.

So we can think of the primes as being the "atoms" of the number system and the composite numbers as the "molecules" of the number system. Again, I stress, this is *just a helpful analogy*. Please don't get confused and take it literally. I am *not* suggesting that there is any mysterious connection between physical atoms and prime numbers. If it confuses you, then it's best to stick to the two previous analogies (building blocks and letters of an alphabet).

The analogy only goes so far, as the primes "bond" with each other through simple multiplication, whereas the bonding of atoms into molecules is much more complicated and subject to variation. Although the word *atom* (actually, it was *atomon*, the plural being *atoma*) was coined by the ancient Greek Democritus[6] to mean "indivisible", it was discovered many centuries later that atoms can be split and that they have component parts (*protons*, *electrons* and *neutrons*). These *subatomic particles* have more recently been split into smaller particles, and using ever larger, more powerful *particle colliders*, scientists continue this process of splitting things into ever smaller things. Primes, on the other hand, really *are* "atoms", in the sense that they are indisputably indivisible (in the "multiplicative" sense we are talking about here[7]) and this basic truth cannot be undermined by any further discoveries.

In this portrayal of prime factorisation, 1 is "the absence of any atoms" (in other words, no molecule at all). We can think of it as the empty space in which our molecules would float. It is there, invisibly implied in any assemblage of atoms (that is, any prime factorisation of a counting number). If asked "what number is represented by empty space in this scheme?", the answer could only be "1".

THE BAFFLING SEQUENCE OF FACTORISATIONS

Let's have another look at some factorisations of counting numbers:

Notice that there isn't a "1" egg. 1 is best understood as "the absence of a factorisation". It might be helpful to think of 1 as the bird which laid all of these eggs! (see page 80)

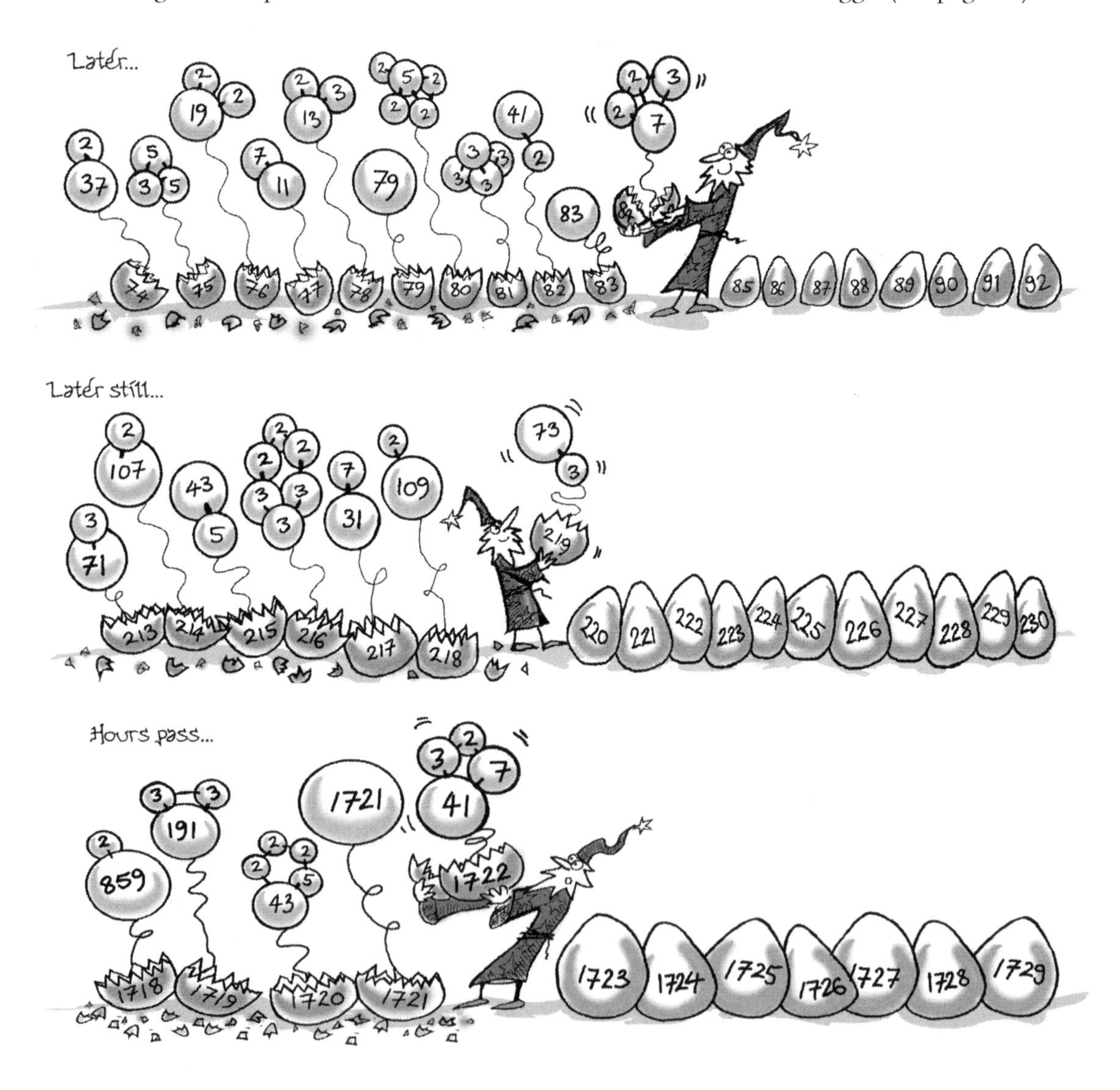

These images might provide some insight into what Ian Stewart, probably the world's most successful author of popular books about mathematics, meant when he wrote

> "*Mathematics is full of surprises. Who would have imagined, for instance, that something as straightforward as the natural numbers* (1, 2, 3, 4, ...) *could give birth to anything so baffling as the prime numbers* (2, 3, 5, 7, 11, ...)*?*" [8]

The bafflement to which he refers runs very deep indeed, as we'll eventually see. But a first taste can most readily be obtained by looking at the last set of illustrations and considering why something as seemingly simple and straightforward as the sequence of factorisations should look like such a jumbled mess.

Ian Stewart's question brings to mind the image of a very polite, mild mannered, predictable suburban mother (embodying the "picket fence" image of the counting numbers included on page 51) giving birth to a "wild child" contrary to all expectations. *Where did* that *come from?*

And the bafflement will only get deeper from now on.

The mathematician, physicist and philosopher Hermann Weyl, in a 1927 book on the philosophy of science, wrote

> "*The mystery that clings to numbers, the magic of numbers, may spring from this very fact, that the intellect, in the form of the number series, creates an infinite manifold of well distinguishable individuals. Even we enlightened scientists can still feel it e.g. in the impenetrable law of the distribution of prime numbers*" [9]

Here "number series" just means the sequence of counting numbers, "manifold" is being used to mean "multitude" [10] and "the distribution of prime numbers" means "how the prime numbers are arranged within the sequence of counting numbers". Translated into plainer language, then, Weyl was saying that the "mystery" or "magic" associated with numbers may arise from the fact that our minds create, in the form of the sequence of counting numbers, an infinite multitude of "well distinguishable individuals". He suggests that "enlightened scientists" can sense the presence of this mystery in the way that the prime numbers fall within the overall sequence of counting numbers.

Take note of how the word "enlightened" is used here – Weyl was perhaps employing a touch of irony, as he's talking about something which supposedly "enlightened scientists" find "impenetrable". The counting numbers are still cloaked in "magic" and "mystery" despite our best attempts to understand them.

So, what Weyl seems to have been getting at is something like this: *When it comes to the deep issues of the number system, even the most brilliant scientists are confronted with the fact that we're not as enlightened as we'd like to think. There's a dark corner which we're still unable to illuminate with our knowledge.*

The mystery, Weyl is suggesting, springs from the fact that, via the simple idea of counting, the human mind is able to conceptualise an infinite sequence of "well distinguishable individuals". Importantly, he also mentions the *distribution of prime numbers* as his chosen example of how this mystery manifests. This "distribution" is something we'll be focusing on extensively in later chapters.

Although Peano's axioms can be used to produce this infinite sequence, the "well distinguishable" nature of the sequence's elements (that is, the counting numbers) is not particularly evident from the construction. Each number after 1 is constructed using the same simple process (adding 1) as used for all the others, giving the impression of uniformity, homogeneity or "mass production". The interesting stuff which we've now started to focus in on only becomes apparent once multiplication has been introduced and, with it, the notion of prime numbers.

chapter 5

a philosophical interlude

You just can't argue with the basic facts about prime numbers. There's never going to be a serious dispute about whether or not 219 is prime. As Hardy put it [1],

> "*317 is a prime, not because we think so, or because our minds are shaped in one way rather than another, but because it is so, because mathematical reality is built that way.*"

There's no way around it. The primes don't change. They transcend history, culture and opinion. If the contents of the physical universe were completely rearranged, or removed and replaced by different contents, *even then* the facts concerning which counting numbers are and aren't prime would remain the same. The only other thing that you could say that about would be the laws of physics. But it's even possible to imagine universes with *entirely different laws of physics* where the system of counting numbers and all of its properties remain unchanged.

One image which occasionally comes up is that of the sequence of primes as a kind of "cosmic code". Here's a good example, from a book about great mathematicians of the past, aimed at non-specialist readers:

> "*Primes are part of the scaffolding of the number system, but the structure of this scaffold is still obscure, still hidden despite the research of centuries. Prime numbers are imbedded throughout the system, yet no pattern in their apparently random appearance can be detected no outward signs distinguish them from nonprimes, and no method exists to tell exactly how many primes occur in a certain number span. They are an infinite code that no one has ever been able to crack.*" (Jane Muir [2])

We'll briefly consider the implications of this metaphor.

The word "code" has a few subtly different meanings. There are legal codes and codes of conduct (these usages are entirely unrelated to our interests). There are computer codes, genetic codes, telephone area codes and product identification codes – here the word is being used to mean any kind of systematically compressed information which can be interpreted within a recognised system (a computer, a biological organism, a telephone network, a commercial database).

But in the popular imagination, "code" is often understood in terms of "secret codes", strings of symbols put together in order to communicate some hidden meaning. This meaning is retrievable only by the intended recipient(s) possessing knowledge of the decoding method. The examples of "codes" mentioned above (computer, genetic, telephone dialling, product ID) aren't *secret* codes. They're not intended to contain *meanings or messages* which can be decoded – they're a different sort of thing.

There are also communication codes, like Morse code and Braille, which are similar in format to secret codes but used in a non-secretive way (the decoding method being publicly available to anyone willing to learn it).

One thing that these various types of "code" have in common is that to the uninitiated viewer, they tend to look like meaningless, unpredictable strings of symbols. The fact that the sequence of prime numbers has a similar feel to this is probably what has led to the emergence of the "code" metaphor.

A secret code must have an "author". So, if we're going to think of the prime number sequence as a code in that sense ("*an infinite code that no one has ever been able to crack*"), then the question arises as to its "authorship". Religious believers might suggest that the author must be "God" (some of them already have, on obscure websites [3]). Those uncomfortable with "God" might suggest "the universe", "the Cosmos" or "the source of creation", if pushed to answer the question. Some people might simply declare the question to be meaningless. (In each case, we'd probably get the same answer if we asked about the "author" responsible for genetic codes.)

I've seen curious websites where earnest, enthusiastic mathematical amateurs struggle to demonstrate some link between prime numbers and DNA, the (genetic) code of life. Nothing convincing has yet been shown, but ideas like this seem to take hold of certain people. Individuals with a poor grasp of even the simplest mathematics are sometimes gripped by the certainty that they will be able to "decode" (or even *have* decoded) the prime numbers. It's as if they're feverishly trying to solve a riddle or decrypt a secret message. There's sometimes an almost religious zeal about it. A similar psychological phenomenon is evident in the title of an extremely confused book published in 1998 – *God's Secret Formula: The Deciphering of the Riddle of the Universe and the Prime Number Code*[4].

If we were to continue with this metaphor and ask who is the intended *recipient* of the message encoded in the prime numbers, the answer would presumably be "conscious lifeforms sufficiently advanced to unravel the theory of prime numbers". What about the *message* hidden by the code? Since the code – if that's what it is – hasn't been deciphered yet, we can't really talk about that, but the adoption of the "code" metaphor hints at an unacknowledged belief in some deep metaphysical message or cosmic truth "hidden" in or behind the sequence of prime numbers (and hence in the number system itself). Some of the quotations I've collected on the subject and scattered throughout this book convey a sense of there being such a "hidden meaning", felt by many within the mathematical community. But we're now talking about *feelings*, very subtle feelings. And feelings don't really come into mathematics[5]. Still, that doesn't mean that we shouldn't carefully examine feelings *about* mathematics. I believe this is a subject worthy of more attention – it will be explored further in Volume 3.

Putting aside any discussion of "authorship" or hidden meanings, we should be able to agree that the sequence of primes, taken as a whole, as a single "item of information", has the following properties:

☆ It's *indisputable*. There can be no serious argument about whether or not 317 is a prime number – this is just about the only thing which *everyone* can agree on.

☆ It's *fundamental*. It doesn't depend on any cultural biases or arbitrary human ideas originating in particular times or places. Rather, it seems to operate at a deeper, more "foundational" level of reality.

☆ It's *nontrivial* (a word mathematicians are fond of – in this setting it would just mean "not simple or predictable", although there's a lot more than that to the term). The sequences $(1, 2, 3, 4, 5, \ldots)$, $(2, 4, 6, 8, 10, \ldots)$, $(1, 2, 1, 2, 1, 2, 1, 2, \ldots)$ and $(1, 1, 1, 1, 1, 1, \ldots)$ could be thought of as "trivial" in this context since they contain no substantial *information* (each can be summarised in just a few words) whereas the sequence of primes clearly does.

☆ It's *communal* (perhaps an unusual choice of word, but it fits), meaning that it's everyone's "property". Conscious lifeforms anywhere in time or space all potentially have/had/will have unrestricted access to it.

Although it would be difficult to back up such a claim with any kind of conventional reasoning, I want to put forward the idea that the sequence of prime numbers is *the most fundamental nontrivial information accessible to any consciousness*[6]. This is information which transcends history, culture, opinions and feelings, as well as anything particular to the human race, the Earth, our solar system, this galaxy or the particular form taken by the physical universe in which we find ourselves. This is why I feel compelled to write about it – faced with the great profusion of beliefs, philosophies and ideologies which I discussed in this book's introduction, it's just about the only thing I know of which I can write about with any certainty!

Even if you don't accept my claims of "most fundamental" and "accessible to any consciousness", the prime numbers certainly comprise some kind of fundamental or universal sequence (or "code", if you like). It's unchanging, accessible everywhere, at all times past, present and future, and (limiting our considerations to human lifeforms for now) *it's accessible to everyone*. I find this curious status of the number system as "common intellectual property for all" rather pleasing. *This stuff applies to you whoever you are.* Everyone has equal access. And in my experience of explaining these ideas to people, I've found that with the right approach, the essence of even the deepest mysteries can be conveyed to anyone who can count, add and multiply and who is prepared to make a small effort to understand.

Truths about the number system provide us with something, perhaps the *only* thing, which everyone can agree about. As I suggested in the introduction, at a time when there seems to be more, and increasingly hostile, disagreement about more issues than ever before, there's something reassuring about this.

Look around a library and consider how many books would remain relevant if they were to travel with you on a voyage into outer space. You'll see that almost everything on the shelves, apart from the maths, physics, chemistry and astronomy sections, is concerned with the goings-on on one particular little planet called *Earth*, and primarily with the affairs of one particular species called *Homo sapiens*. Mathematics stands out in that it *isn't* limited in this way. And physics (of which chemistry and astronomy are just specialised branches) is formulated in terms of mathematics. All of mathematics follows from the system of counting numbers and at the heart of this lies the sequence of primes.

Someone might argue that Shakespeare or the Bible or the poetry of Sappho (or whatever) might remain valid, because you could still experience a deep sense of inner truth or beauty by reading it on your interstellar travels. Indeed, you could, but you must also admit that it's still all about human affairs in one brief epoch of human existence on an ephemeral little sphere in this *huge* space[7].

If we ever encounter conscious, intelligent, communicating lifeforms from beyond this planet, we'll quite possibly have a very different kind of understanding and formulation of any higher mathematics (and, to an even greater extent, physics) from theirs. Still, we might be able to agree on certain fundamental truths – the most fundamental being such facts as that 66 can be divided into the prime factors 2, 3 and 11, and that 67 is indivisible.

In other (somewhat flippant) words, you might be able to chat to an extraterrestrial about prime numbers but quite possibly about almost nothing else. Obviously, there's the problematic issue of needing some kind of common language in which to chat, so this isn't really a serious proposition. However, there *are* people seriously thinking about and working on the matter of extraterrestrial communication.

MESSAGES TO (AND FROM) OUTER SPACE

Because of the absolute, eternal, unchanging, "transcultural" nature of primes, scientists see them as one of the few things we're likely to have in common with any "intelligent life" which might exist in other parts of the universe. If such life does exist, the reasoning goes, it should be capable of deducing the sequence of prime numbers. Being familiar with the sequence, these beings would recognise it in a repeated signal received from across the vastness of space and would realise that some other intelligent lifeforms must have been responsible for the broadcast. This is an idea which has been put forward as a way to signal our presence here on Earth to any extra-terrestrial lifeforms which might happen to be listening.

The idea goes back to at least 1941, when James Jeans, a physicist, astronomer and mathematician, suggested that we could bring ourselves to the attention of intelligent lifeforms on Mars, "*if any such there be*", by using powerful beams of light to flash, in sequence, the first prime numbers: 2, 3, 5, 7, 11, 13, 17, 19, 23, 29, 31,… when the planet was at its closest to Earth [8]. These days, more ambitious schemes set their sights much farther than Mars and use not beams of light but extremely powerful radio signals.

It's of course possible that alien lifeforms exist which deal with "reality" in such a radically different way than we do that they'd have nothing in their awareness or "culture" remotely like our number concepts. But if we're going to go to the trouble of broadcasting a message into space to announce our existence as "intelligent life", then our best bet is to broadcast sequences of prime numbers.

The reverse of this theme appears in the scientist Carl Sagan's 1985 novel *Contact* [9] (later adapted into a Hollywood film), where scientists in New Mexico searching with radio telescopes for signals of intelligent life pick up such a signal from the region of Vega, a star twenty-five light years away. One of the scientists, summarising the startling discovery to his colleagues, points out that "*...no astrophysical process is likely to generate prime numbers.*" Another, briefing the US President's science advisor, further explains: "*It's hard to imagine some radiating plasma or exploding galaxy sending out a regular set of mathematical signals like this. The prime numbers are to attract our attention.*" The point these characters are making is that the source of the signal is almost certainly *conscious* or *intelligent*.

"Project Argus", which features in *Contact*, is clearly based on SETI, an international affiliation of various projects and organisations involved in the *S*earch for *E*xtra-*T*errestrial *I*ntelligence, which has been around since 1960. Since 1985, the SETI Institute has been based in Mountain View, California. For a while, SETI's website was providing answers to frequently asked questions raised by the book and film. I found this claim:

> "*Mathematicians consider prime numbers to be a universal concept. A message containing prime numbers would simply yet unmistakably indicate an intelligent source.*"

The page has since disappeared, perhaps because not everyone agrees with this view. It rests on the assumption that no "natural" astrophysical process could produce such an "artificial" signal, a point I've seen disputed in an interesting on-line piece [10]. The point of contention was that the distinction between *natural* and *artificial* phenomena is (perhaps) a useful one for our limited, Earthbound existences, but once we widen our perspective to include the rest of the universe, it ceases to have any real meaning.

Certainly, the contributor argued, mathematics and science do not recognise such a distinction – it's a *cultural* thing.

That debate mainly concerns messages we might *receive*. At the moment, the point is entirely theoretical since, outside the realms of fiction, no sequences of prime numbers have been picked up from deep space. Since 1960 though, humans have been *transmitting* radio signals into space. The most famous message to be broadcast was beamed from a huge transmitter at Arecibo in Puerto Rico in 1974 (it's now referred to as the *Arecibo message*). It relied on the fact that "intelligent" recipients would be familiar with prime numbers. A repeating signal consisting of 1679 zeros and ones (in the form of electromagnetic pulses), starting "0000001010 1010000000000000101000000101000..." was broadcast. The hope was that intelligent beings elsewhere would notice the repetition and then begin to decode the message by observing that 1679 has the prime factorisation 23×73. If they then thought to convert the 0's and 1's into a 23×73 rectangular grid of black and white squares (or something equivalent), this would then reveal the intended message – a crude

block diagram illustrating roughly what we look like, our population, the relative sizes of the planets in our solar system, the presence of the DNA helix in our biology, its chemical makeup and the size and shape of the dish at Arecibo.

The signal would have reached the Alpha Centauri star system, the nearest place where there could potentially be planetary systems supporting life, in 1979. As I write this in 2008, it's reached to about 200 000 000 000 000 miles from the Earth.

In a lecture given in 2001 at University College Cork in Ireland, philosophy professor Garrett Barden attempted to summarise the information which humans have transmitted into space about ourselves:

> "[*We are*] *bilaterally symmetrical, sexually differentiated bipeds located on one of the outer spirals of the Milky Way, capable of recognising the prime numbers and moved by one extraordinary quality that lasts longer than all our other urges – curiosity.*" [11]

He went on to ponder the wisdom of assuming that other intelligent lifeforms would necessarily recognise prime numbers:

> "*What of the statement that we are all 'capable of recognising the prime numbers'? Not all of us do recognise them. Not all human cultures and civilisations do recognise or have recognised them. The choice of this feature in the description is startling. It is, obviously, a choice made by people within a civilisation where prime numbers are recognised. But the writers need not have chosen them.*"

In fact, even within our own civilisation, there are problems with this approach! As an experiment, astronomer and SETI founder Frank Drake sent a coded message to all the participants who had attended the 1961 Greenbank Conference (the first SETI meeting). It consisted of 1271 digits (zeros and ones only). The idea was that the recipients might recognise that 1271 factorises uniquely as 31×41 and then think to arrange the 1271 digits into a rectangular grid (using, say, black and white squares to represent 0's and 1's) to reveal a hidden message in the form of a visual diagram. Only one person succeeded in decoding the message. As David Darling's "astrobiology" website[12] puts it, "*Given that all involved were of the same species, spoke the same language and were familiar, to some extent, with the workings of Drake's mind, some idea emerges of the problems to be faced in interstellar communication.*"

A similar assumption about the presence of prime numbers necessarily being the work of intelligent life (this time in the context of microbiology rather than astrophysics) has been made by the geneticist Richard Dawkins. In his book *The River out of Eden*, Dawkins attempts to illustrate the precision of DNA codes with a fictional account of a genetic scientist who, while being forced to work against his will for an evil captor, codes a message into a genetically engineered virus. The presence of the coded message is flagged by the encoding of a repeating sequence of the first ten prime numbers. When puzzled microbiologists study the newly appeared virus which has been released into the world (through the self-infected scientist sneezing on his captors), they note this undeniable evidence of external intelligent manipulation and are able to decode the message as the captive scientist intended.

> "*Alerted by the prime numbers – which cannot have arisen spontaneously – somebody tumbles to the idea of deploying code-breaking techniques. From there it would be short work to read the full English text of Professor Crickson's message, sneezed around the world.*"[13]

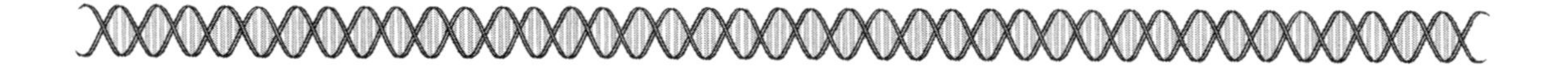

I sometimes hear people argue that number concepts and the number system are "our invention", something which humans have created and which has no existence beyond *our* existence. But the serious thought which has gone into these SETI-type efforts to reach out into the universe (whether or not they're misguided) suggests that if there *is* any nontrivial information available to us which is also accessible to non-human intelligence, it would be basic facts about the counting numbers and their factorisation into primes.

The issue of whether, or in what sense, numbers and other mathematical entities "exist" is a difficult one. I've been drawn into this debate many times and, in my experience, it never really goes anywhere. Serious philosophers continue to discuss the "ontological status" of numbers (the nature of their "existence") in what tends to be inaccessibly difficult and abstract language. Almost no one else gives this matter any real thought. If asked, some people seem to think numbers "live" solely in our inner mental world. Other people see them as being somehow evident in the external physical world. I've gradually come to see them as a kind of bridge or "interface" between the internal and external worlds.

These are issues we'll return to in the third volume. For now, I'd like to stress the "communality" of the number system and what I've already called its "anatomy".

Even if number concepts exist solely within our mental world, they're potentially accessible from within *anyone's* mind. To me, this evokes the image of these things dwelling "at the core of our being", or in that place where all of our "beings" overlap, where all consciousness intersects. We can strip away all the particularities of our individual personalities, cultural identities, ideas and opinions, and we're still left with the same thing.

As a consequence, learning more about the number system (in the right sort of way – probably not in a school-enforced or conventional academic way) could be thought of as "getting to know ourselves a bit better". If we consider the sort of alienation and negative emotional states which the number system produces in so many people (as discussed in Chapter 1), then it seems possible that this exploratory approach might lead us towards a healthier, more "integrated" kind of existence.

Regardless of any other intelligences elsewhere in the universe and our (in)ability to communicate with them, the number system, because of its "universality", acts as a source of continual wonder, arguably the ultimate mental playground. We are only scratching the surface – this thing goes on *forever*, and professional number theorists, those people most familiar with it, are the first to admit how little we really know.

Chapter 6

addition versus multiplication

You've probably been brought up to think of the counting numbers like this...

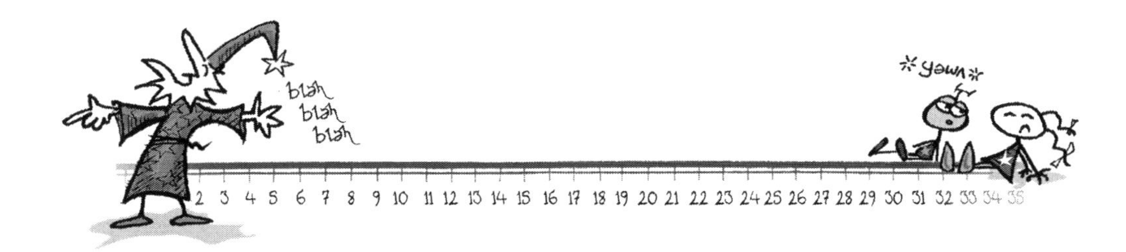

...but I'm going to encourage you to think of them like this:

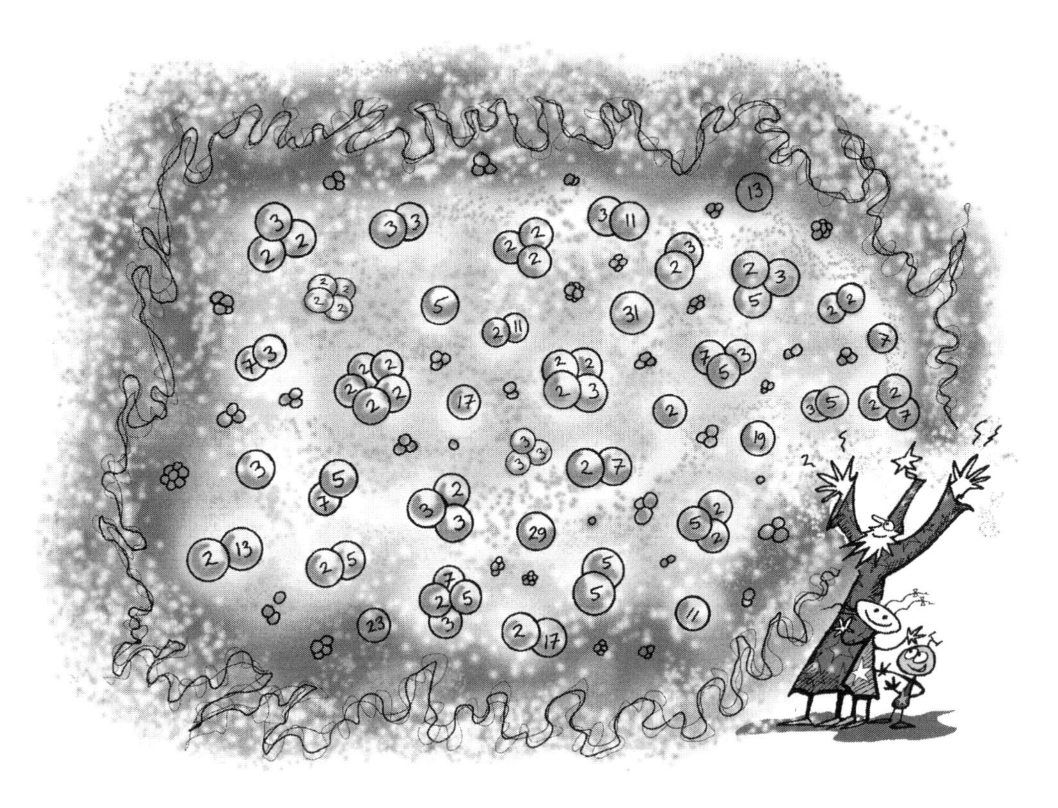

We're going to combine features of the three portrayals of the primes and counting numbers which we saw earlier (building blocks, letters and words, atoms and molecules) to produce a useful image which we can work with alongside the "number line" image.

Imagine you had a big box of table tennis balls, a magic marker and some glue. With these, you could "manufacture" each counting number by working out its prime factors, writing each of these prime numbers on a ball and glueing the balls together to represent the multiplication of prime factors (appropriately repeating repeated factors).

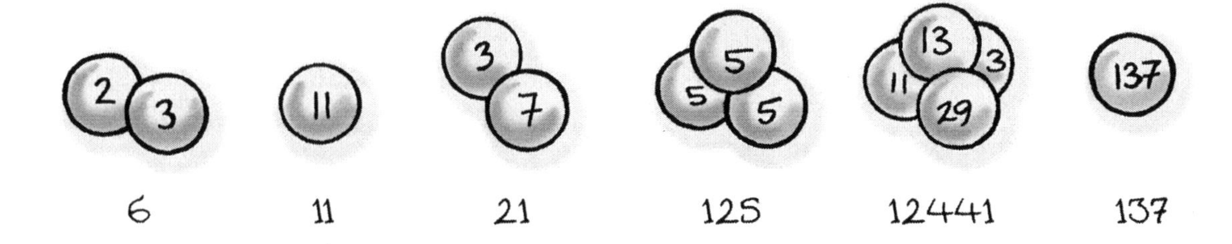

Now, imagine the balls floated in the air and glowed in the dark (why not?). Imagine you had unlimited time, space and materials, that you made representations of *all possible counting numbers* in this way and then threw them all into an infinitely vast dark room, letting them float around, glowing.

Next to the more "familiar" representation of the number system as a sequence of equally-spaced points along a line, keep this rather less "regimented" image in mind. Whenever you see a counting number, try to think not just of an amount (as with a pile of beans) but also of the unique "cluster" of primes which multiply together to produce that counting number.

The counting number 1, in this scheme, can be understood as the space in which the clusters float. If we were going to consider 1 to be a prime number, then we'd be able to construct a whole series of different clusters, like this, to represent the same counting number (42):

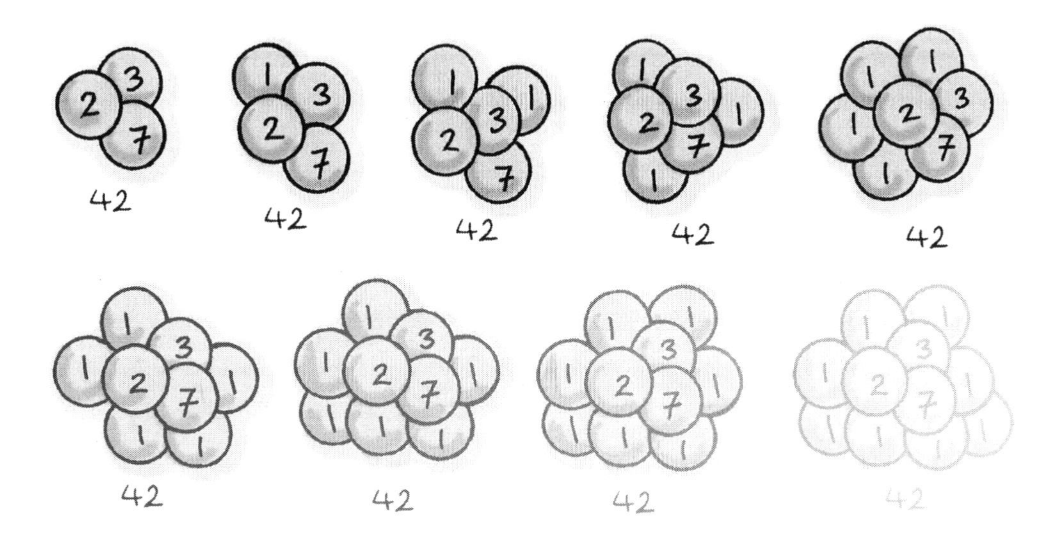

There are infinitely many of these representations (try to imagine planet-sized clusters, made entirely of 1's apart from a single 2, 3 and 7). By ruling out 1, as discussed when exploring the Fundamental Theorem of Arithmetic, we end up with a *unique* cluster representation for each counting number.

It may help to think of a factorisation like $2 \times 3 \times 7$ as being

$$1 \times 2 \times 3 \times 7.$$

We always start with 1 (empty space). We can think of the first prime-ball (2) being "glued" to this emptiness (that would be 1×2). We then glue 3 to that (to give $1 \times 2 \times 3$) and finally glue 7 to *that*.

COMPARING ADDITION AND MULTIPLICATION

Humanity has a fairly thorough understanding of the number system from the point of view of addition (although there's still work to be done relating to the "additive partitions" question discussed back in Chapter 3 – remember, though, that this involves a rather convoluted counting of *ways of combining numbers*).

However, as soon as we consider multiplication, things start to get *much* more complicated and seemingly endless mysteries begin to present themselves.

We'll have a quick look at addition first. To do so, it's most suitable to think of the counting numbers as inhabiting a number line:

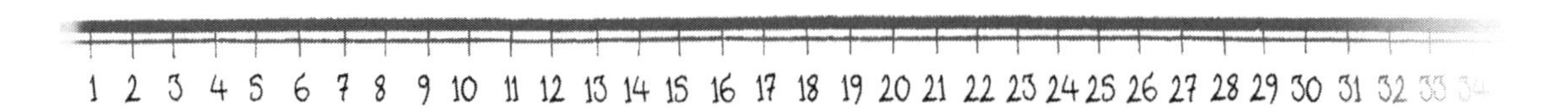

To carry out addition in this setting is entirely straightforward. If we want to perform the addition "12+ 7", we can imagine ourselves starting at the point which would correspond to zero, taking twelve steps to the right, then taking seven more steps to the right. The place you end up will correspond to the number 19.

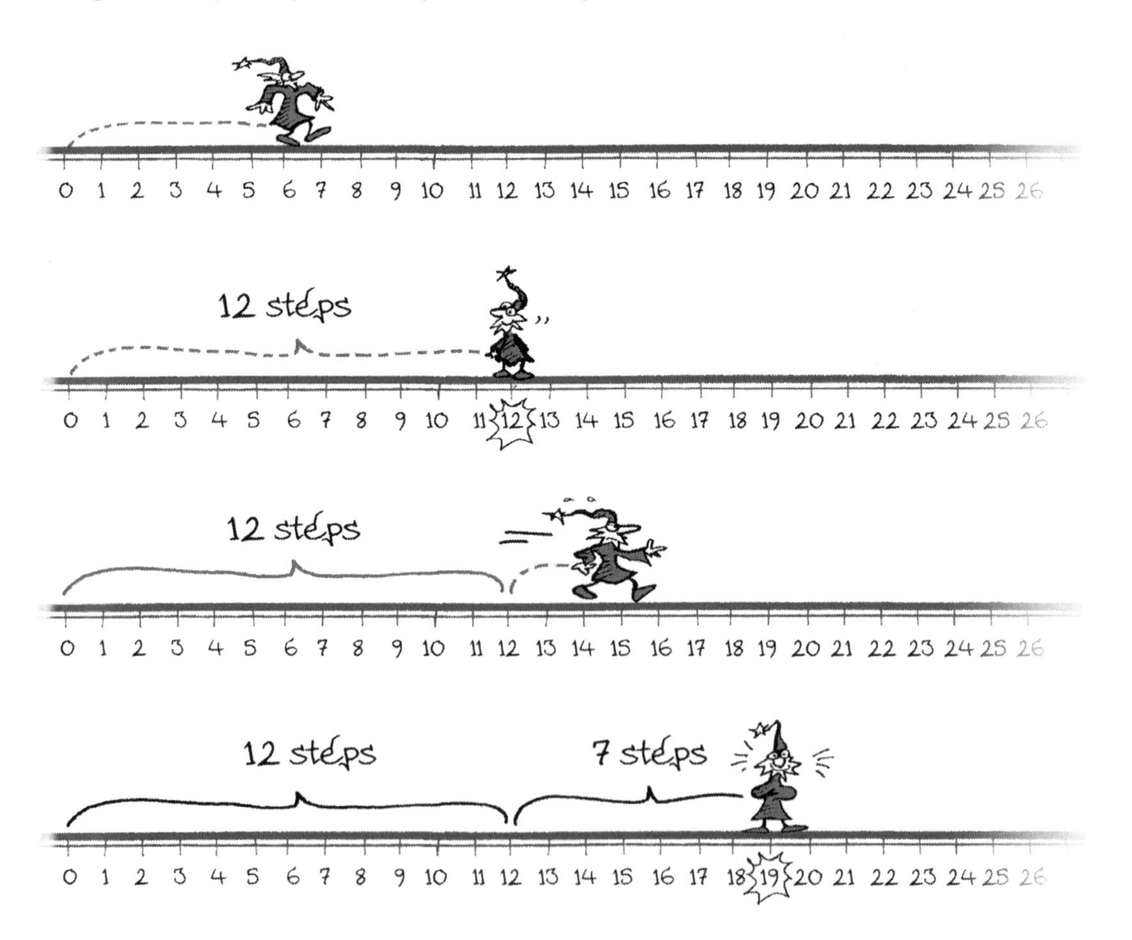

Notice how you could have taken the seven steps first, then the twelve steps – another way of expressing this is "$7 + 12 = 12 + 7$". Mathematicians call this the *commutative property* of addition. It's one of those things kids are made to learn about in school but find really stupid because it seems so obvious (they then forget about it for the rest of their lives). But it's not as obvious as you might think – there are "noncommutative" number systems, where this property which we take for granted doesn't hold! Fortunately, though, we're not going to have to deal with them in this book.

Multiplication isn't so simple. To multiply 7×12 on our number line, we would take twelve steps to the right seven times. Seven times 12.

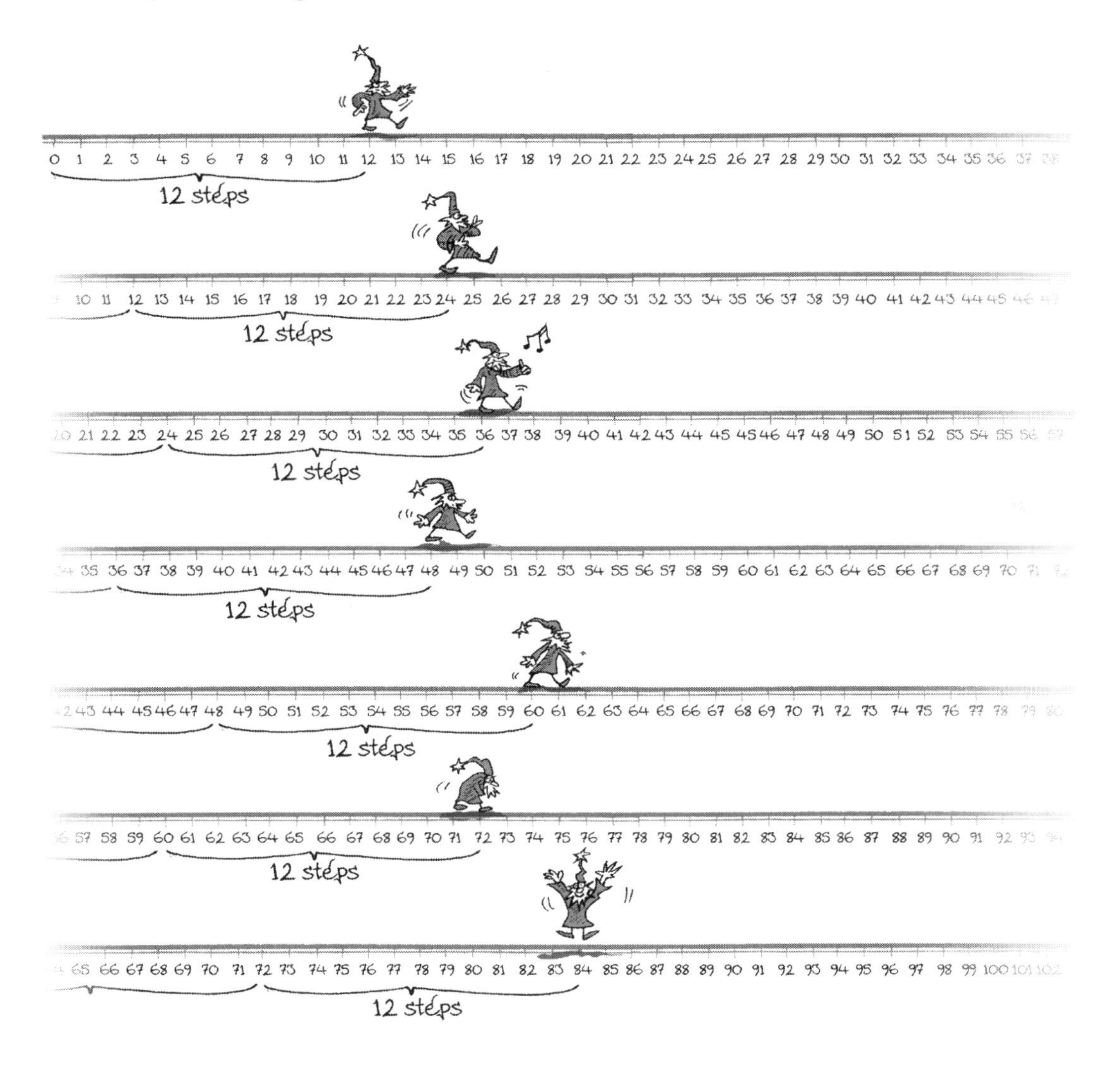

This may not seem much more complicated than addition, but it involves the concept of *repeating something a certain number of <u>times</u>*. People still generally prefer to say "seven *times* twelve" rather than "seven multiplied by twelve". This makes sense if we remind ourselves that

$$84 = 7 \times 12 = 0 + 12 + 12 + 12 + 12 + 12 + 12 + 12.$$

We can see that 84 is 12 added to zero 7 times. In slightly more outdated language, 84 is "7 times 12". Consider the sentence "Seven times she rang the bell above the door." Now consider the sentence "Seven times twelve got added to zero." This is really what "seven times twelve" is short for.

English school children informally call the multiplication tables which they are encouraged to memorise *times tables*.

Like addition, multiplication has a "commutative property". In other words, it doesn't matter in which order you multiply a pair of numbers. 84 is also 12×7 or 12 times 7, "twelve times seven got added to zero":

$$84 = 12 \times 7 = 0 + 7 + 7 + 7 + 7 + 7 + 7 + 7 + 7 + 7 + 7 + 7 + 7$$

Another way we can make sense of 7×12 is to make seven copies of the number line, mark twelve dots on each and then count the dots in the grid that this produces:

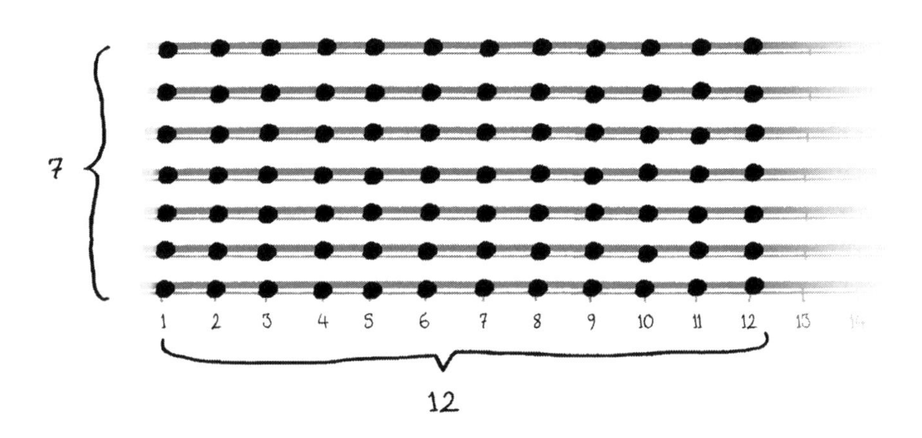

This portrayal helps to illustrate how multiplication relates to the geometric concept of *area*. If we were interested in how much paint we'd need to cover a particular rectangular wall, we could measure the edges of the wall and describe it in terms such as "12 feet × 7 feet". That would be 84 *square feet* of area which we'd need to cover in paint, the square foot being a unit of measurement for measuring area. Looking at the following illustration, we can see how the act of multiplication is involved.

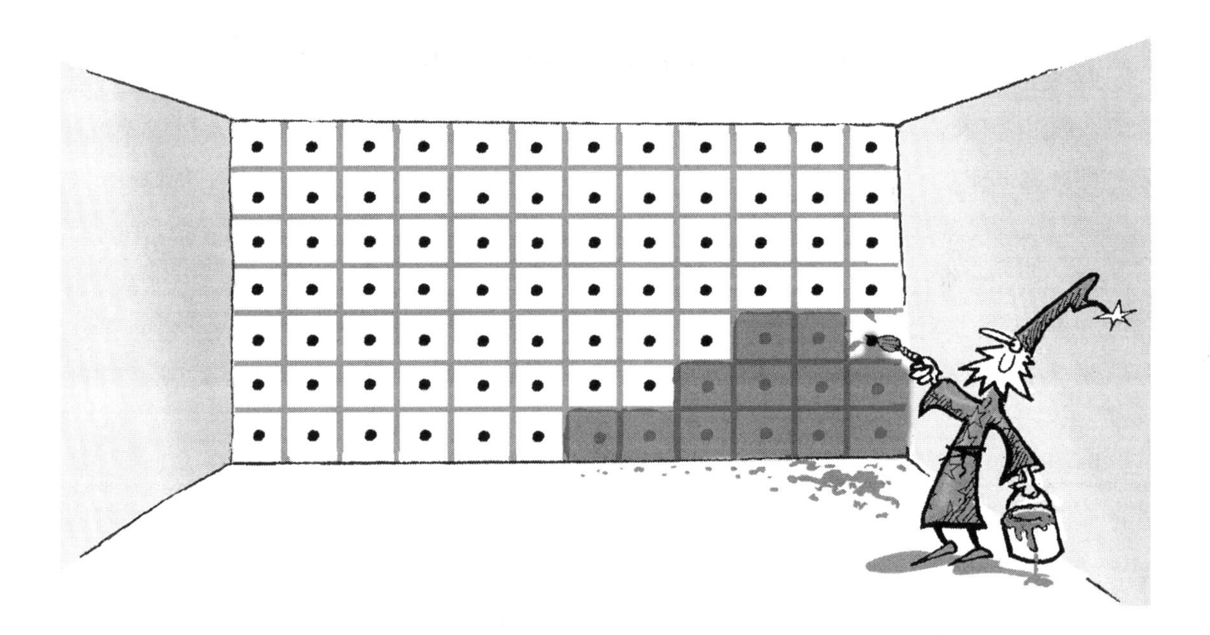

ADDITION AND MULTIPLICATION IN THE "PEANO" AND "CLUSTER" SETTINGS

Within the Peano framework, where numbers are built by adding 1's together, multiplication is clearly a more complicated matter than addition. Addition is well understood, but as the mysteries of the primes are gradually revealed to you, it will become clear that we don't fully understand multiplication or, more precisely, that *we don't understand the relationship between multiplication and addition*. Louis Kauffman, a University of Chicago mathematician whose wide-ranging interests include the foundations of mathematics, makes this point:

"*Multiplication is more complex* [*than addition*]. *When we multiply* 2×3 *we either take two threes and add them together, or we take 3 twos and add these together. In either case we make an operator out of one number and use this operator to reproduce copies of the other number.*"[1]

The word "operator" is used a lot in mathematics, incidentally – its meaning is very precise, but for our purposes it can be thought of roughly as "something that does something to something else".

The point that Kauffman is making is that with multiplication, the two numbers involved are playing different roles. When adding a pair of numbers, both numbers are "doing the same thing" (moving you a certain number of steps along the number line to the right). Here, the roles of the two numbers are very different: one is a motion to the right on the number line, the other is *the amount of times this motion is to be made*. And, interestingly, you can interchange the roles of the numbers and you'll always get the same result.

Think about it this way: 2 and 3 are counting numbers. In the context of "2×3", what are they counting?

The 3 could be counting steps taken along a line, beans placed on a table or marks made on paper. 2 is counting the number of times this action of taking three steps, placing three beans on a table or making three marks is performed. These are very different kinds of counting.

In the case of the 2, the things being counted (something happening three times) all have the number 3 somehow buried inside them. They can all be "segmented" into three similar pieces. These pieces are the things which the 3 is counting (steps, beans, marks), and these are *not* "segmentable". They're basic units, the sorts of things counting numbers are usually thought of as counting. But multiplication suggests *another kind of counting* – "the counting of countings" or counting turned in on itself. This innovation has some very strange consequences, the irregularity of the prime number sequence and the sequence of factorisations being just the beginning.

In the "prime cluster" setting, multiplication is easy! To multiply any two counting numbers together, we just glue their corresponding clusters together to form a bigger cluster.

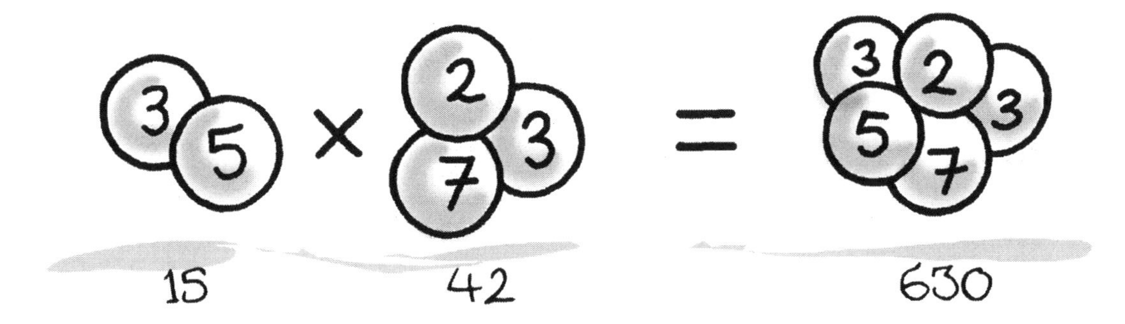

Addition, on the other hand, becomes much more complicated in this setting. The prime factors of counting numbers are not particularly relevant when you're adding the numbers together.

Take a pair of typical two-prime clusters and try adding them:

The result is a single prime. There's nothing about the factorisations $2 \times 5 = 10$ or $3 \times 7 = 21$ that suggests that this should happen.

SHIFTING OUR PERCEPTION OF THE COUNTING NUMBERS

So, we have two competing portrayals of the number system: the rather monotonous

Peano-constructed "line of dots" and the more picturesque dark room full of floating, glowing clusters of table tennis balls.

In the first scenario, addition is a simple process and multiplication is a more complicated, less obvious matter. In the second setting, multiplication is utterly simple, but addition is not at all obvious. Both pictures are valuable, depending on which aspects of the number system we're interested in.

Let's now consider the difference in character between the Peano (addition-based) and "cluster" (multiplication-based) approaches to building the number system. The first characterises numbers as being built up by adding 1's. That's how we were originally taught about numbers, so it's an approach which is familiar to everyone who ever learnt to count. The second approach characterises numbers as being built by multiplying primes together. Although this approach is just as valid, there's very little awareness of it – it comes as a surprise when explained to the great majority of people.

The reason for this bias is that the modern Westernised world has based itself on the use of numbers as *quantities*, as discussed in Chapter 1. People are concerned with acquiring, exchanging and measuring "amounts of stuff". This all comes down to the "bigger-smaller" comparison which corresponds to positions on the (Peano-related) number line. When it comes to capitalism and material goods, a number farther to the right is more desirable than a number farther to the left. Admittedly, it's hard to imagine a culture or civilisation whose dominant relationship with the number system was instead based on prime factorisation! But just because we use and think about numbers in the particular way that we do *doesn't* mean that the first view is somehow more correct, important or relevant than the second. As we journey deeper into the heart of the number system, we'll begin to see that it's the *interaction between the two approaches* which is the true source of mystery.

One aim of this book is to try to balance your view and experience of numbers by focusing on prime factorisation as an accessible, indisputable and yet utterly mysterious aspect of the number system. Through developing an awareness of the fact that every counting number can be "cracked open" to reveal an "internal structure" (its prime factors), they cease to be dull, homogeneous entities, distinguished only by their positions in the sequence. Rather than a row of cereal boxes on an infinitely long supermarket shelf, the counting numbers begin to resemble a world of radically different creatures, each with its own "anatomy" and story to tell.

An interesting exercise is to try to keep this image of an "internal structure" in mind when going about your everyday life, seeing numbers on bus timetables, in newspaper articles, as prices, phone numbers, *etc*. It's not a *practical* way of thinking about numbers: when comparing prices, we need to be able to think about numbers in terms of "bigger" and "smaller", not in terms of their prime factorisations. The "additive" view of number is far more practical in almost every situation because so much of human activity has come to be structured around the use of numbers as quantities. Although of no practical use, doing this may help you to relate to numbers in a more complete way.

As we've seen, if you were to walk along the number line and construct the cluster for each counting number in succession, you'd find that the number of balls per cluster will vary wildly. There's no rhyme or reason in it. Single-ball clusters (primes) show up sporadically along the line, but that's only part of the story. This irregularity, which is experienced both in the arrangement of primes among the counting numbers and in the variation of numbers-of-factors as we proceed along the number line, is the most obvious, "visible" face of the *bafflement* which was earlier expressed by Ian Stewart (and which has been similarly expressed by various other commentators[2]).

The Peano construction gives us the number line and there's little mystery in evidence there – the behaviour is the same all the way along[3]. If you add 3 to 5 or you add 3 to 6 411 238, the same procedure is involved, the same thing happens:

you shift along 3 units to the right. 6 411 238 only differs from 6 411 237 in that it's "one bigger" or "one unit farther down the line".

Given the choice of working in the "Peano factory" or working in the "cluster factory", with the job of systematically building the sequence of counting numbers, you can guess which one most people would choose.

However, there's a subtlety which we must not overlook. Although successively producing 1, 2, 3, 4, 5, 6, 7, 8,... by building their clusters may be more varied and interesting than just stacking 1's together, *it relies on there being a recognised sequence already in place*. We can assemble all possible clusters, all possible combinations of prime factors in our factory, but there's no obvious sense of "this cluster comes after that cluster", so the idea of assembling them *in succession* requires the assumption that the Peano approach is somehow present in what's going on. The clusters are able to slot perfectly into the sequence, but the clusters themselves will not *give us* the sequence. Knowledge of that sequence already implies that the Peano scenario has been established. So the mystery of how the primes arrange themselves within the sequence of counting numbers and the wider mystery of how all the possible factorisations arrange themselves in sequence cannot be understood in terms of either the "cluster" approach or the "Peano" approach alone. It must be understood in terms of how the two approaches interrelate.

We've already observed that the additive approach is much more in tune with the quantitative/materialistic worldview described in Chapter 1. The game we're encouraged to play is to get as far down the economic number line as possible (ideally farther than anyone else). It's not that economists and busynesspeople don't need to use multiplication. They obviously do – for example, multiplying the profit from a unit of production by the number of units sold – but prime factorisation doesn't enter the picture. In terms of money, a capitalist sees no real difference between £5 000 000 000 and £5 000 000 002. A number theorist, on the other hand, when confronted with these numbers, would see wildly different factorisations. To put it simply, the addition-based approach to numbers ties in much more closely with the whole project of "shifting units".

THE "TENSION" BETWEEN ADDITION AND MULTIPLICATION

Let's return to a remark quoted earlier:

> *"To some extent the beauty of number theory seems to be related to the contradiction between the simplicity of the integers and the complicated structure of the primes, their building blocks. This has always attracted people."* [4]

This was included in some lecture notes by mathematical physicist Andreas Knauf in 1998. He's suggesting that what people have found beautiful about number theory through the centuries is the contrast between (a) how simple the counting numbers seem when looked at in terms of addition, and (b) how complicated the counting numbers seem when looked at in terms of multiplication (or, equivalently, prime factorisation). After all, individual prime numbers don't really *have* a "structure", so when Knauf says "structure of the primes", he's referring to the structure of the entire set of primes, taken as a whole. That *does* have a complicated structure, as we shall see. That structure only reveals itself when you view the counting numbers in terms of multiplication.

Most number theorists would wholeheartedly agree with what Knauf is expressing. The most interesting thing about his statement, I find, is the use of the word "contradiction". This is similar to Ian Stewart's (prime-number-related) remark quoted in Chapter 4, "*Mathematics is full of surprises.*" There is no "contradiction" *in* the number system, just as there are no "surprises" *in* mathematics. The "contradiction", if we're going to call it that, is in *our minds and perceptions,* as are any "surprises". The word *contradiction* properly belongs to the study of logic, but we shouldn't get too caught up in precise definitions or the use of language in this casual remark. Andreas Knauf was not attempting to state a mathematical truth but rather *to express a feeling*. He is not alone in feeling this – later we'll be encountering several other, similar, statements from mathematicians and physicists.

One interpretation of the view being expressed is that the irregular pattern of primes and factorisations is the result of some kind of "clash", "friction", "collision" or "tension" between the additive and multiplicative approaches to the counting numbers – the result of addition and multiplication "interacting" within the number system. But we must keep in mind that the irregularity of primes and factorisations *only becomes apparent when the factorisations are strung out along a line, Peano-style*. While they're floating about as clusters in the vast, dark space I described, this irregularity is not in evidence – we have every possible combination of factors represented as a unique cluster but no sense of those clusters *being in any kind of order*.

In the novel *Uncle Petros and Goldbach's Conjecture* which I mentioned earlier, the author makes the following point through his fictional mathematician character Petros Papachristos:

> *"'Multiplication is unnatural in the same sense as addition is natural. It is a contrived, second-order concept, no more really than a series of additions of equal elements. 3 × 5, for example, is nothing more than 5 + 5 + 5'...'If multiplication is unnatural,' he continued, 'more so is the concept of "prime number" that springs directly from it. The extreme difficulty of the basic problems related to the primes is in fact a direct*

outcome of this. The reason there is no visible pattern in their distribution is that the very notion of multiplication – and thus of primes – is unnecessarily complex. This is the basic premise.'"[5]

I'm not sure about "unnatural", the distinction between "natural" and "artificial" being ultimately meaningless, but multiplication certainly feels like something of a *completely* different nature from addition. It's easy to be fooled by the fact that there are these two very similar symbols, "+" and "×" (corresponding to two buttons on a pocket calculator) which are used to combine any pair of numbers to produce another number. The way they are taught can leave you with the impression that these processes or "operations" *are* in the same "category of thing". But, as described by Uncle Petros, multiplication is a "second-order" concept, built out of the "first-order" concept we call addition – multiplication is *repeated addition*, so addition is in some sense simpler or more fundamental.

We should note that Petros' argument must be understood against the background of the Peano construction. If we look at the counting numbers purely in terms of the prime-clusters seen earlier, then it's multiplication which seems natural (just glue the clusters together!) and addition which seems unnatural.

THE GOLDBACH CONJECTURE

The uneasy relationship between addition and multiplication is very nicely illustrated by something called the *Goldbach Conjecture*, one of the most famous unsolved problems in mathematics. Unlike most famous unsolved problems, you don't need an education in higher mathematics to understand what the problem actually is. It's possible to state this one in such simple terms that a child could understand it. It says this:

> *Every even counting number bigger than 2 can be obtained by adding two prime numbers together.*

It's called a "conjecture" because it's widely believed to be true, but no one has yet been able to *prove* that it's true. Mathematics is awash with conjectures put forward by mathematicians who suspect or believe them to be true. Publishing such a conjecture is an invitation to the mathematical community to prove or disprove it.

If we look at the first few even numbers after 2, we find

$$4 = 2 + 2 \qquad 6 = 3 + 3 \qquad 8 = 3 + 5$$

$$10 = 5 + 5 \qquad 12 = 5 + 7 \qquad 14 = 7 + 7 \qquad 16 = 3 + 13$$

$$18 = 7 + 11 \qquad 20 = 3 + 17 \qquad 22 = 11 + 11 \qquad 24 = 7 + 17 \qquad 26 = 13 + 13$$

At the time I'm writing this, all even numbers up to 1 000 000 000 000 000 000 have been checked and found to be expressible as a sum of two primes. This may sound impressive but it's far from conclusive evidence that the Goldbach Conjecture is true. Keep in mind that the counting numbers continue *forever* so there's plenty of time for something to go wrong. Mathematics contains examples of things which appear to follow a pattern up to what would normally be considered *huge* numbers, but which then go on to fail a bit farther down the line. No amount of evidence is good enough when dealing with the infinite. Only *proof* is good enough.

We must accept that we'll never be able to check *all* even numbers bigger than 2. This is why mathematicians are seeking a *proof*, that is, a chain of logical statements which lead to the undeniable conclusion that given any even number bigger than 2, it can be written as the sum of two prime numbers.

Notice how this conjecture relates addition and multiplication. Prime numbers, as we have seen, arise in the multiplication-based view of the number system. They're things which you normally think about multiplying with each other, not *adding together*. Once you start looking at the issue of adding primes together, things get very messy!

Even answering simple questions like "can you make every even number bigger than 2 by adding a pair of prime numbers?" is beyond the capabilities of the most advanced mathematicians on the planet. Godfrey Hardy, a hugely accomplished number theorist, described the Goldbach Conjecture as one of the most difficult problems in mathematics[6]. Usually, in mathematics, the more difficult a problem is, the more difficult it is to understand *the question* (let alone find the answer!) this is a complete reversal of that trend.

The Conjecture came about after Christian Goldbach, a Prussian scholar of mathematics and law, wrote a letter in 1742 to the then famous mathematician Leonhard Euler, asking whether every counting number bigger than 1 could be expressed as *a sum of no more than three primes*. Euler wrote back, saying that it was a difficult problem to which he didn't know the answer. He also pointed out that the problem was mathematically equivalent to the question of whether every even number other than 2 can be expressed as a sum of two primes.

Here, "mathematically equivalent" means that Euler was able to prove that if the answer to either question is "yes", then the answer to the other is also "yes" and similarly, if the answer to either question is "no", then the answer to the other is also "no". Mathematical equivalence is an important concept we'll need later.

When the English translation of *Uncle Petros and Goldbach's Conjecture* was published, publishers Faber & Faber announced a $1 000 000 prize for anyone who could prove the Conjecture. This generated even more interest among amateur mathematicians with whom it was already a popular pursuit (because it's so easy to understand the question). As of 2010, there are several supposed "proofs" of the Goldbach Conjecture posted on the Web, but none has yet withstood the scrutiny of the mathematical community and achieved any kind of acceptance.

Here's something to give you an idea of how slow progress has been with this problem (keep in mind that some of the best mathematicians of every generation have been working on it since the 1740s): at a conference in Russia in 1963, a couple of little-known mathematicians called A.A. Shanin and T.A. Sheptitskaya presented a proof that every even number bigger than 2 can be written as the sum of no more than *20 000 000 000* prime numbers! There have since been considerable improvements (that is, reductions of this unhelpfully large number), but at the moment, the best result we've got is that every even number bigger than 2 is the sum of at most *6* primes[7].

chapter 7
an infinity of primes

The Fundamental Theorem of Arithmetic tells us that there is a unique way to build any counting number by multiplying prime numbers together. I hope that you've come to accept that this is true, and that this truth is about as definite as anything that can ever be talked about or written down.

An important question which we still have to address is whether this building-of-all-possible-counting-numbers can be accomplished *with a finite amount of primes*. I'm using the word "finite" here in the sense that the English alphabet contains a finite collection of letters, whereas the number system does not contain a finite amount of counting numbers. So far, nothing we've seen immediately rules out this possibility. Could it be that the occurrence of primes in the number system eventually stops at some "biggest prime number"?

Some people, vaguely recalling minor news stories about the discovery of a "biggest prime number" get confused about this. I occasionally get asked if there is indeed a single biggest prime number.

Most people instinctively suspect not. Although they couldn't say why, it seems to them that the primes must continue occurring forever.

And they're right.

The discoveries occasionally reported in the news refer to the biggest prime *which anyone has yet found*, not the biggest prime which can possibly exist.

There's a largely invisible, international scene of people (almost all enthusiasts rather than professional mathematicians) busily employing the most powerful computers and sophisticated computational techniques available to find ever larger prime numbers. At the time I'm writing this, the biggest *known* prime number is 12 978 189 digits long. That would take almost 7000 of these pages to write down in this typeface (that's a book about 18 times as thick as this volume)[1]. The number turns out to be what you get if you multiply 43 112 609 2's together and then subtract 1 (so it's the counting number which immediately precedes a number whose cluster contains nothing but 2's – 43 112 609 of them).

Primes built like that (multiplying a load of 2's and then subtracting 1) are of particular interest – they're called *Mersenne primes*. Special prime-hunting techniques have been developed for them, and usually, when a new "biggest yet" prime is found, it's a Mersenne prime.

There are rare occasions when specific examples of "very large" prime numbers are helpful in obscure corners of mathematical research. But the bulk of the activity of "prime hunting" seems to be motivated by a view of prime numbers as *isolated specimens* to be collected, as well as a fixation with quantities, particularly, quantities as *large as possible* – it's the sort of mindset that underlies the popularity of the *Guinness Book of World Records*. We won't concern ourselves with the world of "prime hunting" in this book. Searching for a new "biggest prime" is a very different kind of activity from the one we'll be engaging in for the remaining chapters – that is, exploring how the primes are arranged (distributed) throughout the number system.

But we haven't answered the original question – why can't there be a biggest possible prime number?

This was first proved by Euclid around 300 BCE, while he was (like Eratosthenes, mentioned earlier) based in Alexandria. This achievement marked the beginning of the theoretical investigation of prime numbers which continues to this day.

The proof relies on the FTA being true. More correctly, it relies on part of the FTA being true. We can think of the FTA as having two parts: one says that a prime factorisation exists for every counting number (the "existence" part) and the other says that these prime factorisations are *unique* (the "uniqueness" part). Euclid only needed the "existence" part of the FTA, which was fortunate for him, as that was something he was capable of proving (that every counting number has a prime factorisation – an outline of the reasoning was given back on pages 91–92). The uniqueness part had to wait another couple of thousand years for mathematics to develop to the point where Carl Gauss could come along and prove it (in 1798[2]).

The FTA is what it says it is – fundamental, or *foundational* to the number system. Euclid's proof that there's no biggest prime is built on part of this foundation.

People tend not to be that surprised by *what Euclid proved* – it seems somehow reasonable or appropriate to them that the primes should go on forever. If anything, they're surprised by *the fact that you can prove* something like this concerning something as erratic as the primes, and something which goes on *forever*, stretching off towards infinity. And then, once they've seen it, they're very often surprised by *just how simple the proof is.*

A little was said about the nature of mathematical proof in Chapter 4. A proof involves a chain of logical deductions which shows how whatever it is you're trying to prove inevitably follows from "first principles". These first principles are called *axioms* in mathematics. They're assumed to be true – you can't prove or disprove them, they're just *given*. To prove any other mathematical statement, you must deduce its truth from the truth of the axioms. Peano's axioms cover the basic properties of counting numbers, but as we branch out into more and more abstract mathematics, further axioms are required. As they cannot be proved or disproved, axioms are effectively "adopted" by the mathematical community via a wide-ranging consensus which settles into place over decades or centuries. In the "foundations of mathematics", that area of study where mathematics, logic and philosophy overlap, small numbers of philosophically-minded mathematicians study the properties of various "axiomatic systems" and how changing the axioms changes what is and isn't provable. This is all very far removed from the activity of most mathematical research (which takes place unquestioningly within an agreed axiomatic system). And it's even further removed from the rest of human activity!

Mathematical proof tends to be presented to students as something entirely rigid, unquestionable, "black and white". However, subtle issues remain about the most appropriate choice of axioms, as well as even more subtle matters. Despite what's often said or implied, the boundaries here do involve a certain amount of "fuzziness" – it's not all as clear cut as some people would like to believe. This is a fascinating topic but not one which we'll be able to explore further here. We'll just have to accept that within the worldwide community of mathematicians there's an extremely broad consensus as to what constitutes a proof, and that this seems to work very well. Certainly, you'll get a much more complete sense of agreement on this matter than you would among academics on any other topic.

Euclid's is one of the simplest, "cleanest", most beautiful *and* most ancient of proofs, so it's an excellent starting point if you've never seen a mathematical proof before.

When I claim that the prime numbers go on forever, this has the same intended meaning as each of the following statements:

☆ "There's an infinite amount of primes."

☆ "There's no biggest prime number."

☆ "Given any prime, there's always another prime bigger than it."

☆ "If your reasoning leads you to the conclusion that there's a finite amount of primes, then you must have gone wrong at some point."

Since Euclid's time, people have come up with a number of different proofs of this fact. If you're interested (and have some familiarity with basic mathematical notation), you can read about these in Appendix 3. The version we're going to look at here is undoubtedly the most famous and most commonly encountered, essentially the same as that which Euclid wrote down around 300 BCE.

Mathematicians would normally express this proof in the form of a single sentence, being able to compress a lot of meaning into a small number of words and symbols through a shared familiarity with the precise language and concepts involved. My explanation will be considerably longer, as I am "decompressing" the proof, trying to be as thorough as possible in making sure that you follow the argument.

A "PROOF BY CONTRADICTION"

We're going to prove it like this: we'll (perhaps confusingly) start by assuming the opposite of what we're trying to prove. Let's assume that the primes *don't* go on forever.

The idea now is to show that we must have been wrong in making this assumption.

We do this by starting from the assumption and going on to deduce something which is clearly false. There are many different strategies for putting together a proof – this particular strategy is called *proof by contradiction*. It's a technique widely used in mathematics, but I must stress that not all proofs are like this. Other approaches involve chains of logical deduction of entirely different kinds (the sketch of part of a proof of the FTA in Chapter 4 being an example). Other types of proof include *direct proof*, *proof by induction*, *proof by construction*, *proof by transposition*, *proof by exhaustion*, *combinatorial proof* and *probabilistic proof*.

Our strategy, then, is to begin by assuming that there's a finite number of primes or, equivalently, that there is a largest prime. If we accept this assumption, then we can multiply all of the primes that exist together (one of each) to give one monstrously huge composite number.

It doesn't matter how long our supposedly complete list of prime numbers goes on. If it eventually stops, then we can (at least in theory) multiply all of the primes together to give a single counting number. This number might be bigger than anything we could realistically write down or work with, but that doesn't matter for the purposes of this proof since we won't ever need to.

Now, whatever this gigantic number is, we can certainly add 1 to it. This is a basic property of the counting numbers which we earlier saw embodied as the second Peano axiom. This is the property of the counting numbers which guarantees that "there's always a next one".

This new number, the result of multiplying all the primes together and then adding 1, presents us with a problem.

Here's the problem. We have accepted the truth of the Fundamental Theorem of Arithmetic, that every counting number is either

(a) a prime number, or
(b) can be built by multiplying two or more prime numbers together.

This must apply to our new number.

We can immediately rule out possibility (a). Our number can't be prime, because it will be considerably bigger than any of the numbers in our "complete list of primes" from which it was built. If the list really *is* complete and this number's not in the list, then it can't be a prime number.

Here's a quick example of how this works. Suppose we were very lazy or not very good at counting, and we naively took "2, 3, 5, 7, 11, 13, 17, 19, 23" to be the complete list of prime numbers (so we'd believe 23 to be the "biggest prime").

In reality, we know that 23 *isn't* the largest prime, and anything like a complete list would be a lot longer than this, but for the purposes of simplicity, this list will do.

We multiply together: $2 \times 3 \times 5 \times 7 \times 11 \times 13 \times 17 \times 19 \times 23 = 223\,092\,870$.

Now we add 1: $223\,092\,870 + 1 = 223\,092\,871$.

Clearly, 223 092 871 is not in our "complete list of prime numbers" – it's far too big. The result of multiplying a list of counting numbers bigger than 1 together is *always* going to be bigger than any of the individual numbers in the list.

So we're left with possibility (b), that our number can be built by multiplying prime numbers together.

We've already seen that when a counting number is built by multiplying prime numbers together, it must be divisible by each of these prime numbers.

For example, $1276 = 2 \times 2 \times 11 \times 29$. This is how we build 1276 out of primes. We can picture 1276 as a prime cluster made of four balls. And 1276 is divisible by each of 2, 11 and 29:

$$1276 = 2 \times 638$$
$$1276 = 11 \times 116$$
$$1276 = 29 \times 44$$

In other words, a pile of 1276 beans can be separated into two equal piles (of 638), eleven equal piles (of 116) or twenty-nine equal piles (of 44). 2, 11 and 29 are all divisors of 1276.

But *none* of the primes in our supposedly "complete list" – 2, 3, 5, 7, 11, 13, 17, 19, 23 – can be a divisor of our number 223 092 871. What you'll find when you try to divide by any prime number in the list is that *you will always have 1 left over*. Try it. Bringing about exactly this situation was the whole point of adding the 1 to 223 092 870 in the first place.

$223092871 = 2 \times 111546435 + 1$:

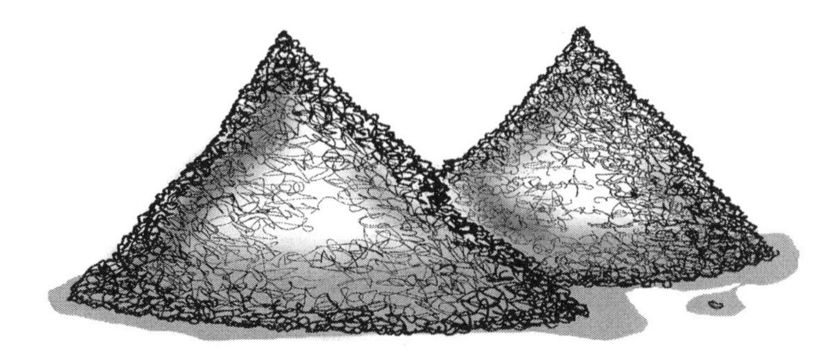

$223092871 = 3 \times 74364290 + 1$:

$223092871 = 5 \times 44618574 + 1$:

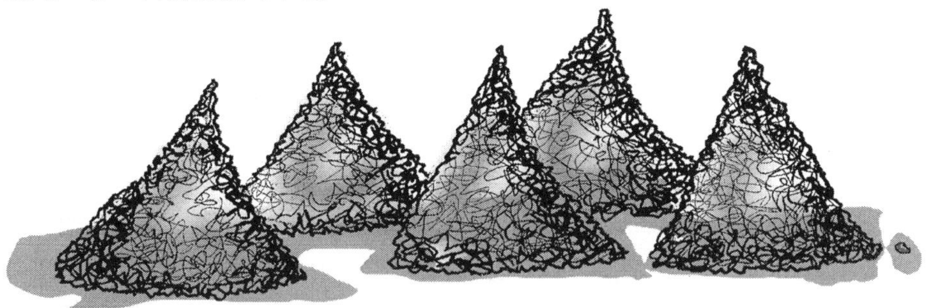

$223092871 = 7 \times 31870410 + 1$:

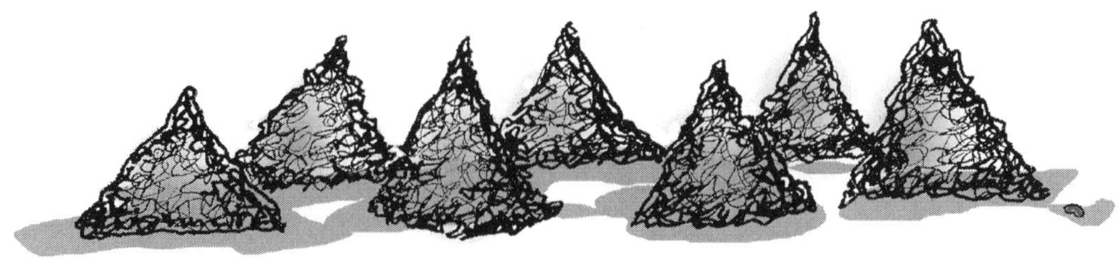

...and similarly with all of the other primes up to and including 23.

If it can't be divided by any of the primes in the list, then it should be clear that our specially manufactured number 223 092 871 can't be built by multiplying any combination of those primes together. Think about it. If it *could* be built in that way, it *would* be divisible by one or more of the primes in our list. And it isn't. So it can't be built in this way.

Summary: From our supposed "complete list of primes" (which could have stopped at 23 or any other prime number we wanted, regardless of size), we produce a counting number, and even though it may be too big to write down, we know with absolute certainty that

(1) it's not in our list of primes, so, assuming our list is complete, it can't be a prime number, and

(2) it can't be built by multiplying primes together, assuming our list is complete.

Because we accept the truth of the Fundamental Theorem of Arithmetic, we know that this situation is impossible. A counting number must either be prime or split up into primes – there's no third option. So we've deduced something which is certainly false.

How did this happen? If you correctly apply logic (as we have) and deduce something which is clearly false, then *you must have made a false assumption at some point*.

As our only assumption was that there's a biggest prime number (or, equivalently, that the primes eventually stop occurring), we can safely conclude that this must have been the false assumption.

This is a classic proof by contradiction. We assume the opposite of what we want to prove, then reason until we arrive at a contradiction from which we conclude that the assumption must be false. And if the assumption is false, then the *opposite* of the assumption must be *true*.

In this way, we can conclude with absolute certainty that there can be no biggest prime number. In other words, the primes must go on forever.

Euclid's proof that the primes continue on forever was summed up by Don Zagier in a simple, highly accessible, sentence:

> "*If there were only finitely many primes, then by multiplying them all together and adding 1, one would get a number which is not divisible by any prime at all, and that is impossible.*"[3]

Uncle Petros' fictional nephew grasped this easily and was most impressed:

> "*'That's fantastic, Uncle', I said, exhilarated by the ingeniousness of the proof. 'It's so simple!'*
>
> *'Yes,' he sighed, 'so simple, yet no one had thought of it before Euclid. Consider the lesson behind this: sometimes things appear simple only in retrospect.'*"[4]

If, unlike him, you *didn't* find the proof simple – if you found yourself getting bogged down, your eyes glazing over, skipping words or lines – then go back over it a couple of times. It's worth making the effort. Take it one sentence at a time, see if you agree or disagree with what's being said. I'm quite sure that you'll get it soon enough, and that you'll agree that each step of the proof is logically sound.

We should take note of the fact that the proof of this theorem (for that's what the proven statement "there is no biggest prime number" can now be called) relies on the truth of another theorem – the FTA. We needed to use the fact that every counting number has a prime factorisation[5]. This building of theorems upon theorems is very common in mathematics. It's the standard way in which new mathematical truths are deduced. The truth of this theorem ("the infinitude of primes", we might call it) can now be used in the proofs of more complicated theorems. The truth of those theorems can then be used in proofs of even more elaborate theorems. And so on.

"BIG" NUMBERS

A very important observation which I've hinted at, but not yet stated openly, is this:

There's no such thing as a "big number".

An obvious consequence of this is that there's no such thing as a "big prime number". My point here is that numbers are only "big" *relative to each other*. To most people, 10 000 000 is a "big number", as they rarely deal in numbers of that size, financially or otherwise. Most of us rarely have to deal with 10 000 000 people, objects or units of currency. If you're an astronomer, geologist, economist or microbiologist, 10 000 000 probably doesn't seem that large, but 10 000 000 000 000 might.

Ultimately, whatever you think of as "big" can be dwarfed by a much bigger number. Just multiply it by itself, for example.

You might be thinking 10 000 000 000 000 *is* a big number. Well, 10 000 000 000 000 times 10 000 000 000 000 is 100 000 000 000 000 000 000 000 000, and that number is *10 000 000 000 000 times bigger* than 10 000 000 000 000.

Do you think 10 000 000 000 000 is a big number? Here, a mark has been made where your number, which you thought was "big", lives on the number line. If I multiply it by itself to get a new number, my number would be something like 81 250 000 miles off the page, to the right. A ray of light would take over seven minutes to reach it.

Size only has a *relative* meaning when we're dealing with counting numbers. If someone finds a prime number with over nine million digits, that's a "big" number relative to ordinary Earthbound concerns, so it's seen as an impressive feat. It's probably a bigger number than any we're ever going to have to concern ourselves with. *But it isn't "big" in any absolute sense*.

If we continue thinking along these lines, we eventually come to realise that the range of numbers we deal with in our Earthbound reality is utterly miniscule – "infinitely small", in fact. It must be. For at any point in history, there was a largest number which had been directly thought of, referred to, stated, written, or in some other way expressed. Now, imagine marking that number on a number line. Where on the line would we find the result of multiplying the number by itself? In any portrayal of the number line which included this new, considerably larger number, the original number would be so close to zero as to be indistinguishable from it.

There's no avoiding it. The numbers just keep on going. Unlike anything else we'll encounter in our material lives, we are here dealing with something genuinely *infinite*.

For a moment, let's put aside our familiar image of the (horizontal) number line and instead think of a (vertical) "number pond", an infinitely deep pond where the surface represents 1 and the numbers continue on down forever. We can imagine people fishing in the pond, lowering their fishing lines into deeper and deeper water, seeking to catch increasingly large prime number "fish". Some have even devised net-like "sieves" for this purpose.

How deep are they really fishing, those people I described earlier, employing the latest computer technology to catch nine-million digit "fish"? Because the range of numbers humanity deals with will always be finite, we're inevitably restricted to an *infinitely thin* slice on the top of the pond – from the surface down to whichever "biggest" number anyone has had any dealings with.

So they're not really fishing at all, they're just scratching the surface, probing an (infinitely thin) scum on the surface of the number pond – and they always will be.

Despite the fact that the nicely manageable "clusters" we looked at earlier could all be fitted onto a page, the vast, vast majority of the counting numbers (more than 99.9999999999999999999999999999999999999999999% of them[6] – and that would be true if I filled the rest of this book, or a whole library of books, with 9's) would produce clusters too big to be physically represented in *any* way. Picture huge clusters, planet-sized, galaxy-sized, too big to fit inside the universe (and beyond). There comes a point on the number line where there's no longer enough time, space or resources available to ever (even theoretically) "write down" or in any other way physically express the numbers that follow.

No one will ever see even 0.000000000000000000000000000000001% of the number system. Nowhere near. Again, keeping that "1" at the end, I could fill this book with 0's and that statement would still be just as true. Most of the number system is just "out there", "doing its thing", well beyond the reach of our minds or any machines we could ever devise.

Using ever-faster computers to search for record-breaking primes can be compared to astronomers using ever-larger telescopes to look for distant galaxies, *etc*. However, there is a difference. At the moment, mainstream science tends not to support the view that physical space extends infinitely, although various cosmological theories continue to be debated – but the number system *certainly* does. So it might be meaningful to talk about "vast distances" in astronomy, as there could possibly be an absolute measure (the total size of the universe) to which these distances might favourably compare. But in the number system there can be no absolute size against which we can compare a counting number and declare it to be "big". Any number that has ever been declared "big" could equally well (in fact, more accurately) be described as "miniscule".

Unlike the flattering image of space travelling, mountain climbing, deep sea exploring, prime hunting humanity as a mythological hero on a quest to the outer regions, this awareness makes us feel tiny and insignificant. It's humbling.

FINAL THOUGHTS ON EUCLID'S PROOF

We should note that Euclid's proof relies on the Peano-based idea that given any number, you can "just add 1". The cluster-based approach doesn't involve any assumptions about what primes are available, but from those that are, it can be used to build new counting numbers endlessly (the amount of available "primes" could be finite or infinite). The act of "adding 1" is alien to the cluster-based approach – given a factorisation, say $33 = 3 \times 11$, adding 1 to get 34 yields the factorisation 2×17 and it's not immediately clear that "2×17" should follow "3×11". Introducing the idea of *adding 1* brings together addition and multiplication in a way which *forces* there to be an infinite range of primes to build clusters from. Euclid's proof involves a fusion of the two approaches (cluster-based and Peano/just-add-1). It's closely tied up with how the mysterious "interface" between addition and multiplication works.

It's remarkable that we're able to prove a (nontrivial) statement about something being *infinite*. Here we are, destined to be stuck forever in the "pond scum" of the number system, and yet we can be *certain* that these special counting numbers, the primes (which we've managed to acquire but a modest collection of), go on *forever.* The number system, by its (axiomatically defined) nature, has certain properties and, from these properties, we can logically deduce more properties. This is one of them: the prime numbers continue on forever. And that's the WAY IT IS.

chapter 8

patterns and formulas

"Prime numbers have always fascinated mathematicians, professional and amateur alike. They appear... seemingly at random, and yet not quite: there seems to be some order or pattern, just a little below the surface, just a little out of reach." (U. Dudley [1])

When talking casually with people about prime numbers, these are the two questions which come up most often:

☆ Is there a formula for producing prime numbers?

☆ Is there a pattern in the sequence of primes?

Both questions are rooted in this more subtle question:

"Is there *order* in the sequence of primes?"

Here, the word *order* is intended to mean "comprehensible arrangement", the opposite of "disorder". *Disorder* is related to the problematic idea of *randomness*, which will also be discussed in connection with the prime numbers in Volume 3.

But what people usually want to know when they ask these questions is either...

Given a list of all primes up to a certain point, is there a simple way to determine the next one in the sequence?

...or else...

Is there a simple procedure which can be used to crank out primes endlessly, at a fairly steady rate?

The simple answer to these (related) questions is "no". However, the situation is far from simple, as we shall see, partly because words such as *formula*, *pattern* and *order* do not have absolute, fixed meanings.

IS THERE A FORMULA?

By "formula", people usually mean "a string of mathematical symbols which can be used to solve a particular problem". In a more general sense though, the word is similar to "recipe" – a set of instructions which, if followed correctly, lead to a desired result.

A simple example would be a *conversion formula*. To convert a temperature from Celsius to Fahrenheit, you use the following conversion formula: multiply by 9, divide by 5, then add 32. That reads a bit like a recipe. As a string of symbols, this formula would be written "$F = \frac{9}{5}C + 32$". They mean exactly the same thing. You start with a temperature in degrees Celsius, follow the instructions, and arrive at the desired result – the equivalent temperature in degrees Fahrenheit.

A system was described back in Chapter 3 for producing a list of primes. It was first presented as a visualisation involving a rope bridge with numbered planks. This was then revealed to be a picturesque reworking of the *Sieve of Eratosthenes*, an ancient procedure which involves systematically removing all composite numbers from the sequence of counting numbers.

The Sieve of Eratosthenes could be described as a "formula" since it can be systematically applied to find the, say, 117th prime number. This would be a formula in the "recipe" sense, though, because there's no compact string of mathematical symbols to describe the procedure. However, it turns out that the sieve starts to slow down considerably as we seek increasingly large primes. The sieve provides an example of how primes can be systematically produced, but the real question is *how efficient the system is*. At the moment, apart from a few refinements of the original sieve idea, there isn't a substantially faster way to produce an exact list of primes up to a certain point. If we had a "simple formula" of the kind most people are asking about, it could be used to program a reasonably modest computer to churn out primes in rapid succession, quickly surpassing anything now possible.

There exist some "almost formulas" which appear to generate primes unfailingly but then, after a while, do fail. Here's one which Leonhard Euler discovered:

Take a counting number, multiply it by itself, subtract it from the result and then add 41.

Here are some examples:

$$1\times1-1+41=41,\quad 2\times2-2+41=43,\quad 3\times3-3+41=47$$
$$4\times4-4+41=53,\quad 5\times5-5+41=61,\quad 6\times6-6+41=71$$

This produces primes all the way up to $40\times40-40+41=1601$ (prime), but it then stops working. $41\times41-41+41=1681$, but that's just 41×41, clearly *not* a prime number.

Here's another version of the same idea: Take a counting number, multiply it by itself, *add* the number to the result and then add 41. This works up to $39\times39+39+41=1601$ (prime), failing for $40\times40+40+41$ which equals $1681=41\times41$ (composite).

There are other examples of these deceptive "formulas" which produce primes up to a certain point. It's this *style* of simple formula which people have in mind when they ask the question, rather than a sieve-style recipe for finding prime numbers. But it's a mathematical certainty that no formula like this could ever be used to produce primes indefinitely, so the mathematical community gave up the search for anything like one long ago. There are other, much more complicated formulas which *do* produce the entire collection of primes. Despite being of some minor theoretical interest, they are of no practical use to prime hunters, or anyone else, being so absurdly time-consuming to apply that they're effectively useless. The physicist Manfred Schroeder has described them as "*hilarious at best and distinguished by total impracticality*"[2]. Interested (and more mathematically confident) readers can find out more in Appendix 4.

In his thoroughly comprehensive *New Book of Prime Number Records*, a book far more interesting than its title suggests, Paolo Ribenboim authoritatively states: "*...there is no reasonable formula or function representing primes.*"[3] The key word here is "reasonable".

WHAT ABOUT PATTERNS?

What about a *pattern*? Is there a pattern in the sequence of prime numbers? That depends on what you mean by "pattern". In one sense, the sequence of primes *is* a pattern.

Of the many possible definitions of pattern, the one which is probably most useful here is *a recurrent set of features or characteristics*.

Wallpaper patterns come to mind. Recurrence or repetition is the key here. When we find a pattern in a sequence of numbers, we're able to predict what's going to happen next. Here are some examples:

1, 2, 1, 2, 1, 2, 1, 2, 1, 2, 1, 2, 1, 2, 1, 2, 1, 2, 1, 2, 1, 2, 1, 2, 1, 2, 1, 2,...

7, 14, 21, 28, 35, 42, 49, 56, 63, 70, 77, 84, 91, 98, 105, 112, 119, 126,...

11, 6, 28, 0, 9, 5, 5, 3, 11, 6, 28, 0, 9, 5, 5, 3, 11, 6, 28, 0, 9, 5, 5, 3, 11,...

4, 12, 36, 108, 324, 972, 2916, 8748, 26 244, 78 732, 236 196, 708 588,...

1, 1, 2, 1, 2, 3, 1, 2, 3, 4, 1, 2, 3, 4, 5, 1, 2, 3, 4, 5, 6, 1, 2, 3, 4, 5, 6, 7,...

1, 1, 2, 3, 5, 8, 13, 21, 34, 55, 89, 144, 233, 377, 610, 987, 1597, 2584,...

In the first four sequences, we either have a "block" of numbers repeating endlessly, or else each number is generated from the previous number by repeatedly applying a fixed procedure (add 7 or multiply by 3, for example). The last example is the least obvious, but there is a pattern – after the first pair of 1's, each number is the sum of the previous two numbers in the sequence. This is the famous *Fibonacci sequence*, which has been found to show up all over the biological world[4].

In figures such as these...

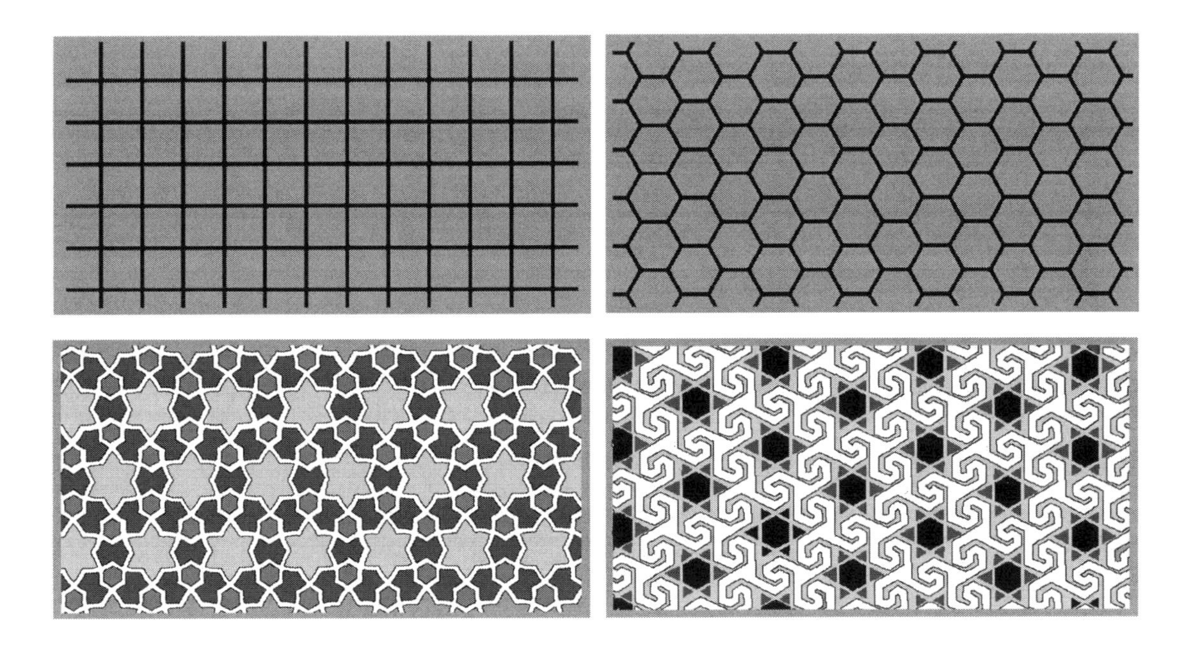

...there's a pattern which you can immediately detect and, in your imagination, extend forever in all directions.

All of the examples of patterns which I've just given are "perfect" patterns, patterns whose behaviour farther out along the sequence or farther out into the visual plane can be predicted with absolute accuracy. They exist in the world of mathematics – of pure number and geometry. Geometry, incidentally, has been described as "the unfolding of number in space" (whereas music has been described as "the unfolding of number in time").

We also find patterns like these, or at least approximations of them, in the "real world".

Looking at a decorative tiled floor or repeating wallpaper pattern, you can also imagine the pattern being extended forever in all directions. We are still essentially in the world of geometry here as these physical patterns were consciously, intentionally produced by geometrically-aware humans as representations of some endlessly repeating form in two-dimensional space.

Moving away from the human world, we still find countless "patterns" of one sort or another. There are patterns in seashells, leaves, animal markings, rock formations, honeycombs, cloud formations and the grooves left in the sand by a receding tide. If you look closely, though, there's always a certain amount of randomness or deviation from the pure geometric patterns which these imperfect physical patterns are suggesting.

Turning from geometry to look at complex physical processes, we notice patterns in people's behaviour – the order in which they do certain things or how they react when a certain type of situation arises. But people can be thoroughly unpredictable creatures at times, so even if we're very familiar with someone's pattern of behaviour, we can still sometimes make a totally wrong prediction. Here, we are very far from the pure patterns of number and geometry.

The development and behaviour of plants and animals follow certain patterns. Given an egg, we might consider what it's going to hatch into: what it will look like, what it will eat, how big it will be. Given a wildflower seed we might similarly wonder what it's going to grow into, how tall it will be, what colour its flowers will be. If we recognise the egg or seed from some previous experience, then we know more-or-less what it's going to produce this time. Still, we can never rule out mutations, "freaks" and exceptions – deviations from the well-established biological pattern.

We're aware of patterns in the weather which allow us to predict what it will do next. But, as everyone knows, even the most highly skilled meteorologists can be completely wrong at times.

The seasons are a kind of pattern, in the sense that we can expect spring to follow winter, *etc*. Although noticeably less so than in the not-that-distant past (rather worryingly), we can still predict with a certain amount of accuracy when certain flowers will bloom, migrating birds will return or heavy rains will come.

In these naturally-occurring physical examples, unlike the "perfect" mathematical patterns in the sequences we looked at, we cannot achieve 100% accurate prediction. The best we can do is to take some measurements, collect some data and carry out some statistical analysis. Based on that, we can then predict, within a certain range of accuracy, what will happen beyond that part of the pattern which we can already perceive in time or space. We'll be correct 80% or 95% or 99.9% of the time, depending not just on how well we understand the pattern but also on how wildly it varies (which is related to how much randomness or "deviation" is involved).

When weather forecasters say "70% chance of rain", they're effectively saying that 70% of recorded situations with "similar" conditions have been followed by rain. Their assertion is based on the idea that there *are* patterns in the weather which allow (within the span of a few days) reasonably accurate prediction.

The main thing I want you to keep in mind from this rather extensive discussion of "patterns" is the major distinction between the "perfect" mathematical patterns of number and geometry, and the "imperfect" patterns which occur in the physical world.

SO, IS THERE A PATTERN IN THE PRIMES?

Now, back to the prime numbers. As each prime can only appear once in the sequence

$$2, 3, 5, 7, 11, 13, 17, 19, 23, 29, 31, 37, 41, 43, 47, 53, \ldots,$$

any "pattern" that there might be could not possibly involve blocks of numbers repeating. Looking at the first few primes marked on a number line...

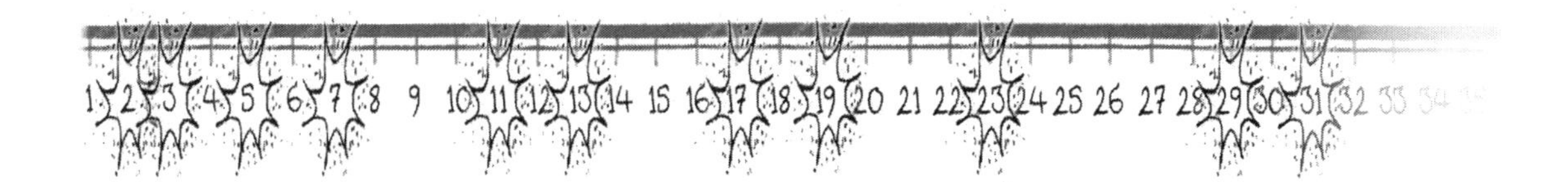

...some enthusiastic novices refuse to accept that there's not a relatively simple pattern in the primes and think that *they* (despite their lack of mathematical awareness) are going to be the first person to find the pattern, to "solve the mystery" or "crack the code". The lack of any obvious pattern seems to attract people's interest and to stir up deep feelings within them. There's an unease, evident in some of the quotations I've collected, and certain types of people feel challenged to resolve this. This is perhaps understandable if we consider just how much of our mental activity involves *pattern recognition* – recognising someone's face or voice, recognising letters and words on a page, recognising a song on the radio, *etc*. Something seemingly patternless, yet at the same time so *fundamental*, leaves some of us feeling somewhat lost or uneasy.

Some people think that they can spot the beginnings of a pattern in the *spacings between the primes*. In particular, the way that 5 and 7, 11 and 13, 17 and 19 space themselves suggests a pattern to some. However, an examination of a more extensive list of primes quickly dispels any such wishful thinking. Some fairly simple mathematics can be used to rule out the possibility of any "spatially repeating" pattern like this.

When people ask about a "pattern" in the primes, I think they generally mean a situation where you can predict exactly what will happen next based on what's gone before. This would involve some kind of repetition or behaviour which could be predicted by repeatedly applying a simple rule. But if such behaviour were present, we'd be able to express the pattern as a "reasonable formula", to use Paolo Ribenboim's words. And, as Ribenboim has explained, there is no such formula. In the sense most people intend, then, we can say that it is not possible to find a "pattern" in the prime numbers.

But here's a surprise: if we widen the notion of "pattern" so that it also includes such things as "weather patterns", then there *is* a pattern in the prime numbers. This is quite a long story, which we'll come to soon. First, though, there are a couple of related issues which it would be helpful to look at.

Another question which people tend to ask, related to the "formula" and "pattern" questions, is "How do you recognise a prime number?" A mathematician would ask "How do you test for primality – is there a simple method?". The first thing to realise is that if there *were* a simple method, then there wouldn't be such huge computational efforts going on to find primes. We *can* systematically test a number for "primality" (see Appendix 5 if you're interested), but the bigger the prime, the longer this will take, and very soon we are beyond the range of our computing technology. This is closely related to the way in which the Sieve of Eratosthenes (which can itself be used, inefficiently, to test for primality) slows down as we deal with ever larger numbers.

The lack of obvious pattern or formula means that, on first inspection, the sequence of prime numbers appears not to be "orderly". In other words, it appears to be *disorderly*. This somehow bothers or surprises many people. Ian Stewart was expressing this kind of view when he wrote:

> *"Mathematics is full of surprises. Who would have imagined, for instance, that something as straightforward as the natural numbers (1, 2, 3, 4, ...) could give birth to anything so baffling as the prime numbers (2, 3, 5, 7, 11, ...)?"* [5]

We'll encounter a lot more of this bafflement and surprise as we go on. It will become something of a theme throughout the three volumes.

MATHEMATICIANS' FEELINGS ABOUT THE "DISORDER" IN THE PRIMES

Until the 1790s, there's no record of anyone noticing any kind of pattern in the prime numbers. In 1751, Leonard Euler, at that time the most accomplished mathematician ever to have lived, wrote (overly pessimistically),

> "*Mathematicians have tried in vain to this day to discover some order in the sequence of prime numbers, and we have reason to believe that it is a mystery into which the mind will never penetrate.*"[6]

In *Uncle Petros and Goldbach's Conjecture*, author A. Doxiadis expresses this view:

> "*The seeming absence of any ascertained organizing principle in the distribution or the succession of the primes had bedevilled mathematicians for centuries and given Number Theory much of its fascination. Here was a great mystery indeed, worthy of the most exalted intelligence: since the primes are the building blocks of the* [*counting numbers*] *and the* [*counting numbers are*] *the basis of our logical understanding of the cosmos, how is it possible that their form is not determined by law? Why isn't 'divine geometry' apparent in their case?*"[7]

In crudely religious terms, this is effectively saying "*God put this number system together, so why does its internal structure look like such a complete mess?*"

Notice that Doxiadis used the words "had bedevilled", *not* "has bedevilled". As we'll go on to see, certain mathematical discoveries in the 19th century *did* reveal a sort of "organizing principle". However, the same sorts of "bedevilment", "fascination", "mystery" and sense of "how is it possible...?" which earlier mathematicians reported are still regularly experienced by people first examining the sequence of prime numbers. Everyone who learns about the primes (beyond the basic idea of what they are) is in some sense reliving humanity's collective process of discovery and passage through a series of stages from the darkness of ignorance into the light of awareness. I've found it highly instructive to observe the passage of individuals through these stages and how they react to their newly acquired knowledge.

Our collective experience of the number system and psychological orientation with respect to the counting numbers are apparently such that we initially feel there *should* be a pattern in the sequence of primes.

There's a tension between the facts (a) that the primes are timeless, indisputable, transcendent and built into the very fabric of reality, and (b) that they somehow manage to appear "disorderly" to us, almost to look *wrong*. People don't necessarily *consciously* think about it in this way, but reactions can be partly due to unconscious perceptions and attitudes. It's also worth keeping in mind that ideas about numbers (and reality in general) have changed over time, and we've all been conditioned to think in terms of the current intellectual "fashion".

In the last quoted passage, Doxiadis provides an excellent impression of how the behaviour of the primes seemed to mathematicians up until the end of the 1700s. Around that time, a pattern *was* observed – but, perhaps surprisingly, it was of the "imperfect" variety (that is, more like a "weather pattern" than a "wallpaper pattern"). To see it, we'll need to make a subtle adjustment to the way we think about primes.

THE "DISTRIBUTION" OF PRIME NUMBERS

Note that Euler referred to "the sequence of prime numbers" and Doxiadis to "the distribution or the succession of the primes". The key is to look not at individual prime numbers in isolation but rather to see the whole "scatter", "splatter" or *distribution* of primes throughout the number system. We're going to shift our attention away from the significance of primes as individual entities and towards the totality of primes as a single entity "living" inside the system of counting numbers.

If we mark out the counting numbers along a number line and highlight the locations of primes, we see that there's an overall "shape" or "form" which is produced by the sequence as a whole (it's an infinitely long shape, of course). It's this which is meant by "distribution of primes" and it's this which mathematicians have been trying to make sense of since the late 18th century.

To some people, the word "distribution" brings to mind images of vehicles delivering (or "distributing") commodities from factories and warehouses to retail outlets. This is a rather different usage of the word – it's describing the active *process* of the commodities being "distributed". The usage we have in mind instead concerns the "passive" reality of how something has *been* distributed. If you look at a book describing the birds, insects, fungi or wildflowers of the British Isles, say, you'll find little maps showing blotches in different parts of the country, indicating where you're most likely to find a particular species. Such a map shows the *distribution* of a species across the British Isles (the actual processes responsible for this distribution don't come into it). The "distribution of prime numbers" is *this* kind of distribution.

The distribution of goshawks in the British Isles.

Of course, these maps can only ever be approximate. The actual geographical distribution of ox-eye daisies or wrens is continually changing and is far too intricate to ever be completely known or described. The distribution of primes, on the other hand, is unchanging. But, because it's infinite, it, too, can never be known or described in its entirety (at least not in any immediately obvious sense). So the same kind of approximate "mapping" turns out to be appropriate for initially describing the distribution of primes, as we shall go on to see.

This approach stands in contrast to the previously discussed (and much more widely publicised) "prime hunting" which exploits the power of networked computers and various innovative computational shortcuts in a neverending quest to find the next "biggest prime number ever found". Prime hunting will always be restricted to the infinitesimal "edge" of the number system – the "pond scum" in my earlier fishing metaphor – so it tells us very little about the *distribution* of primes, a far more important (and mysterious!) matter.

Think of stamp or coin collecting – the motivation behind prime hunting has something in common with these. The difference between prime hunting and prime number theory is perhaps comparable to the difference between stamp collecting and studying in detail how every aspect of the international network of postal systems works. Collecting stamps might give you a few small clues if you were engaged in such a study, but there's a lot more to the postal system than the diversity of postage stamps! There's a similar difference/relationship between hunting animals and studying in detail their biology, social behaviour and the entire ecosystem of which they are a part. They're loosely related (knowing a bit about bears would certainly be helpful if you're hunting them), but they're *very* different activities.

We've seen that the sequence of primes continues indefinitely and that, at least on the surface, it displays an unpredictable or irregular quality which seems somehow at odds with our expectations of how something as fundamental as the number system ought to behave. Rather than thinking of individual primes as distinct collectible items (our collection of which is limited, but gradually expanding), we'll try to think of the collection of primes as an *interdependent whole*. Yes, we can pick out and consider individual prime numbers but they only have a reality as part of a collection, within the context of the entire number system.

The "prime hunting" view of prime numbers would be something like a bucketful of pebbles. They're seen as distinct, inanimate objects which we can individually discover and add to our collection. The bigger picture of which they're part is less important than the fact that they're all primes, that they all belong in the bucket.

I'd encourage you to think of the primes more as "living things", as interdependent "social beings", something like a school of fish. The population of these fish would be infinite, of course, and the infinite ocean that they're swimming around in would represent the whole number system. Without this ocean, the fish would cease to have a meaningful existence. Also, they form a social group: each is an autonomous individual, but they've evolved as social creatures, so there's something like a "group mind" in operation, coordinating their behaviour. If you watch the movements of a school of fish (or some flocks of birds), the sudden, precisely timed, collectively executed changes of direction can leave you with the feeling that the individual creatures are acting with something like a single will.

Yes, the number 31 is prime. That's indisputable and we don't need to worry about the entire sequence of primes to establish it as a fact. But "31" only has meaning within the context of the entire number system. Not only that, it's just much more fruitful for obtaining a deeper understanding to see 31 as part of an infinite "school" or "flock" of primes, which have a "collective behaviour".

LOOKING FOR THE RIGHT KIND OF PATTERN

Historically, the primes have frustrated pattern-seeking humans, as they seem to do what they want, to pop up whenever they feel like it. As Don Zagier put it,

"[D]espite their simple definition and role as the building blocks of the [*counting numbers*]*, the prime numbers…grow like weeds among* [*them*]*, seeming to obey no other law than that of chance, and nobody can predict where the next one will sprout."* [8]

The enthusiasm for seeking "big" primes is perhaps partly driven by an unconscious hope that if we could see just a bit farther out along the sequence of primes, a pattern might become visible.

The problem is that anyone motivated like this is looking in the wrong way – looking for the wrong kind of pattern.

We've discussed two kinds of patterns: "perfect" patterns involving numbers and geometry, and "imperfect" patterns which occur in the physical world (like weather patterns or the behavioural patterns of animals), where we can predict what will happen, with limited accuracy, by applying statistical thinking.

Clearly, the primes live in the "perfect" world of number. In fact, they inhabit the very *heart* of that world. Naturally, we would expect any "pattern" which they might contain or produce to be of the "perfect" kind. But, remarkably, the opposite is true.

The pattern discovered in the sequence of primes at the end of the 18tfh century has the "imperfect" or "statistical" quality possessed by real-world, natural, social and organic patterns.

The pattern in the prime numbers concerns their distribution. In other words, it describes some aspect of their *collective behaviour*.

As our "school of fish" metaphor suggests, individual primes can't exist in isolation. They interrelate at some level, somehow "cooperating" in order to obey the observed "social" pattern, collectively, even though an individual prime/fish seen in isolation may appear to be "doing its own thing".

You'll surely agree that the movements of an individual willow warbler have no bearing whatsoever on the map of the distribution of willow warblers in the British Isles. Bearing this in mind, consider the following statement:

> "*Some order begins to emerge from this chaos when the primes are considered not in their individuality but in the aggregate; one considers the social statistics of the primes and not the eccentricities of the individuals.*"[9]

Even if you're feeling comfortable with the idea of the prime numbers as a collective, or as interdependent, almost "social", entities, the phrase "social statistics of the primes" may still seem a bit unusual. An explanation is in order.

Imagine that a group of researchers had gathered large amounts of data on a small island nation with only one urban centre, where a major temple stands. Suppose that they concluded from their statistical analysis that there was a relationship between the distance someone lived from the city and the probability that they had ever visited the temple. They could express this relationship as a mathematical formula and deduce, say, that someone who lived within 10 miles of the city had a 64.8% probability, someone who lived within 20 miles had a 42.5% probability, and someone who lived within 50 miles had an 11.4% probability. Given a random selection of 1000 people living within 20 miles of the city, we would expect about 425 (42.5% of 1000) of these people to have visited the temple. The significant word here is "about". No amount of statistical data will allow us to make an *exact* prediction about the 1000 people (who were chosen at random) – we'll always have to allow for a margin of error. Still, statistics often works very effectively in situations like this (opinion polls predicting election results, *etc.*) *How* and *why* it works are subtle and fascinating questions, but we won't be able to explore them properly in these books.

We can think of the kinds of predictions that can be made by collecting and analysing data like this as being based on *pattern detection*, the pattern being of the "imperfect" variety. Suppose that an organisation carries out an opinion poll, asking 10000 randomly selected people who they'll vote for. 2380 say they'll vote for a particular candidate. Only a small sample of the population has been surveyed, but the pattern detected here is that something like 23.8% of voters intend to vote for this candidate. It could turn out to be wildly inaccurate, of course, but based on statistical experience, this figure is expected to be close to the actual figure announced once the votes have been counted.

This is the kind of pattern we can detect in the primes.

Here's the basic idea. The pattern is expressed as a formula which allows us to predict (within a certain range of accuracy) *the amount of prime numbers less than some given counting number*. This prediction can be guaranteed to achieve 90%, 95%, 99%, 99.99999999% (or whatever-percent-you-like) accuracy by requiring that the given counting number is big enough. Unlike the islanders or voters studied by the statisticians, an infinite supply of counting numbers and primes is available, and it turns out that the accuracy of our prediction can be made as close to 100% as we like, provided that we're willing to look at big enough chunks of the number line.

Because of these considerations involving ranges of accuracy, this pattern which was detected in the prime numbers at the end of the 18th century could be said to concern their collective "statistical behaviour".

Although Euclid proved that the supply of primes will never run out, we've already made the informal observation that the primes certainly *seem* to "thin out". The nature of this pattern tells us that they *do* thin out and, beyond this, it tells us about the *rate at which they thin out*. This is much like the way in which the imagined researchers' observations allow them to deduce the rate at which the proportion of people who have visited the island city temple diminishes as we travel away from the city.

I'm very happy to announce that the pattern in the primes is most easily explained using *spirals*. A simple, poetic summary would be that *the prime numbers thin out at a rate closely related to the way in which a snailshell uncoils*.

For this reason, we must now make a brief diversion, looking into the properties of a particularly beautiful family of spirals.

Chapter 9

Spirals

There are several different types of spirals, only one of which we'll be concerned with. There are crude spirals like these...

...and there are regular, geometric spirals like these (called *Archimedean spirals*)...

…and then there are these:

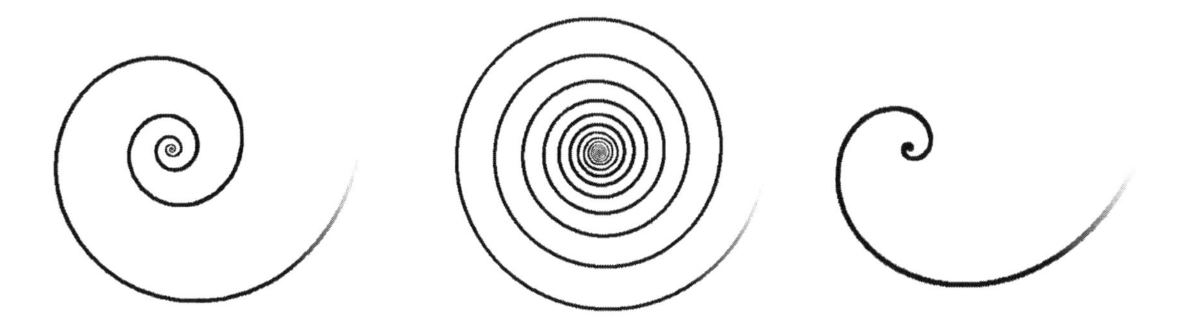

This last type can be seen in snailshells and some seashells, and they're distinguished from other types of spirals by a number of interesting properties. One is that you can zoom in at (or out from) their centre and they'll continue to look exactly the same, except for having been rotated – that's why a spiral like this drawn on a spinning disc creates a strong impression of inward or outward movement. Another key property is that if you draw a straight line through the centre, all of the crossings which the spiral makes with the line (and there will be infinitely many of these) are at the same angle.

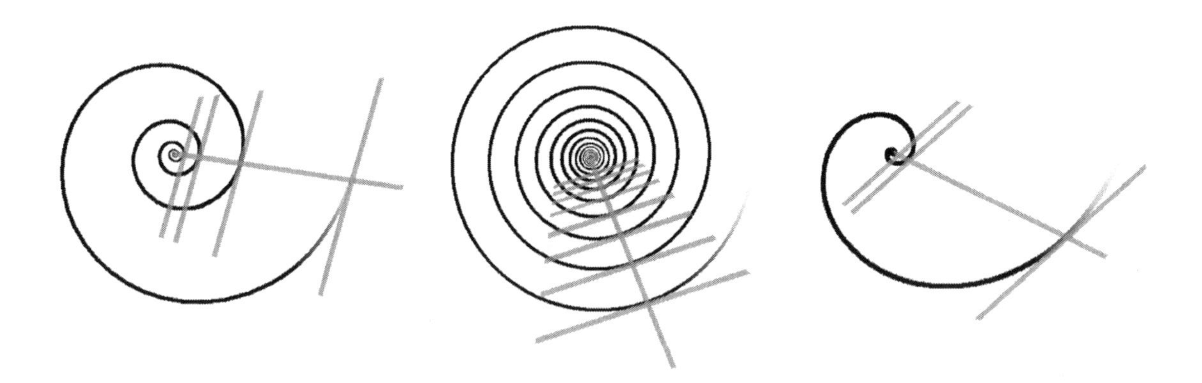

There's a whole "family" of these – they're known as *equiangular spirals* [1]. Due to their many pleasing properties, this kind of spiral was known in centuries past as the *spira mirabilis* ("marvellous spiral").

HOW TO CREATE AN EQUIANGULAR SPIRAL

One way to create a circular path would be to attach a rope to a peg in the ground and walk, keeping the rope taut:

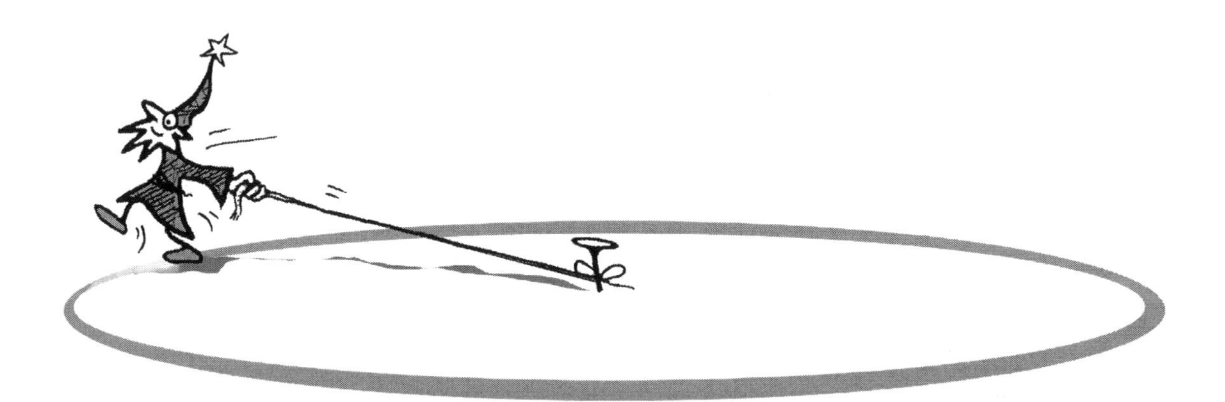

An equiangular spiral path can be created by doing something similar, although slightly more complicated. We're going to use the property that one of these spirals will always cross a line through its centre at the same angle. We attach an *elastic* rope to a peg and attach a handle like this at the other end...

...where we can set the angle of the pointer arrow relative to the handle.

If we keep the handle in line with the elastic rope (and thus with the peg) while walking in the direction of the pointer, we'll do this:

The handle apparatus is designed to keep you walking at a *fixed angle* to the centre as the rope stretches – that's why this will produce an equiangular spiral path.

If you're sitting by a fire on a hilltop at night and a moth flutters into the fire, it will have arrived *not* in a vague line from a distant point as most people would imagine (the moth supposedly being "attracted to the light"). Instead, it will have been following something very close to an equiangular spiral centred on the fire. This is because moths have evolved the ability to navigate using the light of the moon. In order to fly in an approximately straight path, they maintain a *fixed angle* with the moon. This means their path is effectively "spiralling around the moon". But the moon is so far away that even several miles of this spiral path would be indistinguishable from a straight line (much as the spherical surface of the Earth is large enough to seem flat from the localised perspective of its inhabitants). However, if a moth confuses the light of your campfire for the light of the moon, then its path will follow an equiangular spiral centred on the fire, which (if it's unlucky enough to be going the wrong way) will take it to its fiery death[2].

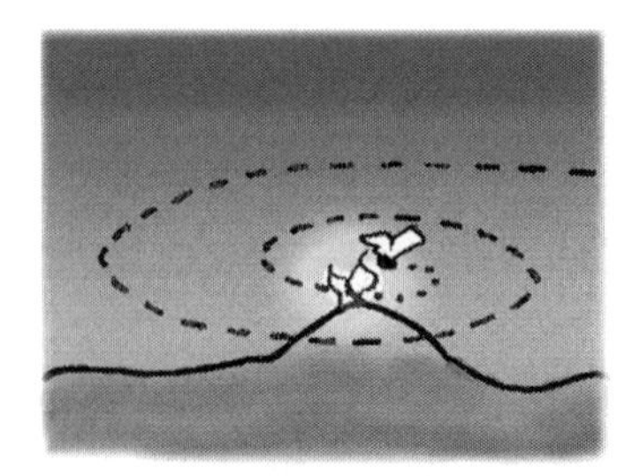

TWO "EXTREME" SPIRALS

You can vary the "tightness of winding" in these spirals as much as you like:

Notice how, as you set the pointer on your handle closer and closer to a right angle (that is, a 90° angle), the spirals become ever more tightly wound:

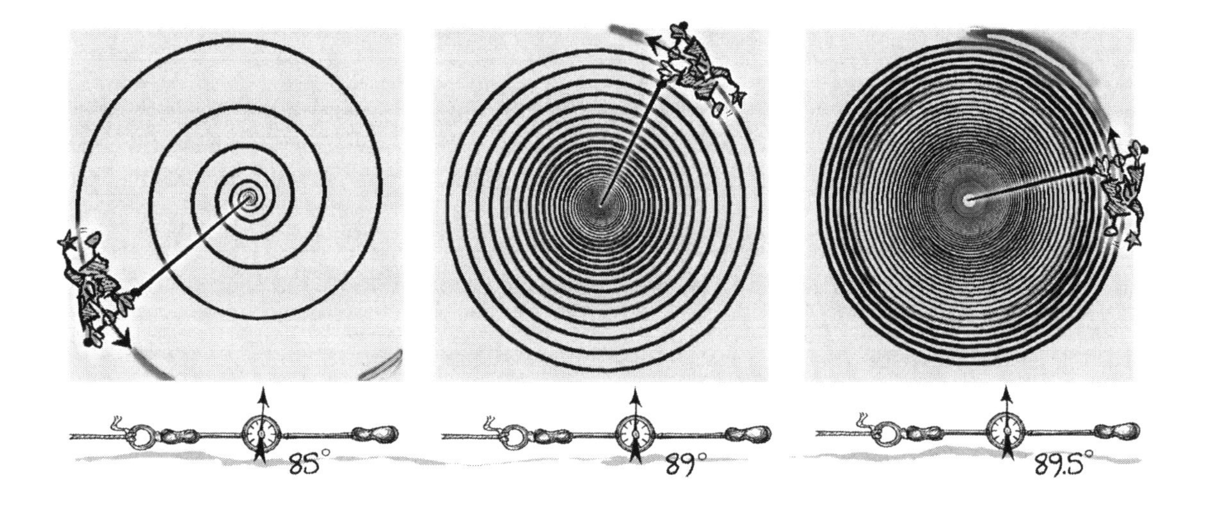

When the angle actually *is* a right angle, you'll end up walking in a circle. In this sense, a circle is just a very particular kind of spiral (mathematicians would call it a "degenerate" spiral in this context). If you set your arrow closer and closer to a 0°

angle, then you'll find that the spirals become increasingly "loose", like this:

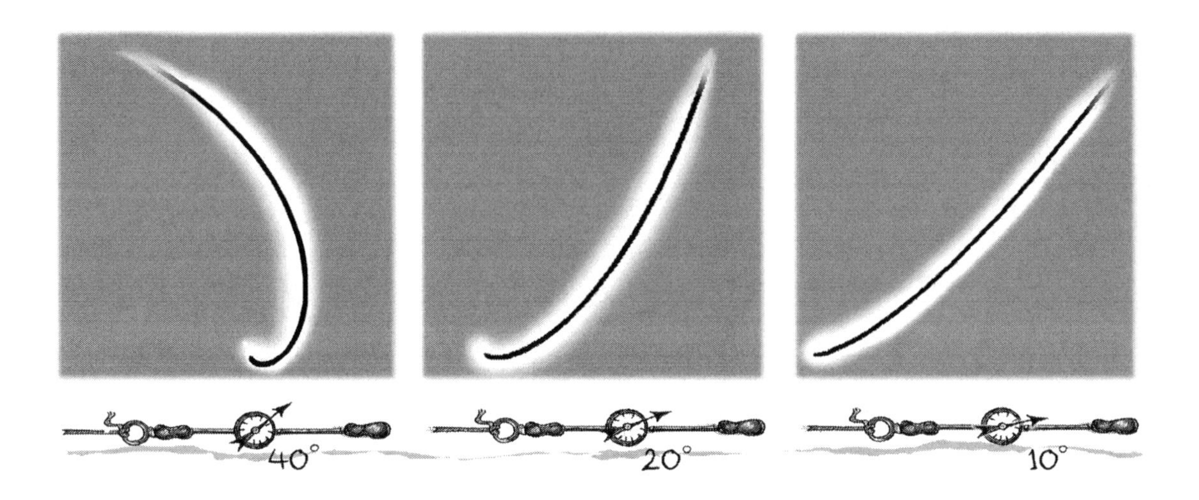

If the angle actually *becomes* 0°, then we end up walking out along a straight line. So, in this sense, a half-line[3] is another kind of ("degenerate") equiangular spiral.

COUNTING COILS

Let's now choose a less extreme example of an equiangular spiral and shade in a circle centred at the spiral's centre:

We'll use the radius of this circle as our "yardstick" or "unit". Not yet having any sense of distance here, we can simply declare that the radius of the circle is 1.

Given any number bigger than 1, using the same centre, we can draw a circle with this number as its radius – in other words, the chosen number equals the distance (number of units) from the centre out to the new circle.

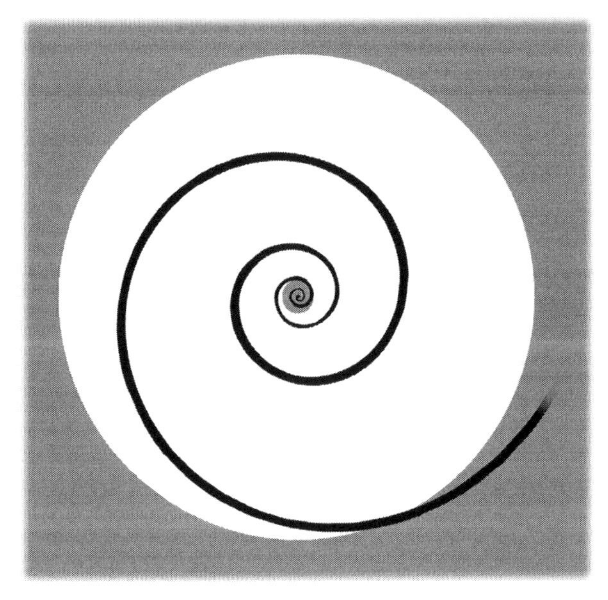

Small radius: 1. Large radius (chosen number here): 16.

Having drawn this circle, rather than shading in its interior, we're going to shade in all the space *outside* it. The white bit that's left, what I'm going to call the *visible region*, is shaped like the iris of an eye.

Now, suppose we wanted to know *how many coils* of the spiral appear in this visible region. We'll have to stop and consider how you would measure this. First, let's draw a line from the centre out through the point where the spiral emerges from the shaded-in circle. We'll use this line as a reference from which to measure.

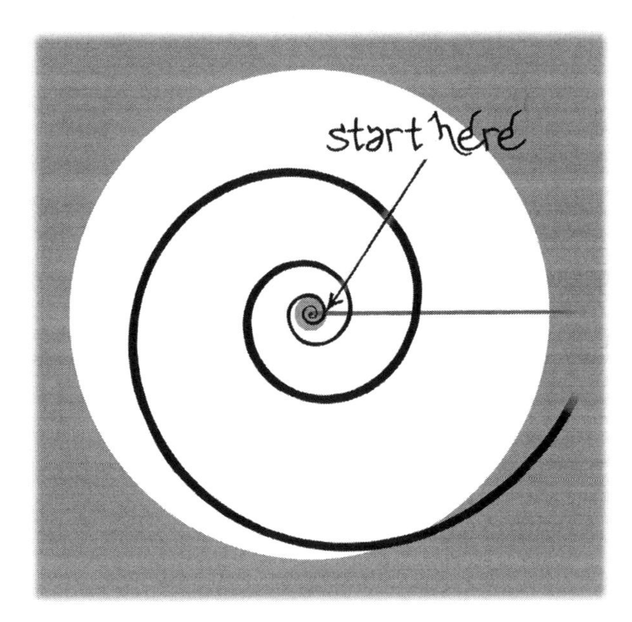

Let your eye follow the spiral out from its "starting point" on the edge of the small shaded-in circle and count how many times it crosses this line. It's two. But there's a bit of spiral left over. That could be counted as part of a coil but not a whole one. How would we measure that? Think of the face of a clock. We talk about "half an hour" or "quarter of an hour", and the distances travelled by the minute hand around the dial (halfway around, a quarter of the way around) correspond to these lengths of time. We don't usually think of 20 minutes as "a third of an hour", but it clearly is, and 20 minutes involves the minute hand travelling 1/3 of the way around the dial. So, if we look back at our leftover piece of spiral, we can see that it travels a bit more than three quarters of the way around. That would be about 0.8 (or 80%) of a coil, as three quarters would be 0.75 (or 75%) of a coil. In this way, we can say that there's a bit more than two-and-three-quarter coils in the visible region, or about 2.8 coils.

The number of coils is what is known as the *logarithm* of the number we started with (the radius of the bigger circle). Logarithms have a particularly bad reputation, having terrified generations of young people forced to learn about them. When taught in schools, rather than using a simple explanation involving spirals, for some reason, they're usually presented in terms of tedious definitions, rules and formulas – and *why* you're learning about these mysterious and confusing things is never made clear. In the past, pupils were expected to work with logarithms by looking things up in hideously dull "logarithm tables" (books full of columns of numbers). These days, electronic calculators tend to be used. Either way, a great many people have left school having *learnt about logarithms* but *understood nothing of what they really are*. All that remains is an unpleasant feeling. Spirals, on the other hand (especially equiangular ones), seem to be universally popular. Modern Western people, in my experience, tend to find spirals pleasing and, far beyond that, they appear to have had a kind of religious or spiritual significance at various times in all parts of the world[4].

One simple (and spiral-free) explanation of what logarithms do is that they "measure the size of numbers". This might sound a bit strange, as a number is already a measure of its own size. But consider the way people talk about things like a "six figure salary". The *number of digits* tells you something about the size of the number. Every time you go up by a digit, the numbers you're talking about have increased tenfold. Logarithms measure sizes of numbers in a way which is very closely related to this.

Equiangular spirals are very often called *logarithmic spirals*.

But (you might protest) there's a whole *family* of these logarithmic spirals, and the number of coils in the visible region is going to depend on which spiral we're using. The more "tightly wound" the spiral, the more coils we would see in the visible region. This is true. Just as there's a range of possible logarithmic spirals for us to work with, there's a range of different logarithms. When talking about logarithms, you must specify a *base*. Any number bigger than 1 can be a logarithm base. "Base 10" logarithms are the ones most commonly taught in schools as they're the ones which have traditionally been used in engineering and banking. But 10 just happens to be a convenient base for calculating [5] – it has no special status, and any other base is just as valid.

Each choice of base gives a different logarithmic spiral. It works as follows. Look to see how far out along the line the spiral makes its first crossing. That's the base. Recall that "1" here is the distance from the centre to the edge of the shaded disc (the inside edge of the visible region). So, here's what the "base-10 spiral" looks like:

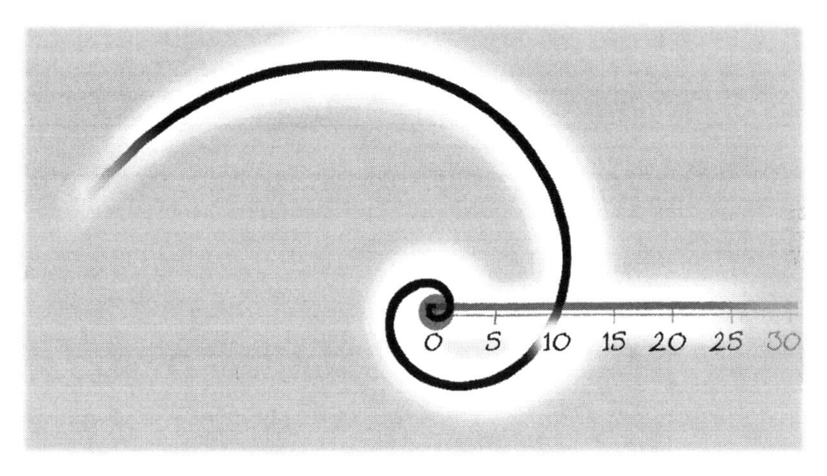

Notice how the first spiral crossing beyond the shaded disc (of radius 1) is 10 times the distance from the centre of the spiral to the edge of this disc. The next crossings will be at 100, 1000, 10000, 100000 and so on.

As you can see, this isn't as tightly wound as the last spiral we saw (which looks like, let's say, a "base-2.7 spiral"). So, for any given number which we use to create our "visible region", the amount of coils visible (that is, the logarithm of the given number) will be smaller with this "base-10" spiral. The more tightly wound the spiral, the more coils you'll see, and so the bigger the logarithm of a given number will be.

THE "GOLDEN SPIRAL"

People sometimes ask if the logarithmic spiral is related to the *golden mean* or *golden ratio*. If you're not familiar with this, it's a number slightly bigger than 1.618, with some very special properties. The most widely known of these is that a rectangle whose sides are 1 and the golden mean...

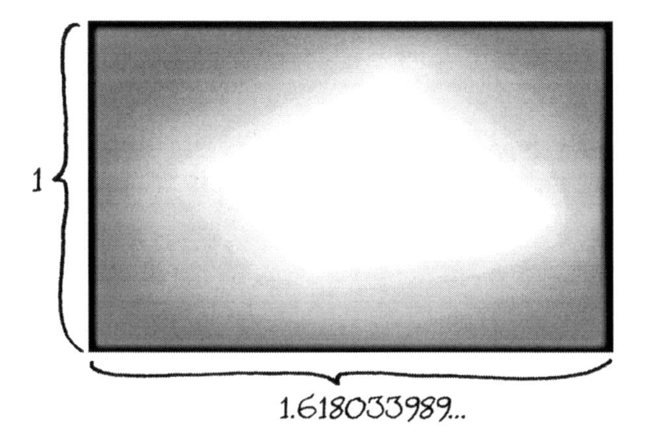

...can be broken down like this:

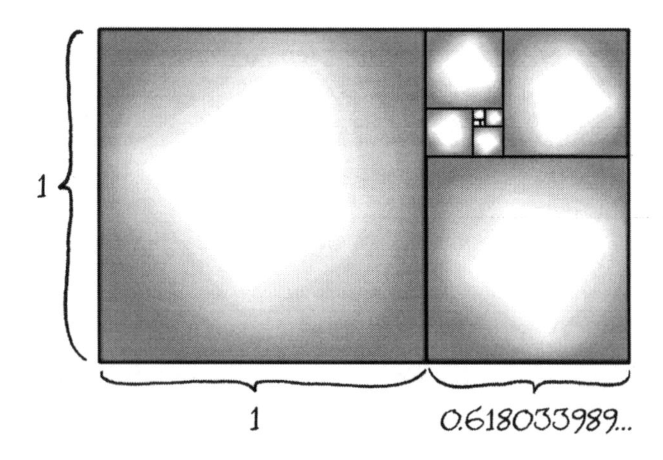

By cutting a square from one end of the rectangle, you're left with a copy of the same rectangle, just smaller, and turned on its side. This continues on down, forever, closing in on the point where these two sloping lines cross:

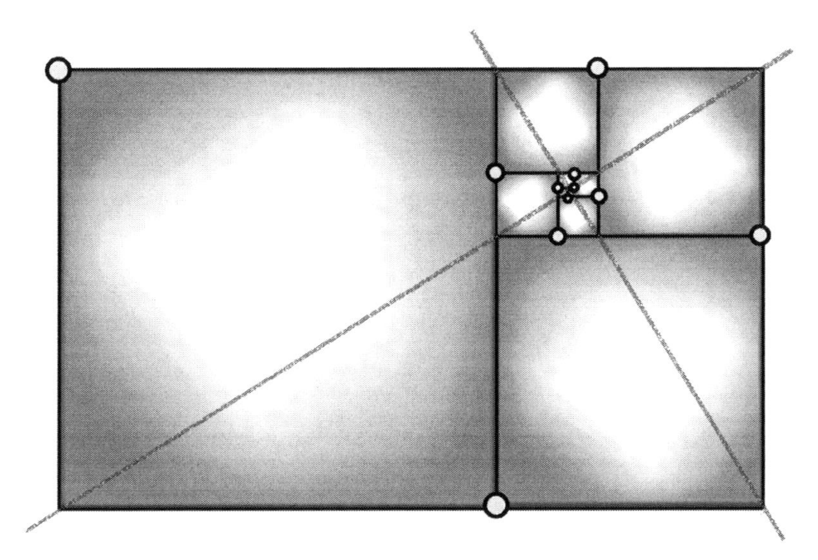

The golden mean *does* relate to logarithmic spirals in a way, because if you mark corners as I've done above, you'll find that they fall precisely on a logarithmic spiral:

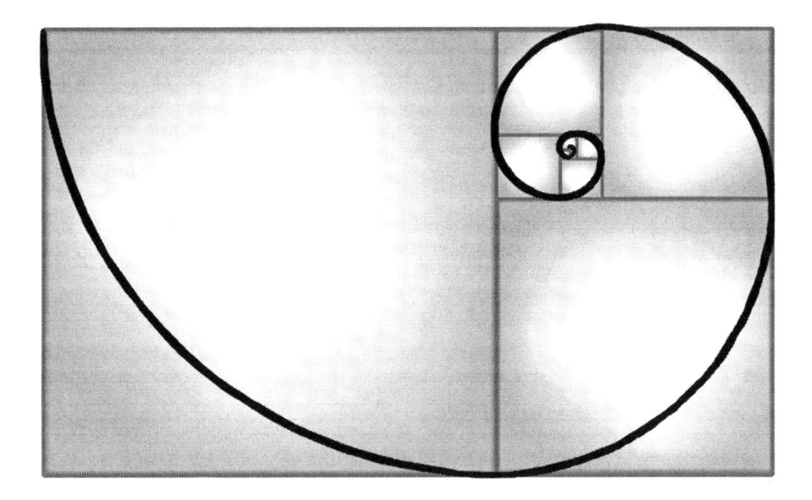

This has a base of about 6.85[6]. As beautiful as it is, it's not directly related to our exploration of the number system, so that's the last we'll see of the golden mean.

AN EXTRAORDINARY NUMBER

There is one very special base for logarithms which we'll need in order to understand the distribution of prime numbers. It's so special, in fact, that logarithms of this base are usually called *natural logarithms*. There is a mathematical sense in which they're the most "natural" kind of logarithm to work with. Some calculators have both a "log" button and a "ln" button. "log" gives the base-10 logarithm of whatever number you enter, while "ln" gives its natural logarithm.

But what's the base of this "natural" logarithm?

To answer that question, we'll first need to consider a mathematical riddle of sorts.

Suppose we have a road, with a stone marking zero and "milestones" corresponding to the counting numbers (they just have to be equally spaced)...

...and suppose there's a ladybird travelling at 1 "mile" per hour (the word "mile" could be replaced here by "foot", "metre", "yard" or "furlong" – this won't affect the outcome).

When the ladybird reaches 1, being in a bit of a hurry, she quickly drinks a magical elixir which allows her to speed up according to the rule "*wherever she is, that's how fast she's going*". So, by the time she gets to the "2" milestone, she's travelling at exactly 2 miles per hour. By the time she gets to the "10" milestone, she's going 10 miles an hour. And so on.

When she's between milestones, we can still describe her location. Here, for example, she's 5.75 miles from the zero-stone, so she's travelling at 5.75 miles per hour:

It's a continual process of acceleration. By the time she's reached the "100" milestone, she'll be moving at 100 miles per hour. By the time she's gone 10000 miles, she'll have reached a speed of 10000 miles per hour. So there's some serious acceleration going on here. To put it simply, the farther she goes, the faster she gets, and the faster she gets, the farther she goes.

Had she arrived at the "1" milestone and *not* partaken of the elixir, she would have continued to travel at a mere 1 mile per hour. After one hour, she would be at the "2" milestone, after two hours she'd be at the "3" milestone, and after nine hours she'd be at the "10" milestone. Predicting her progress would be easy. *With* the elixir, though, it's a very different story. After one hour, she'll have to have gone farther than the "2" milestone *because she's accelerating*. But how far will she have gone?

I sometimes ask people this question – where will she be one hour after drinking the elixir? It requires no real mathematical knowledge to understand the question, and it's interesting to see what people guess. I've heard all kinds of things: "three?", "three and a half?", "about 40?", "millions and millions!". Some people pick up on the fact that whatever the answer is, it will be a *significant* number. And it certainly is.

It turns out that the ladybird will be 2.7182... miles from the zero stone after one hour. After two hours, she'll be 7.3890... miles. After three hours, she'll be 20.0855...

It also turns out that 2.7182... × 2.7182... = 7.3890...

...and that 2.7182... × 2.7182... × 2.7182... = 20.0855...[7]

So, if you want to work out how far the ladybird has gone from hour to hour, you just keep multiplying by 2.7182...

This distance grows very rapidly. After nine hours, rather than being at the "10" milestone (as in elixir-free situation), the ladybird is over 8000 miles from the zero stone. And after 24 hours, rather than being at the "25" stone, she'll have gone more than 25 000 000 000 miles. You can check this on a calculator – just keep multiplying those 2.7182...'s.

But where did this 2.7182... number come from?

The number is known as "e", and mathematicians and scientists of all kinds use it regularly. One of best known places where it occurs is in the growth of populations, particularly populations which grow at a rate related to their size (imagine a colony of bacteria such that, the bigger it gets, the faster it grows). Formulas relating to population growth almost always include at least one "*e*". But it also shows up in a great diversity of other contexts, from astronomy to sociology to chemistry to particle physics.

There's another number called "pi" which you've probably heard of (it's written as "π", the Greek equivalent of the letter "p"). If you consider the simple question "*How many copies of the diameter of a circle does it takes to wrap around that circle*?"...

...the answer turns out to be 3.1415926..., a curious number that looks almost as if it's been chosen at random, as if "God" had thrown a dart at the number line. And yet *it's built directly into the structure of reality – which seems a bit weird*. The number *e* is very much like this, it's just a lot less well known. The explanation involving the ladybird is something most intelligent youngsters could understand, and yet, if mentioned at all in a school education, *e* is defined in an awkward way which makes very little sense to most people, providing no insight at all as to "where it comes from"[8].

If we now mark out where the ladybird will be after 1, 2, 3, 4 and 5 hours, we see this:

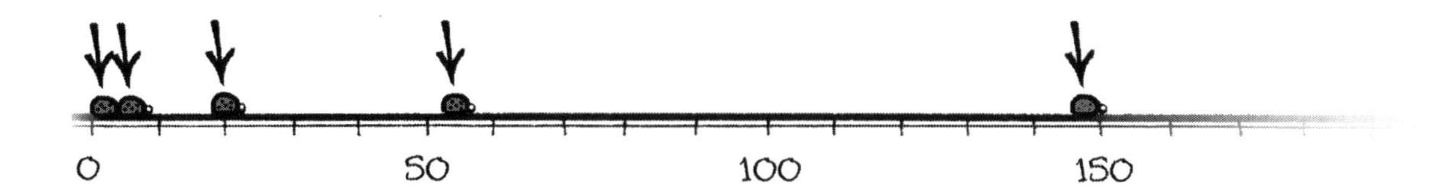

These points all lie on a particular logarithmic spiral with its centre located at 0:

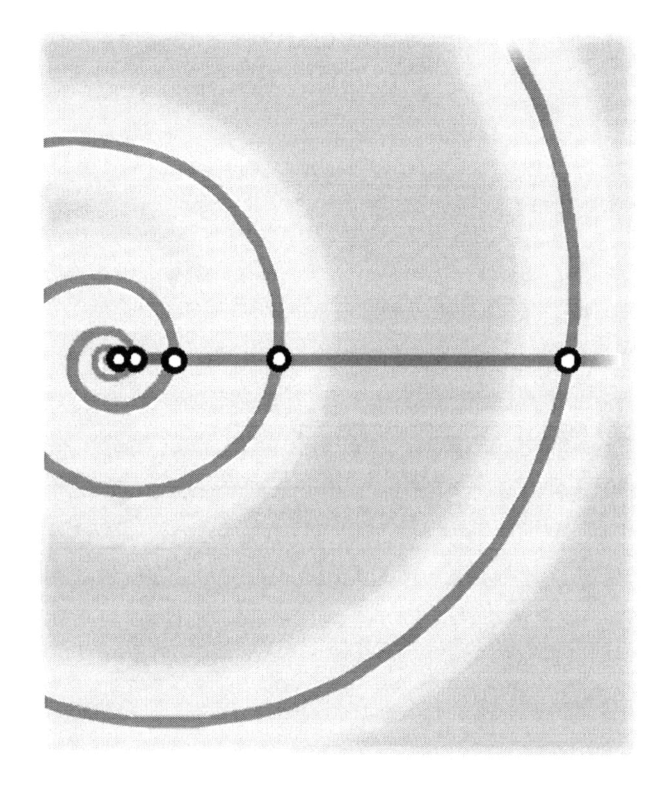

The dots marked here correspond to the positions of the ladybirds shown on the number line above.

If we shade in the circle centred at 0 with radius 1, we're left with (most of) a spiral which first crosses the "road" at 2.7182... (or e). This is the spiral, then, which corresponds to "base-e", or "natural" logarithms, as they're more commonly known. So, to find the natural logarithm of a number, mark that number on the "road", draw a circle centred at zero which passes through it, and then shade in everything outside the circle. The number of coils of the "base-e spiral" inside the visible region is the natural logarithm of your number.

What about the inside of the small, shaded-in circle? The spiral obviously continues spiralling inwardly around zero as you'd imagine, but how does the ladybird scenario tie in with this? Easily, it turns out. Suppose that immediately after she gets to the "1" milestone and drinks the elixir, you pick her up and turn her around so that she's facing 0. She starts travelling towards 0 at 1 mile per hour (she's at 1, so her speed is 1) and as she moves through locations corresponding to numbers less than 1, her speed begins to drop. When she gets to 0.9, her speed has reduced slightly to 0.9 miles per hour. By the time she arrives at 0.5, she's approaching 0 at just half a mile per hour. At 0.1, her speed is a mere 0.1 miles per hour. It turns out that, like this, she'll *never* reach 0 (although she'll come as close as you like, if you wait long enough). Marking her positions after $1, 2, 3, 4, \ldots$ hours, we see this:

These are just the crossings of the "base-*e*" spiral with the road between 0 and 1.

If you take the logarithm of a number between 0 and 1, your answer will be a negative number. Perhaps you can see why. If not, don't worry – it doesn't matter for our purposes.

If you were to pick up the "elixered" ladybird and put her at 0, but pointing towards 1, she'd then be going at a speed of 0 miles per hour. In other words, she'd be at rest and would stay at 0 forever. But if you tapped her, however slightly, in the direction of 1 (say you pushed her to 0.000000000000000000000001), she would start moving, slowly (at the ridiculous pace of 0.000000000000000000000001 miles per hour in our case), but very gradually accelerating. *Eventually*, she'd reach 1 and then (without need for further elixir) continue on beyond it with exactly the same type of acceleration we saw earlier.

Of all the possible types of logarithm, it's only the natural logarithm associated with this "base-*e* spiral" which shows up with any regularity in the mathematical sciences. And it's this one which we'll need in order to make sense of the pattern in the sequence of prime numbers which I referred to at the end of the previous chapter.

All equiangular spirals share certain nice properties, so, you might be wondering, what's so special about this *particular* logarithmic spiral? There are numerous properties which distinguish it, but perhaps the easiest to describe is as follows. Imagine you are travelling around the spiral at "one coil per hour" (this would mean you're accelerating in terms of your actual motion along the spiral, as each successive coil is bigger than the previous one – think about it). The speed at which you're moving away from the centre will be *the same as your distance from the centre*. So, if you're maintaining a speed around the spiral of one coil per hour, when you're at the unique point on the spiral which is at a distance of 23 miles from the centre, then, at that moment, you'll be moving away from the centre at 23 miles per hour. No other base of logarithmic spiral has this property.

HOW TO CREATE THIS PARTICULAR SPIRAL

You might be wondering how we would construct this "base-*e* spiral" using the "elastic rope trick" seen on pages 169–170. The question comes down to the angle at which we set the pointer on our handle.

It should be the same as the angle at which the spiral crosses the horizontal line (see the picture on page 182). We just set the pointer to this angle and start walking.

Looking at the spiral, we can see that it should be something like this...

...but how, exactly, do we find that angle? As it happens, there's a remarkably easy way to do this.

We start with a circle of radius 1, then "open out" the circle and flatten it, to produce something with a length that turns out to be 2 times that number π ("pi") mentioned earlier. It's about 6.28. Stand that upright and draw a diagonal line as shown:

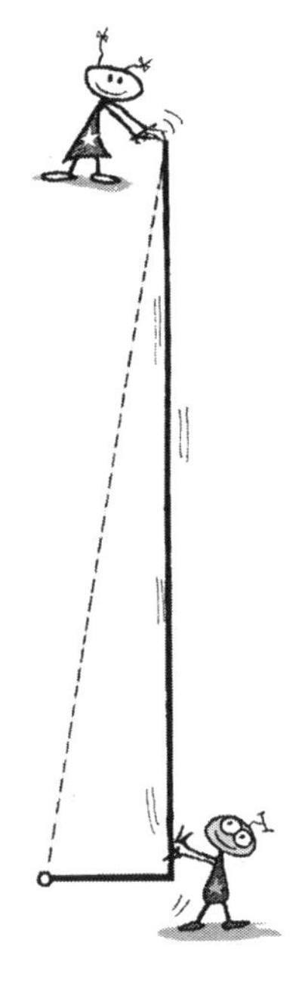

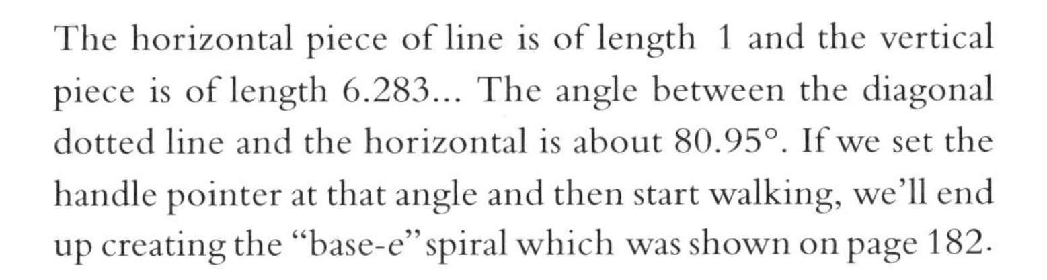

The horizontal piece of line is of length 1 and the vertical piece is of length 6.283... The angle between the diagonal dotted line and the horizontal is about 80.95°. If we set the handle pointer at that angle and then start walking, we'll end up creating the "base-e" spiral which was shown on page 182.

That will produce the "base-e" spiral. So e and π are related (and there are other ways in which they are related – everything in mathematics is ultimately related, and usually in more than one way).

chapter 10
the distribution

Spirals, logarithms, *e* – all very interesting...but wasn't this book supposed to be about prime numbers?

Indeed it was, and it is.

The last we heard about them was something involving an "imperfect", "statistical" "weather pattern"-kind-of-pattern which describes the prime numbers' *distribution*. It was this, I claimed at the end of Chapter 8, which related to logarithmic spirals (or to "the way in which a snailshell uncoils", to put it a bit more poetically) – that's why we needed to make that diversion. It's related, in particular, to the "base-*e* spiral", which is why we've focused on that one specifically.

As I've explained, the seemingly wild, unpredictable nature of the prime numbers can be understood as the result of the interaction (or clash) between addition and multiplication. The primes' role as factors clearly relates to multiplication, while their various positions on the number line are more closely related to addition – think of how we add 1's to get from a number to its successor.

To most people familiar with them, logarithms are associated with sophisticated applications in engineering, economics, physics or mathematics. But regardless of any such "usefulness", as we'll next see, logarithms also *relate directly to the interaction of addition and multiplication* and that's perhaps the simplest explanation as to why they show up in the (unlikely, to many people) context of prime number theory.

If you have two numbers, you can find their logarithms as I've described:

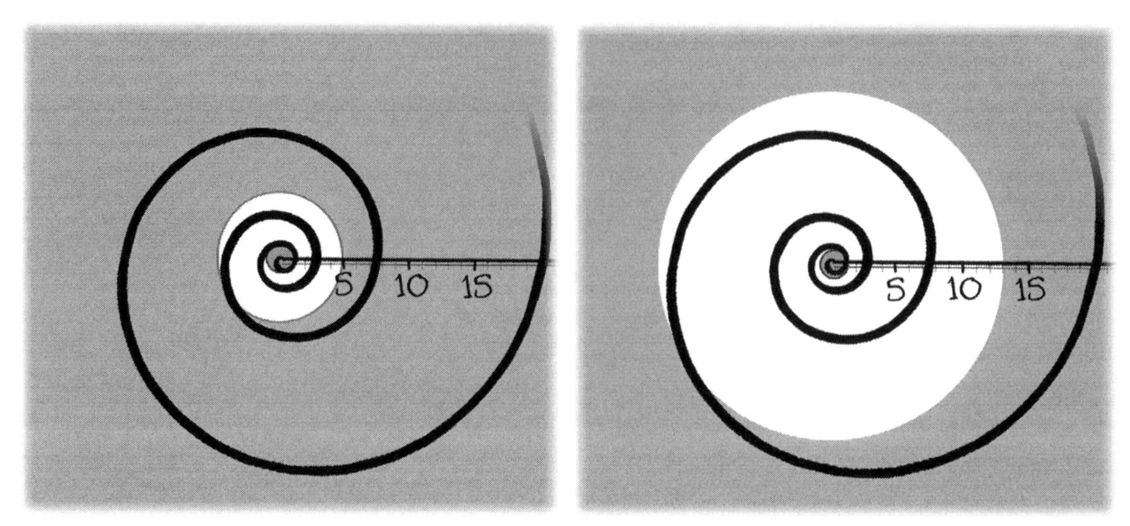

In this case, the chosen numbers are 5 and 13. Counting coils in the "visible" (white) regions, we can see that their logarithms are roughly 1.6 and 2.6, respectively.

If you now *multiply* the two chosen numbers, you'll find that the logarithm of the result is the same as what you get if you *add* the two original logarithms (see image below). In this way, logarithms (of any base) relate multiplication and addition.

Multiplying 5 and 13, we get 65 (so the inner circle is now too small to see). Counting coils, we get about 4.2.

There's another way to illustrate this. Given any number, there will be a "distance from centre" that you would be if you had travelled around that many coils (from the "starting point" of the logarithmic spiral – the point where it emerges from the shaded-in circle).

Given any two numbers of coils, we can find the two distances-from-centre, as illustrated in the next image:

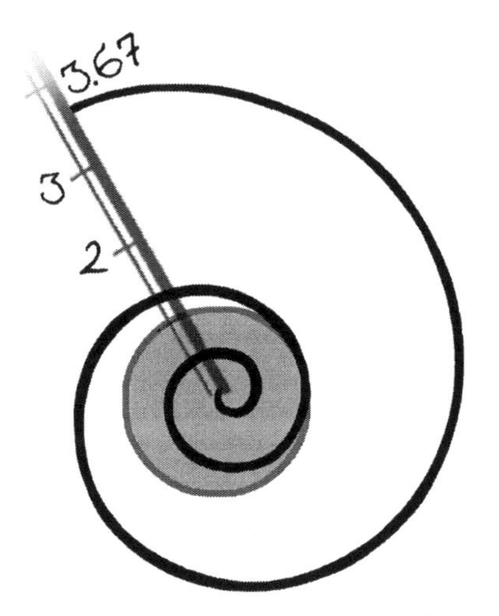

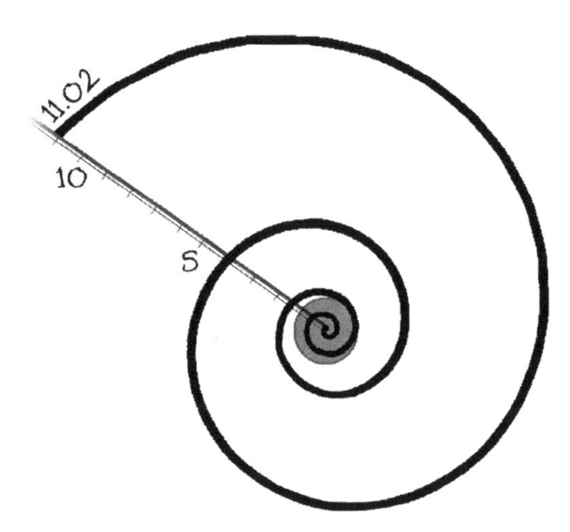

Here, the chosen numbers of coils to be traversed from the edge of the disc are 1.3 (which takes us about 3.67 units from the centre) and 2.4 (which takes us about 11.02 units from the centre).

If we *add* the two numbers of coils and then travel the total number of coils around the spiral from the starting point, our distance from the centre will then be the same as what we would get if we *multiplied* the two distances-from-centre.

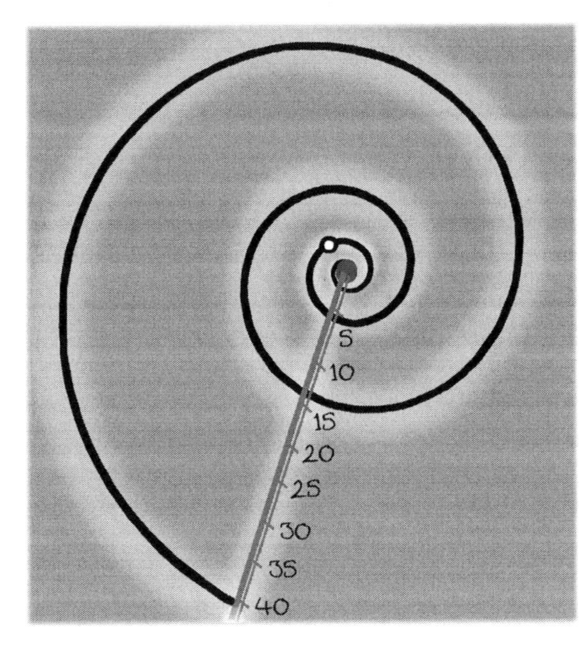

Total coils = 1.3 + 2.4 = 3.7. The white dot marks where the 1.3 coils end and the 2.4 begin. Final distance from centre = 3.67 × 11.02 = 40.44.

And prime numbers? They can be understood both as factors and as the results of adding together amounts of 1's. So, the relationship between multiplication and addition is linked to the distribution of the prime numbers within the counting numbers. But we've just seen that it's also linked to the properties of logarithmic spirals. Because of this, the "imperfect" pattern in the prime numbers can be expressed in a way which directly involves logarithmic spirals.

COUNTING PRIMES

One way to work with the distribution of primes is to keep track of how many of them have appeared up to any given counting number. If you have exact information of this kind, then you can easily see where the primes are. For example, the fact that there are 563 primes less than 4093 but 564 less than 4094 means that 4093 must be prime. Whenever the count jumps by 1, a prime has just appeared.

Rather than counting how many primes there are *less than* a given number, we could count how many primes there are *up to and including* that number. There's very little difference (literally – the two ways of counting can never produce numbers that differ by more than 1). There are 563 primes up to and including 4092, there are 564 primes up to and including 4093 and there are also 564 primes up to and including 4094. The count jumps when we reach a prime, rather than at the number which immediately follows (as in the first approach). It has become common practice to count primes up to and including, rather than less than, given numbers. We'll stick with this approach.

In the 19th century, mathematicians began working with something called a *prime counting function*. As far as we're concerned, this can be thought of as a kind of infinite table like this:

counting number	1	2	3	4	5	6	7	8	9	10	11	12	…
amount of primes up to and including that number	0	1	2	2	3	3	4	4	4	4	5	5	…

counting number	22	23	24	25	26	…	106	107	108	109	110	…
amount of primes up to and including that number	8	9	9	9	9	…	27	28	28	29	29	…

counting number	6187	6188	6189	6190	6191	…
amount of primes up to and including that number	804	804	804	804	804	…

Here's a visualisation to help you see how a table like this would be put together. Imagine that you are walking along the number line, this time in the form of a sequence of numbered paving slabs. Each time you step on a prime-numbered slab, it lights up. When that happens, you add 1 to your count. You start from slab 1 on the number line with your count set at 0. At each counting-number slab, you record in your table the current count of how many primes you've stepped on.

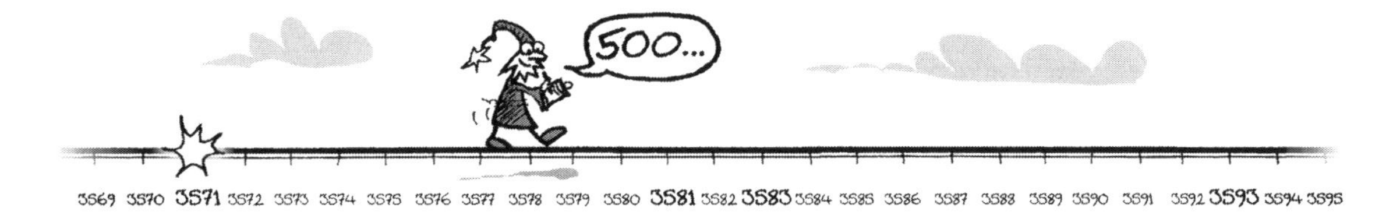

There are 16 prime numbers up to 56. There are 146 primes up to 846, and 500 up to 3578.

A PATTERN IN THE PRIME COUNT

The pattern found in the sequence of primes at the end of the 18th century, the one which relates to spirals, describes a relationship between the numbers on the upper row and the numbers on the lower row of our table – an *approximate* relationship. And it's quite simple, really: you simply take a number from the upper row and *divide it by its own logarithm*. That will give you an approximation for the corresponding number below it. In other words:

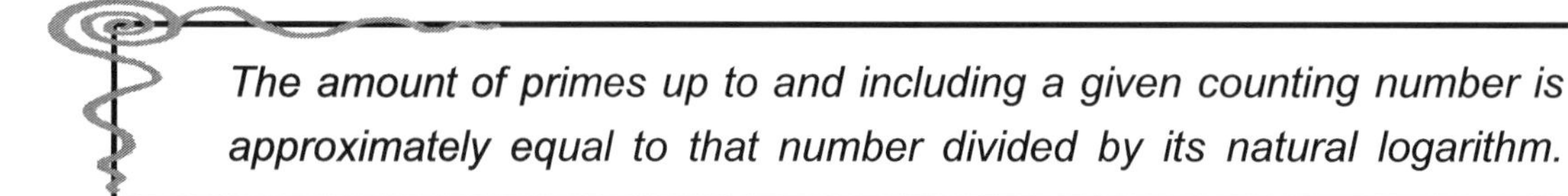

The amount of primes up to and including a given counting number is approximately equal to that number divided by its natural logarithm.

Here's an example. We pick 258. We want to know how many prime numbers there are up to and including 258. First, then, we need to find its natural logarithm:

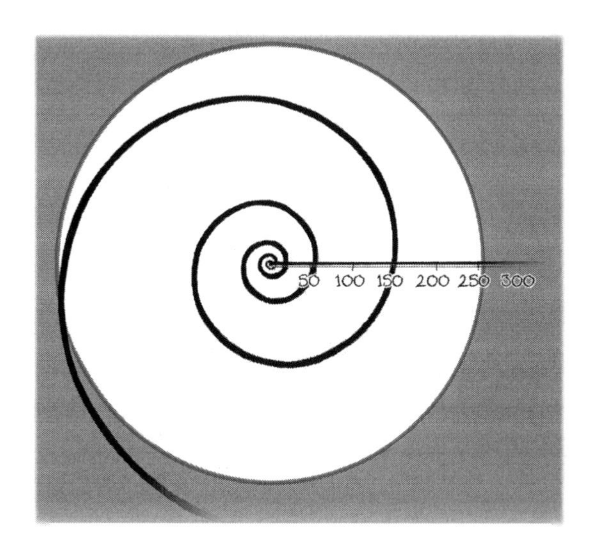

The outer edge of the visible region is at a distance of 258 from the centre and its inner edge is too close to 0 to be visible here. There are approximately 5.553 coils contained within it (although not all visible here), so the logarithm of 258 is approximately 5.553. If we divide 258 by this, we get about 46.462, which means that there are about 46 primes up to and including 258. The actual number is 55.

That makes this estimate about 84% accurate. That's not *too* bad, but it's not very impressive either. Applying this method to ever larger numbers, the accuracy gets closer and closer to 100%. For many smaller numbers, though, it's not so good[1].

Here's an even simpler (but less accurate) approximation. First, we draw a number line extending out from the centre of the "base-*e*" spiral, then we refer back to the earlier visualisation which involved the glowing paving slabs. Suppose that, as well as keeping an accurate count of the primes you've stepped on, you also keep track of *how many times you've crossed the spiral*. That number is never going to differ by more than 1 from the logarithm of the number you're standing at because the relevant number of coils is going to be equal to the number of times the spiral has crossed the line, plus (usually) an extra bit.

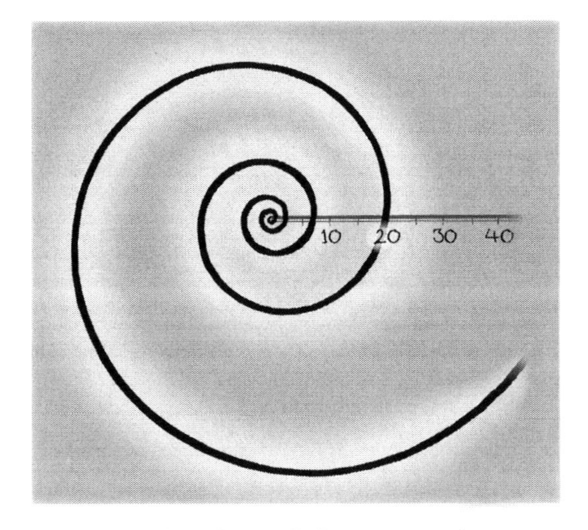

So, dividing the number you're standing at by the number of spiral crossings produces something reasonably close to what you'd get if you divided it by its logarithm.

If we're looking at a "base-e" spiral, the first crossing of the number line after 1 occurs around 2.7, the second occurs around 7.4 and the third occurs just after 20.

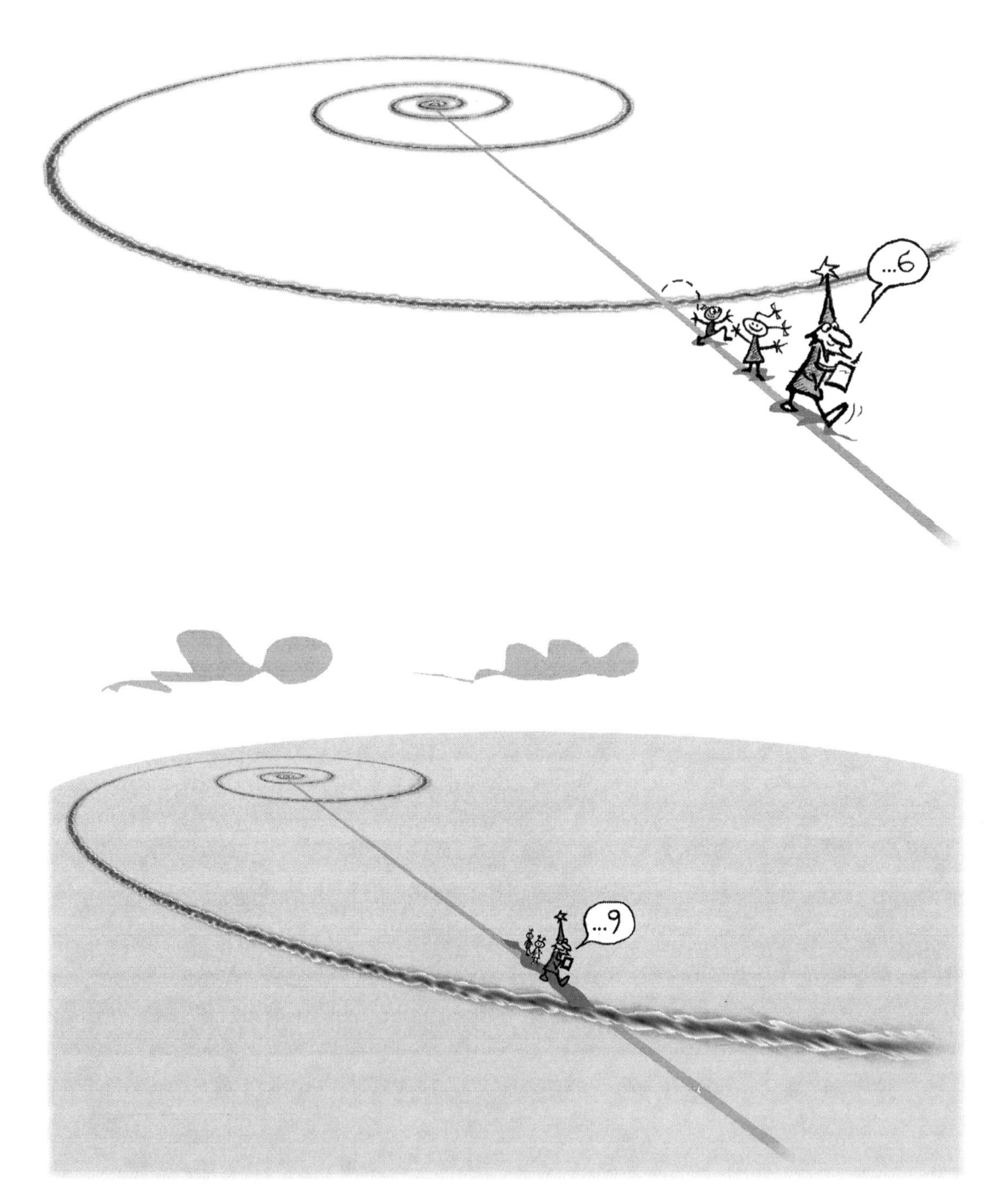

Between the sixth crossing of the spiral, at around 403, and seventh, at around 1097, we divide by 6 to get a rough approximation of the prime count. Between the eighth crossing, at around 2981, and the ninth, at around 8103, we divide by 8.

We now return to the more accurate approximation which involves dividing by logarithms. When we get out to 3000, the logarithm is almost exactly 8. This means that the approximation we get for the number of primes less than 3000 is almost exactly 3000 divided by 8, or *one-eighth* of 3000. Another way of looking at this is that approximately 1/8 of the counting numbers up to and including 3000 are primes. When we get up to 163 000, the logarithm is almost exactly 12, so, by the same kind of reasoning, we can conclude that approximately 1/12 (one-twelfth) of the counting numbers up to and including 163 000 are primes.

Of course, none of this should be taken to suggest that the primes are in any sense *evenly spaced* throughout these stretches of the number line. No, as we've seen, the primes "bunch up" near the beginning of the counting number sequence, gradually thinning out. We'd expect spacings between successive prime numbers to tend to gradually get larger as we proceed along the number line.

I'm using the words *approximate*, *approximately* and *approximation* a lot, so it's about time that I make clear their exact meanings in this context.

It'll be easiest to think in terms of *percentage error*. As we saw, there are 55 primes up to and including 258 and the estimate we got was 46.462. If we subtract to get the difference, 8.538, divide that by the actual number, 55, and then multiply by 100, we get 15.524. This means that our estimate has an error of about 15.52%. Another way of saying this is (subtracting 15.52 from 100 to get 84.48) that the estimate is about 84.48% accurate. Even if you'd struggle to calculate a percentage error given the task, I hope you'll at least have a *feel* for what "97% accurate" and "3% error" mean and can accept that these things can be calculated in a straightforward way. If you'd like to see it explained, have a look at Appendix 6.

The estimate we've seen for the number of primes up to and including a given counting number is "approximate" in the sense that *the percentage error tends to shrink as the numbers we look at get bigger*. We can make this statement more precise. Suppose you name a percentage error, as small as you like, say 0.0000000000000000000000000001% error. Well, *it has been proved* that there *must* be a counting number beyond which the percentage error of the estimate *forever* stays below 0.0000000000000000000000000001%. This counting number would be *very* large by most people's standards, far larger than anything that's ever encountered in any practical setting, but we can be *certain* that such a number exists. It can even be calculated. The same is true regardless of how tiny you insist on making your percentage error.

So, roughly speaking, *the estimate can be made as accurate as you like if you're prepared to look far enough down the number line*.

The table on the opposite page should give you some feel for how the percentage error shrinks. You'll see that although the error is shrinking, it's not doing so very quickly. However, as a result of what's been proved, we can be sure that:

☆ If we choose our counting number to have more than six digits, then the accuracy will always be at least 90% (that is, less than 10% error).

☆ For numbers with more than ten digits, we can guarantee 95% accuracy (that's less than 5% error).

☆ With more than 65 digits, the accuracy will always be at least 99% (1% error).

Mathematically confident readers who'd like to examine the reasoning behind these assertions can find it in Appendix 7.

The important thing to remember here is that although we can never reach 100% accuracy, we can get as close as we want. To guarantee 99.99999% accuracy, we must take our counting number up to at least 6 514 417 digits! Just to write down a number that size in this typeface would require a book nine times the size of this one[2].

counting number	number of primes up to and including it	its logarithm	estimated number of primes up to and including it	percentage error
1	0	0.0000		
2	1	0.6931	2.8853	188.54%
3	2	1.0986	2.7307	36.54%
4	2	1.3862	2.8530	44.27%
5	3	1.6094	3.1066	3.56%
6	3	1.7917	3.3486	11.62%
7	4	1.9459	3.5972	10.07%
⋮	⋮	⋮	⋮	⋮
100	25	4.6051	21.7147	13.14%
101	26	4.6151	21.8845	15.83%
102	26	4.6249	22.0541	15.18%
103	27	4.6347	22.2235	17.69%
104	27	4.6443	22.3926	17.06%
⋮	⋮	⋮	⋮	⋮
1000	168	6.9077	144.7658	13.83%
1001	168	6.9087	144.8886	13.76%
1002	168	6.9097	145.0124	13.68%
1003	168	6.9107	145.1361	13.61%
1004	168	6.9117	145.2599	13.53%
⋮	⋮	⋮	⋮	⋮
10000	1229	9.2103	1085.7362	11.66%
10001	1229	9.2104	1085.8329	11.65%
10002	1229	9.2105	1085.9297	11.64%
10003	1229	9.2106	1086.0265	11.63%
10004	1229	9.2107	1086.1233	11.63%
⋮	⋮	⋮	⋮	⋮
100000	9592	11.5129	8685.8896	9.45%
100001	9592	11.5129	8665.9689	9.45%
100002	9592	11.5129	8666.0482	9.44%
100003	9593	11.5130	8666.1275	9.45%
100004	9593	11.5130	8665.2068	9.45%
⋮	⋮	⋮	⋮	⋮

Everything in the last three columns has been rounded off.

The *absolute error*, as opposed to the percentage error, is the difference between the actual number of primes and the estimate. If you go back to the table, you should be able to see that this is going to become enormous (eventually bigger than any number you might name). But the whole idea of *percentage error* is to look at what *proportion* of the actual count this absolute error is. The absolute error may become huge, but the numbers we're dealing with at that point will be considerably *more* huge, so the proportion will be tiny (and it just keeps getting tinier).

Curiously, the fairly simple mathematical truth we've just been looking at, known to mathematicians as the *Prime Number Theorem*…

The amount of primes up to and including a given counting number is approximately equal to that number divided by its natural logarithm.

…is not generally taught to university mathematics students [3]. I don't know why. As you've seen, it's not *that* difficult to understand. But, for some reason, it's not considered worth teaching, so there are many people with maths degrees (even some with doctorates) who have never even *heard* of this.

THE "SOCIAL STATISTICS OF THE PRIMES"

If you don't want to worry about any of the technical details, if the idea of formulas involving logarithms still terrifies you, then the best thing is just to hang onto the basic idea that *the primes thin out at a rate which is closely related to the way a snailshell uncoils*. But illustrated opposite is the way mathematicians express the Prime Number Theorem (we'll call it the "PNT" from now on). If you're interested, the caption explains what each part of the formula means. Don't worry, though, we won't need to use any of this notation.

The "π" here is not related to the number of diameters it takes to wrap around a circle – it's just the Greek letter "p". "$\pi(x)$" means the number of primes up to and including a given number "x". The symbol "~" means "approximately equal to" in the "shrinking error" sense that's been described (the percentage error gradually dwindles to zero). The fraction-like thing to the right of that means "what you get when you divide your given number x by its natural logarithm".

You may be wondering why I'm dwelling on theorems and formulas, after having promised to show you a *pattern*. But I think it's fair to say that what the PNT tells us – what we've just seen involving the approximate rate at which the primes thin out – *is* a kind of "pattern" in the sense of a weather pattern or the migratory pattern of a species of bird. It describes an overall tendency within a range of accuracy, while individual details may fluctuate considerably. It tells us approximately how many primes to expect in a given region of the number system. For example, since it tells us that there are approximately 144.76 primes up to 1000 and approximately 263.13 up to 2000, we can deduce that in the region of the number system between 1000 and 2000, there should be about 118 primes, since $263.13 - 144.76 = 118.37$. The actual number may be quite a bit more or less (in that particular example, it's 135). The PNT just describes a "statistical tendency". When you take a big enough picture, though, and look far into the number system (or deep into the "number pond"), the occurrence of prime numbers can be seen to follow this tendency with a margin of error that shrinks towards zero. It could almost be described as a "collective behavioural pattern", to echo Davis and Hersh's reference to the "social statistics of the primes"[4].

In the 1790s, Carl Gauss (who, you might recall, proved the Fundamental Theorem of Arithmetic) noticed that the primes tend to thin out according to the "logarithmic" rule described by the Prime Number Theorem.

He discovered the pattern "experimentally" at the age of 14 or 15. The young Gauss apparently used to count the primes in a stretch of 10 000 counting numbers whenever he "*had an idle fifteen minutes*"[5]. The primes could be seen to follow this pattern as far as he was able to count (the largest prime known at that time was 2 147 483 647). In the framework of our earlier analogy with astronomical observation, Gauss and his contemporaries were using the "naked eye" (unassisted mind) and were able to "see" as far as ten-digit primes, whereas today, mathematicians have "huge telescopes" (powerful computers) and can "see" out to primes with *nine million* digits.

Around the same time, the mathematician Adrien-Marie Legendre noticed a similar (apparent) pattern. His approach gave a slightly different estimate for the number of primes less than a given number, but it also involved logarithms and had the property that percentage error shrinks towards zero as you consider ever larger numbers. He published a conjecture about this in 1798. Remember that a "conjecture" (like the Goldbach Conjecture which we encountered earlier) is a mathematical statement which has been proposed as *possibly* true but has not yet been proved or disproved. Legendre turned out to be wrong in the details but had the right general idea.

There's been some historical confusion over who arrived at the idea first, since Gauss didn't publish his findings, rather mentioning them many years later in a letter to a friend. But it now seems to have been clearly established that Gauss was being truthful in his retrospective claim.

The historical details of the Prime Number Theorem needn't concern us. With European mathematics developing in the way that it was, someone was bound to have noticed this pattern in the primes eventually. Who actually *did*, historically, is irrelevant from the point of view of this exploration. One of the uglier aspects of the mathematical research scene is the occasional arising of a bitter dispute over who first proved or discovered something. This is unfortunate, and puzzling, especially if we consider the universal, communal nature of the subject matter. Mathematics, in a strange way, unites humanity. And yet, even in connection with this invisible world we all share access to, there are disputes over a kind of "ownership".

About a century later, in 1898, two mathematicians working independently, Jacques Hadamard and Charles de la Vallée-Poussin, *proved* that the prime numbers follow this apparent pattern, the logarithmic thinning out which I've described, where the error shrinks towards zero as we proceed through the counting numbers.

Their proofs, unlike Euclid's proof of the infinitude of primes or the proof of the Fundamental Theorem of Arithmetic (the two proven mathematical theorems which we've already looked at) were *tremendously* complicated, involving intricate mathematical reasoning well beyond anything yet hinted at in this book or anything someone not immersed in the mathematical sciences could ever imagine. The mathematical theory on which their proofs were based wasn't developed until the middle of the 19th century, so it was unavailable to Gauss or Legendre. The best *they* could do was to make a conjecture (only Legendre published his). Once the tools were available, someone was bound to prove it before too long (and in the end, two people did, almost simultaneously).

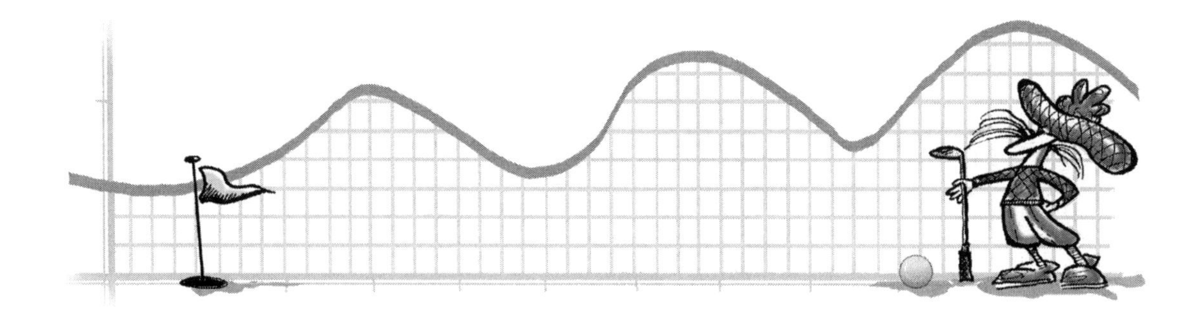

chapter 11

staircases

Here's a way you can visualise the counting of prime numbers which was described in the previous chapter using tables:

At each counting number, a column of stones is conjured into being. If the number is prime (examples here being 2, 3 and 7), then the column contains one more stone than the previous column. If not (the example here is 45), then the column is the same height as the previous column. The space at 1 has *no column at all* – we think of this as "of height zero".

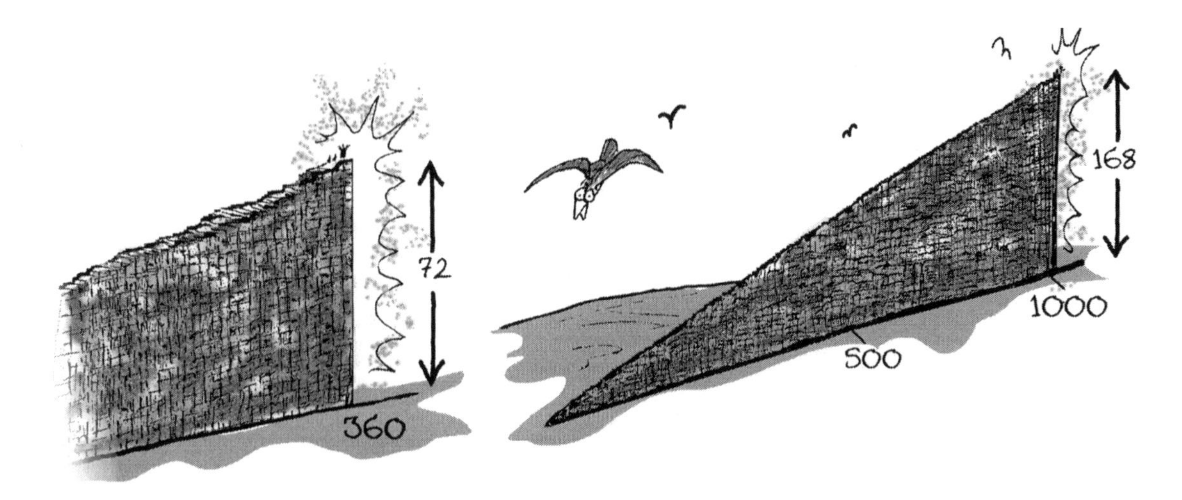

At 360, the height of the staircase is 72 because there are 72 prime numbers less than 360. At 1000, the height of the staircase reaches 168, the number of primes less than 1000. Because a new stone is added each time a prime is found, the height of the staircase "records" the amount of primes which occur up to each number.

If we zoom out far enough from this "prime number staircase", the individual stairs are no longer visible:

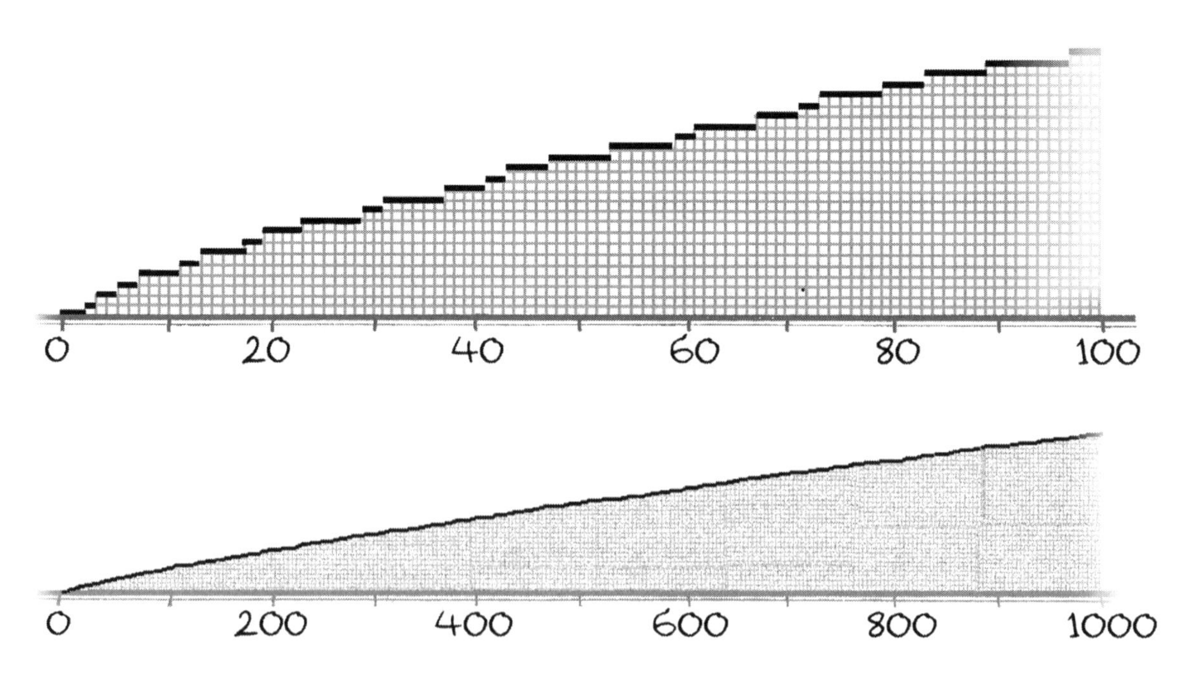

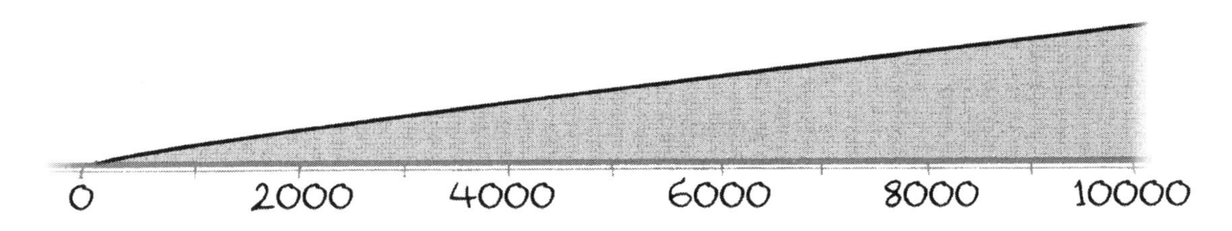

In the 1000 range, you can just about see the individual steps at this sort of scale, but by the time we reach the 10000 range, they're no longer visible.

The staircase, of course, will continue on above the number line forever (and will continue, however gradually, to climb, since new primes are guaranteed to keep occurring). Presented with a large enough chunk of this staircase, we're actually able to *see* evidence of a "tendency towards smoothness" in the irregular sequence of primes. That is, the staircase starts to *look like* it's a continuous curve. If we zoom in to examine a little piece of this seeming curve, though, we'll find that it's made of steps – these reflect the "individual eccentricities" of the primes. Zooming back out, their collective "social tendency" again becomes evident.

This "staircase" is known as the *graph* of the prime counting function. If you're familiar with functions and graphs, this should make sense to you. If you aren't, don't worry – you'll only really need to understand how the staircase is built, and that should be clear from the illustrations we've just seen.

The Prime Number Theorem tells us that the height of the staircase above a given number (that is, the amount of primes up to and including that number) should be approximately equal to the number divided by its own logarithm.

Suppose, then, that we again travel out along the number line, working out this PNT-supplied approximation for each counting number and marking a point at that height above the line, like this:

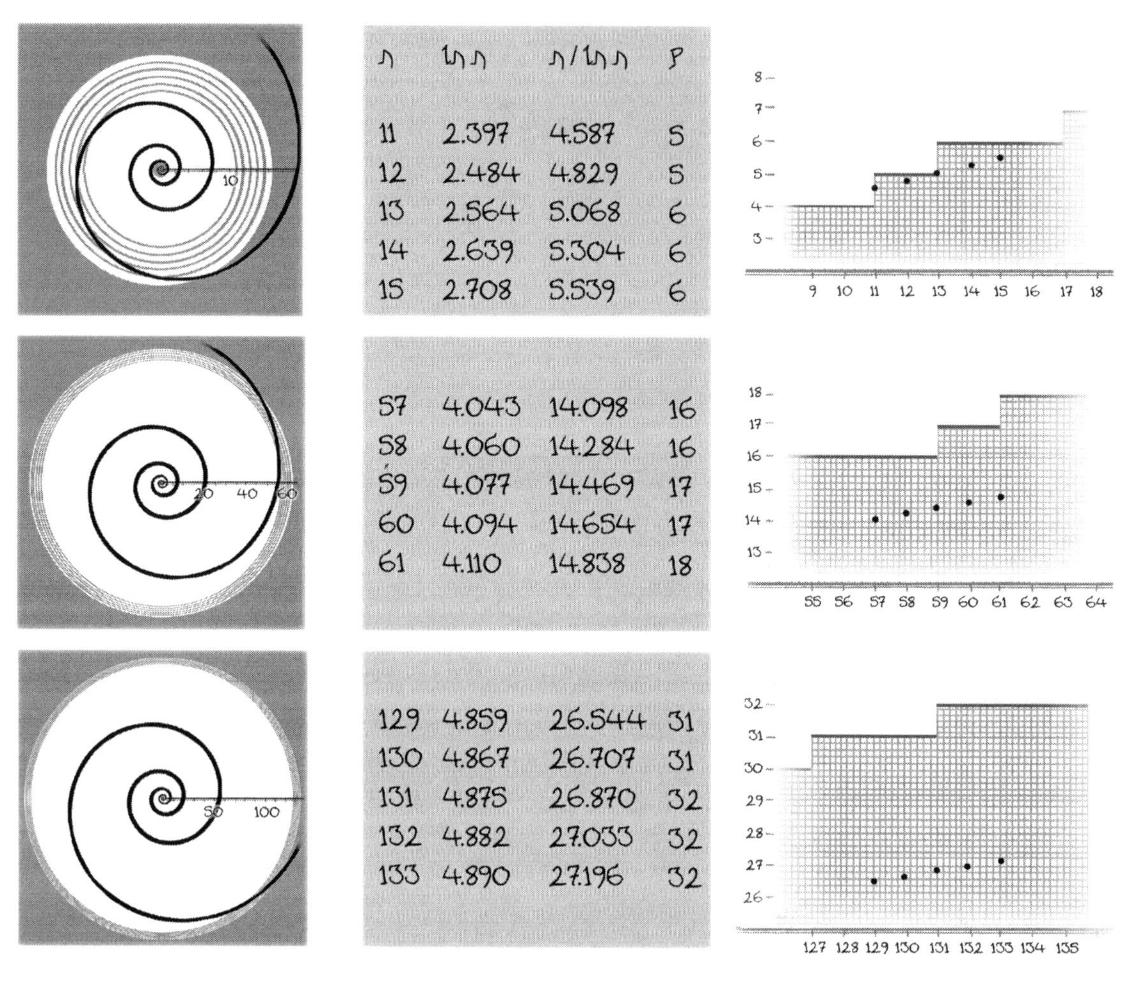

n	ln n	n / ln n	p
11	2.397	4.587	5
12	2.484	4.829	5
13	2.564	5.068	6
14	2.639	5.304	6
15	2.708	5.539	6
57	4.043	14.098	16
58	4.060	14.284	16
59	4.077	14.469	17
60	4.094	14.654	17
61	4.110	14.838	18
129	4.859	26.544	31
130	4.867	26.707	31
131	4.875	26.870	32
132	4.882	27.033	32
133	4.890	27.196	32

Here, "n" is a counting number, "ln n" is its natural logarithm (count coils in the first column) and "$n / \ln n$" is the number divided by its logarithm, which gives the height of the relevant dot in the third column. This approximates the number of primes ("p") up to and including the number.

Floating these dots above all counting numbers up to 100, we see this:

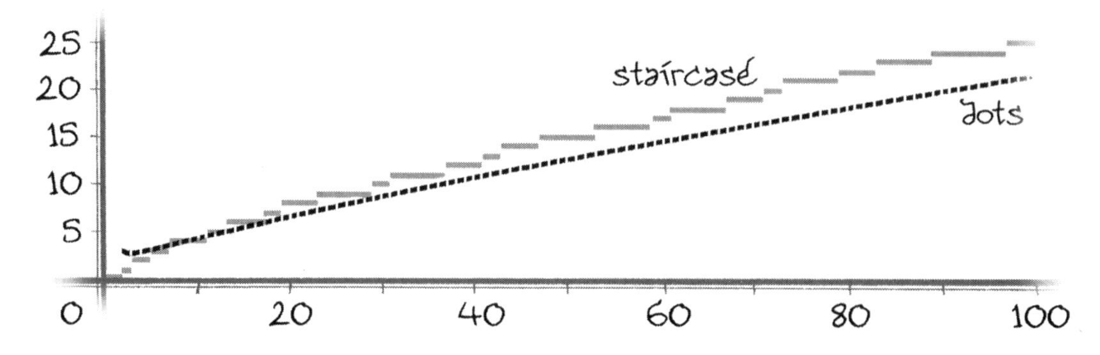

The sequence of dots *also* suggests a smooth curve, one which follows the staircase to some extent, although perhaps not as closely as we might have liked.

The gap between the "curve of dots" and the staircase visually represents the margin of error in the PNT. You may not be convinced that this error is going to shrink towards zero as we get farther and farther down the number line. Remember, though, we're not talking about *absolute error* (that would correspond to this gap). What shrinks towards zero is the *percentage* (or *relative*) *error* and that's based on the size of the gap *relative* to the height of the staircase. The gap will become increasingly huge but the staircase height will always be *much* bigger. We work with proportions (expressing these as percentages) asking, at any counting number, "What proportion (or percentage) of the staircase height is the gap between the staircase and the corresponding dot?" We'll find that *this* shrinks down towards zero as we get into ever more distant realms of the number system.

The dots we've floated above the number line are all located directly above counting numbers. But we can take *any* number bigger than 1 and, whether it's whole or not, divide it by its logarithm and then float a dot above it at that height. If we did this for all numbers bigger than 1, we'd get the smooth curve which our sequence of dots suggested. Similarly, the prime counting function doesn't have to be restricted to counting numbers. We can just as well ask "how many primes are there up to and including 104.75?" as we can about 104. The answer will be the same, of course, 27. The horizontal pieces of the staircase should make sense in this light.

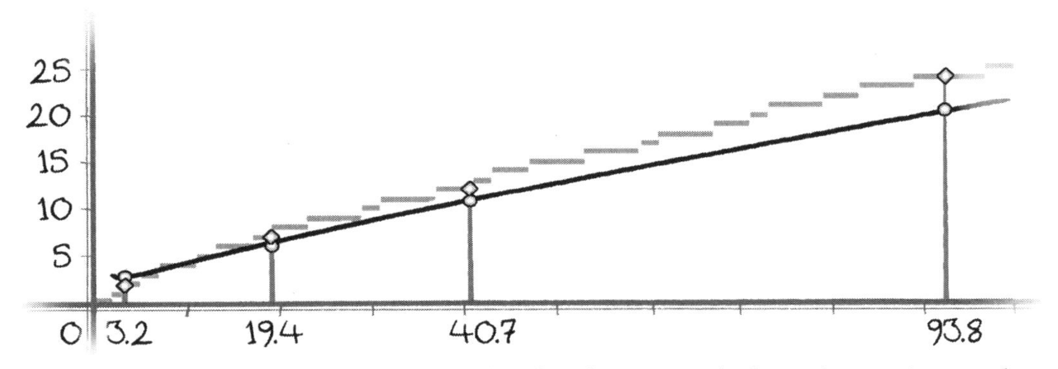

Here, the curve and staircase are compared at four locations which aren't counting numbers.

FUNCTIONS AND GRAPHS

You may have learnt about *functions* and *graphs* as part of your mathematics education, but perhaps not. Or perhaps you've forgotten the details. Not to worry. Although many students these days confuse the two concepts, a function *isn't* the same as a graph. A graph is a kind of visual description of how a particular function works. So – what's a function? You can think of a function as being like a "spell" which can be cast on the number line, causing each point to move instantly to a new location (in a very specific way). For example, if I take every number and multiply it by itself, that's a function:

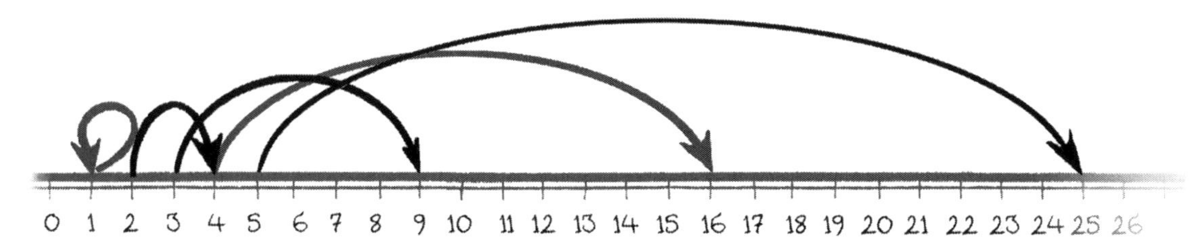

If I add 3 to every number, that's also a function:

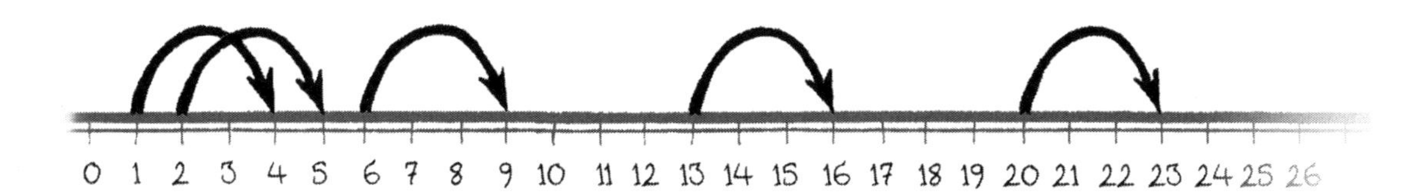

If I multiply every number by itself and subtract 2, that too is a function:

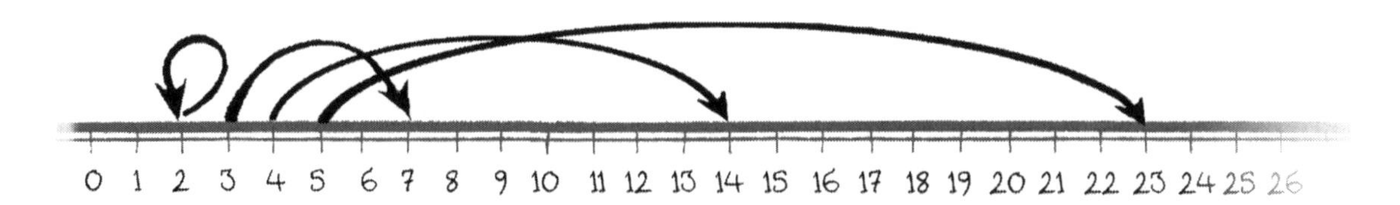

These are all fairly simple examples. Here's one that's a little bit more involved but still a function – divide every number by its natural logarithm:

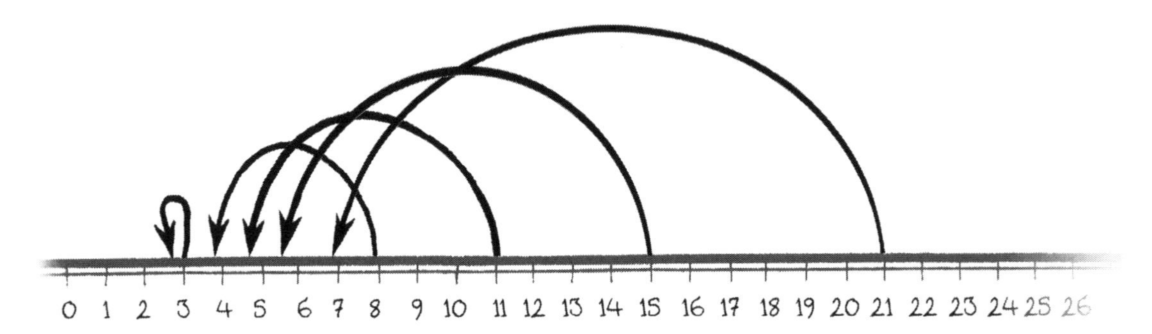

3 divided by its logarithm (1.09) is 2.73; 8 divided by its logarithm (2.08) is 3.85; 11 divided by its logarithm (2.38) is 4.62; 15 divided by its logarithm (2.70) is 5.53, and so on.

If, given a number, we count the amount of primes up to and including that number, we're looking at what we've already called the "prime counting function".

Functions are often explained in terms of "input" and "output". You put a number into the function, the function "does something to it" and spits another[1] number out.

The idea behind the *graph* of a function is that for each number on the number line, you work out "what the function does" to that number (in other words, the number you get as your output) and float a point at that height above the appropriate location on the number line. This creates a 2-dimensional image which allows you to "read" what the "effect" of the function is.

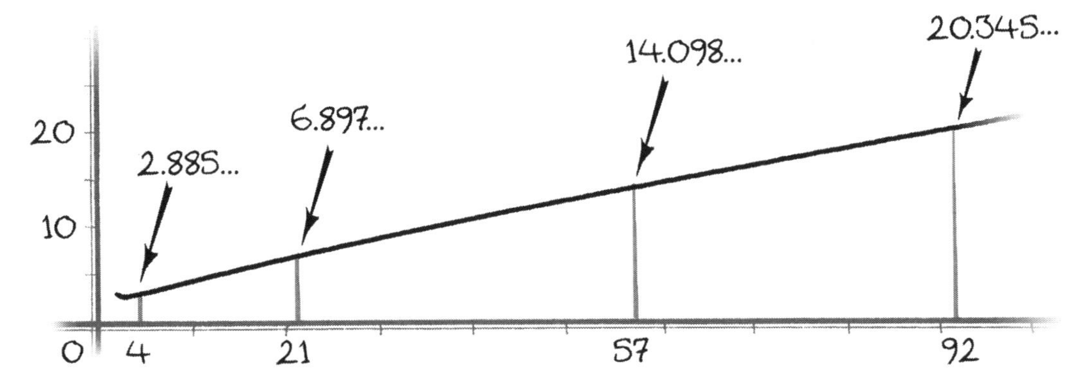

This is the graph of the function that divides each number by its own logarithm.

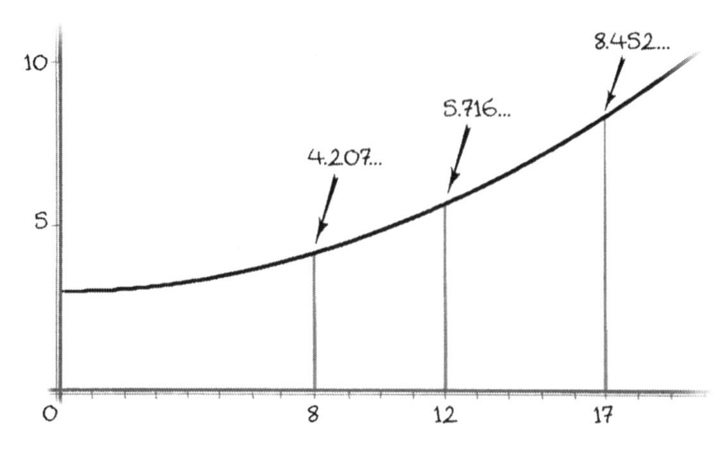

The function graphed here multiplies each number by itself, divides the result by 53 and adds 3.

To read the graph, you choose a number you're interested in, find it on the number line and then see what the height of the graph is there. That's what the function "does" to your chosen number – whether it does it by multiplying the number by itself, adding 3 to it, dividing it by its own logarithm, counting the number of primes up to and including it, or whatever (and the possibilities for defining functions are truly endless). To help make sense of certain important concepts which will be needed in Volumes 2 and 3, we're going to visualise functions in another way. This may seem a bit pointlessly fanciful, but it really will help later:

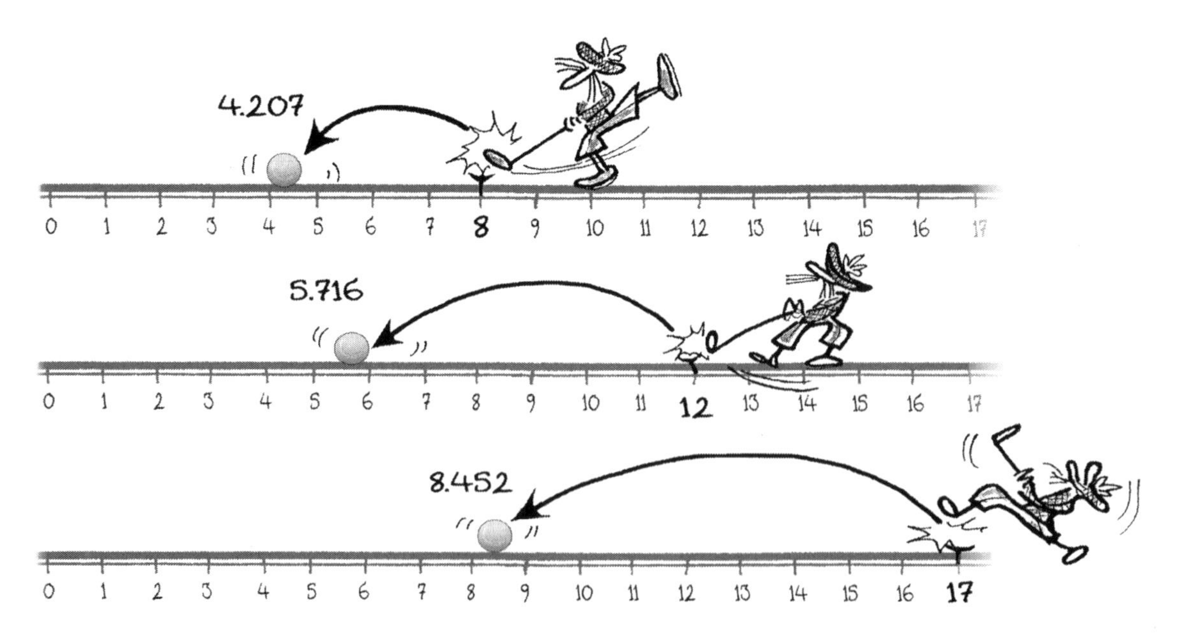

There's one of these "golfing sprites" for every possible function, and each one has complete knowledge of its function. Wherever you choose to place the ball, it knows exactly where to hit it. It's always perfectly accurate. And it is *entirely consistent*. If you place another ball at the same location, it will get hit to the same destination as the first. How does it do it? We can think of the sprite as using its function's graph in order to hit the ball to the correct location. Here's how that works:

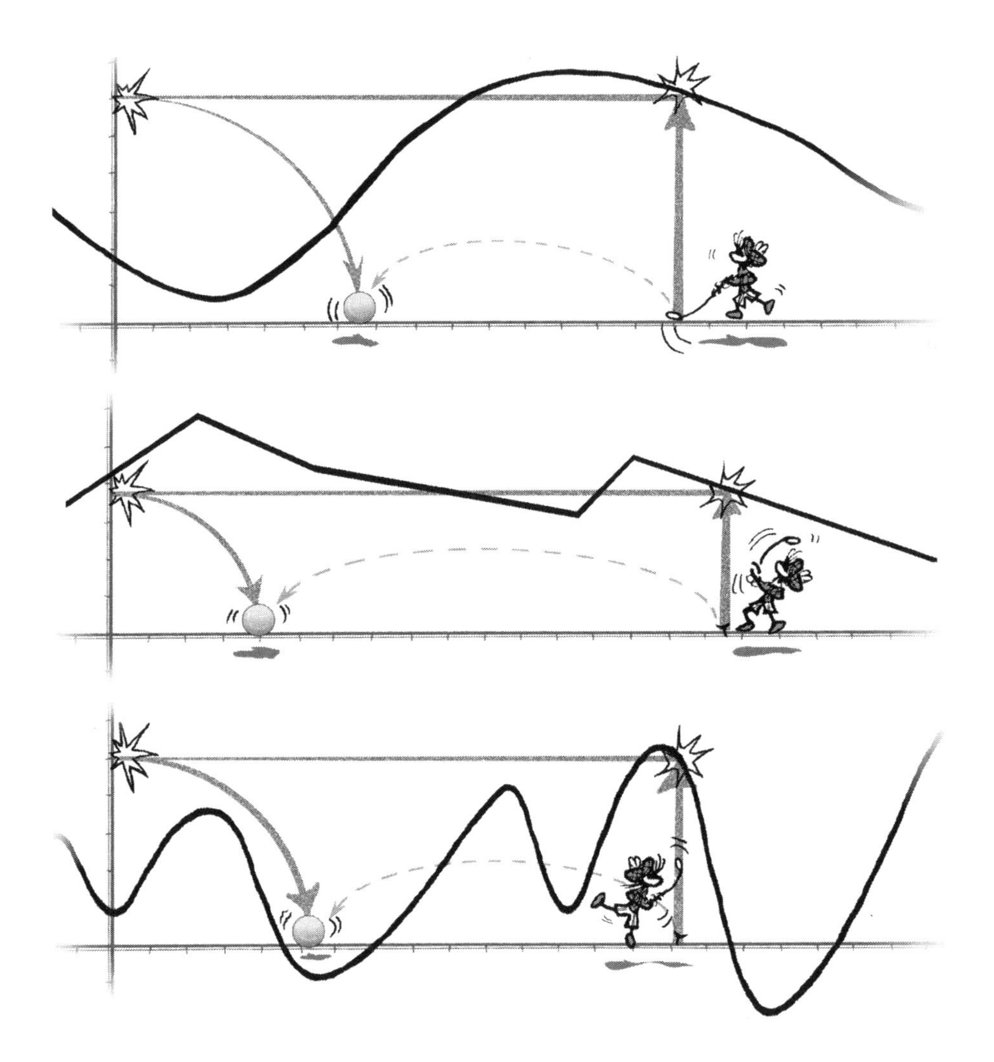

The sprite sends the ball directly up until it meets the graph, at which point it "turns left" and travels horizontally until it hits the vertical line through zero. It then travels a quarter turn clockwise around the zero point. The ball has now returned to the number line and is in the position which the function (corresponding to the graph in question) is supposed to send it to.

Graphs can also dip below the number line (when the "output" of the function in question is a negative number):

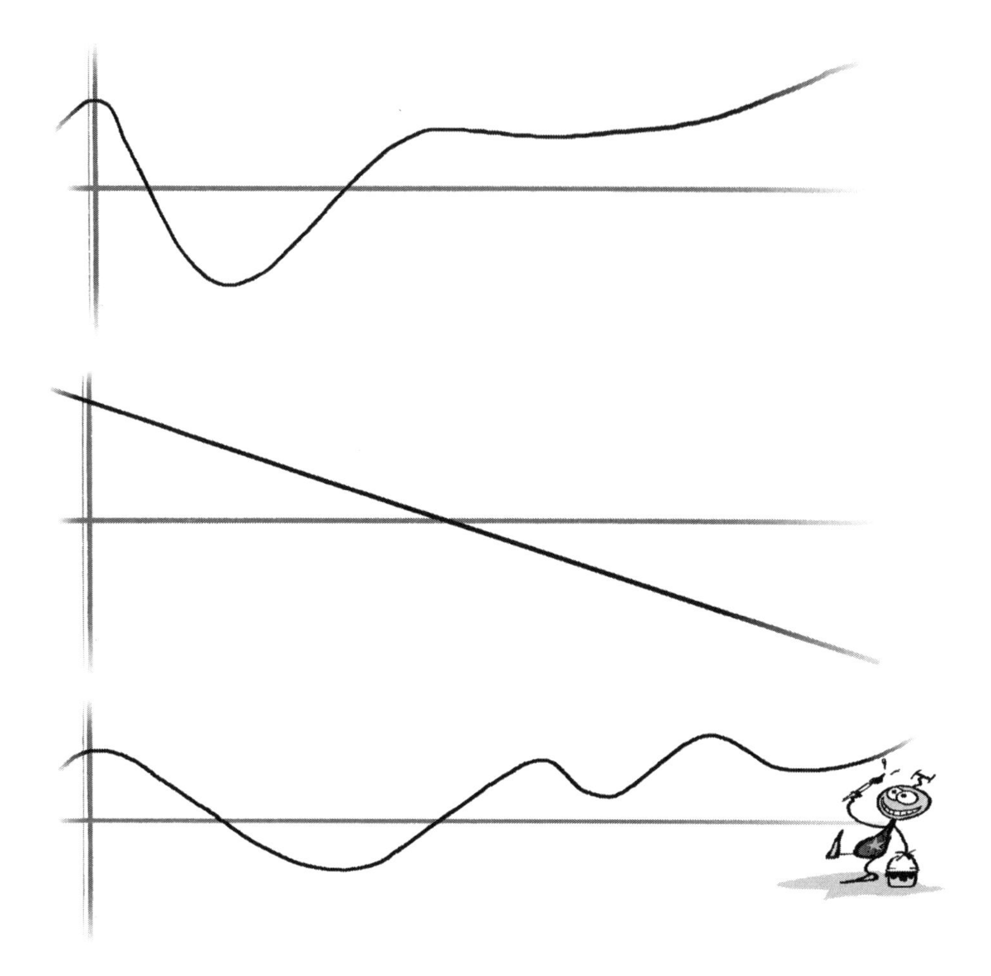

The sprite reads such graphs in almost the same way. It imagines the ball travelling down (rather than up) to meet the graph, then travelling horizontally towards the vertical line through zero, then travelling a quarter-circle clockwise around zero until it meets the number line. In this way, it meets the number line to the left of zero, that is, its final destination will be a negative number. So, if the graph is *above* the number line at a certain number, it means that the function takes that number to a *positive* number; if the graph is *below* the number line there, it means that the function takes it to a *negative* number.

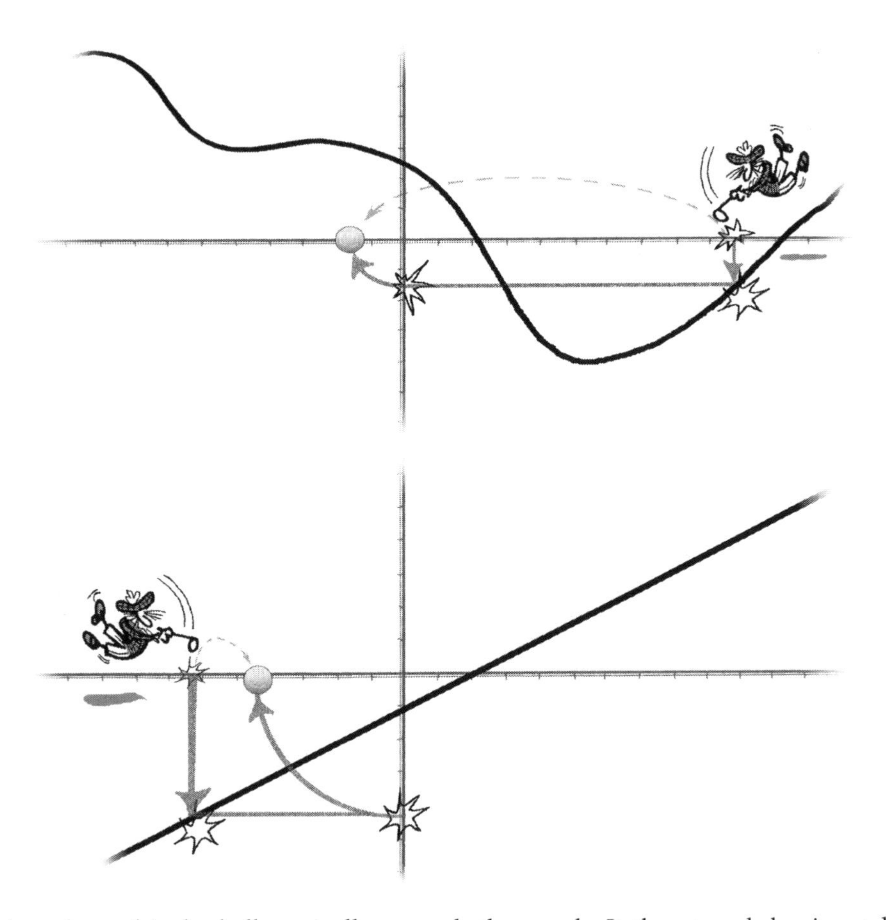

The sprites always hit the ball vertically towards the graph. It then travels horizontally to the vertical line through zero and performs a clockwise quarter turn. So, where the graph dips below the number line, the "output" of the function will always be a negative number.

Where the graph *crosses* the number line, the function takes the number in question to *zero*. When studying a particular function, these numbers are often of great importance – they're called the *zeros* of that function. Notice how the sprite's action works in such a situation. The ball doesn't need to travel up or down to meet the graph, as the ball is already *on* the graph. So the sprite just has to whack the ball horizontally until it reaches the vertical line through zero. This means that it ends up at zero itself. The quarter-turn rotation has no effect – think of a circle of size zero – the purpose of this rotation is to take points from the vertical line around to the number line, and in this case, there's no need.

APPLYING THIS TO PRIMES AND PRIME COUNTING

Let's now apply this way of thinking to the two functions which interest us most right now, that is, the prime counting function (whose graph corresponds to the "staircase" we built earlier) and the "each number divided by its own logarithm" function which the Prime Number Theorem supplies as an approximation for the prime counting function. I'll call the second one the "PNT approximation function".

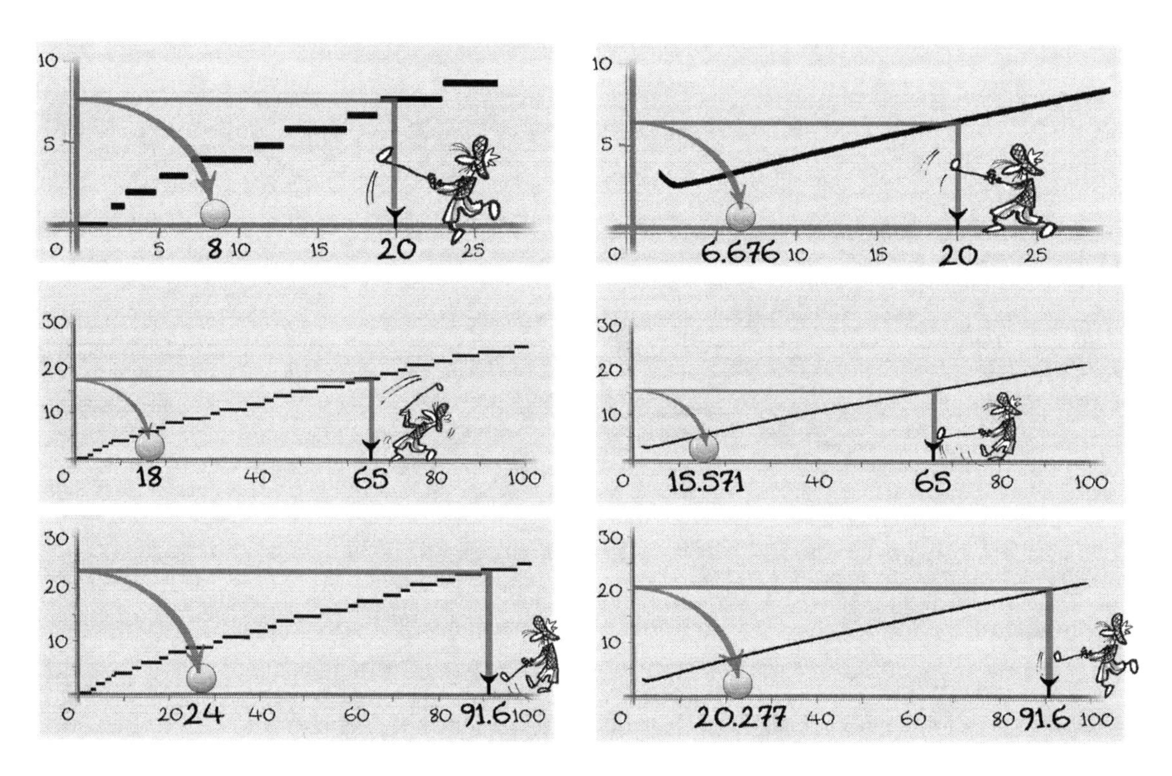

All of the graphs in the left-hand column correspond to the prime counting function. All of the graphs in the right-hand column correspond to the "PNT approximation function".

One thing we can immediately see is that the PNT approximation function's sprite will hit pairs of balls placed very close together to a pair of locations which are also very close together. But the prime counting function's "staircase" sprite will sometimes hit balls placed close together to locations separated by a whole unit – this is because of the "sudden jumps" which occur in the staircase graph when a prime number appears:

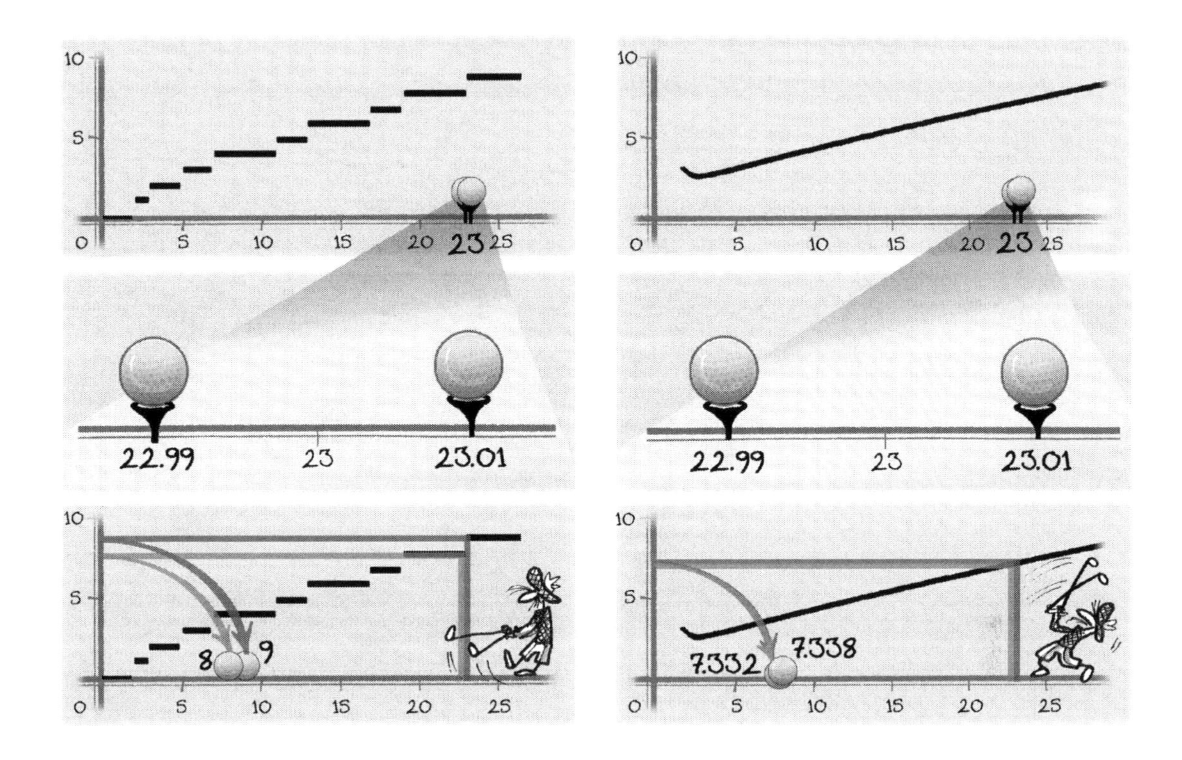

For this reason, we say that the PNT approximation function is *continuous* and the prime counting function is *discontinuous*.

OTHER CONSEQUENCES OF THE PRIME NUMBER THEOREM

The Prime Number Theorem can be used to deduce certain other facts. Suppose we didn't want to know how many primes were less than some given number, but rather we were interested in approximating what, say, the 250th prime is.

Some reasonably simple mathematical manipulations[2] can be used to show that an approximation for this can be given by simply multiplying 250 by its logarithm. That gives $250 \times 5.521\ldots$, or 1380.365... In fact, the 250th prime is 1583, so that's a fairly reasonable approximation, although not that impressive (12.8% error). But if we were to do this for the thousandth prime, the millionth prime and so on, we'd again find the percentage error shrinking towards zero.

There's an improved version of this. It says that to get an estimate for, say, the 250th prime, you take the logarithm of 250 (which is about 5.521), then take the logarithm of *that*, which is about 1.709...

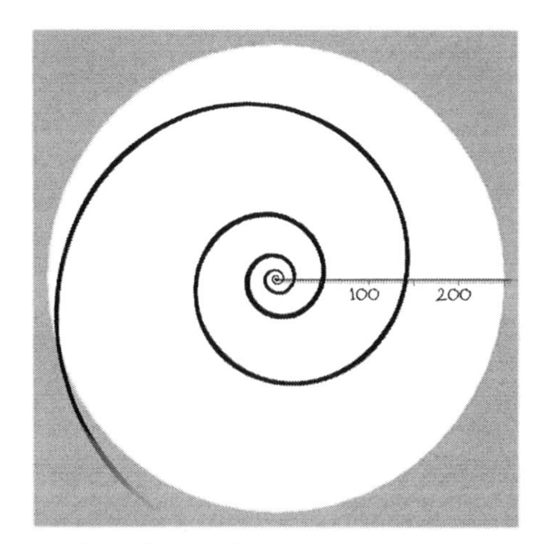

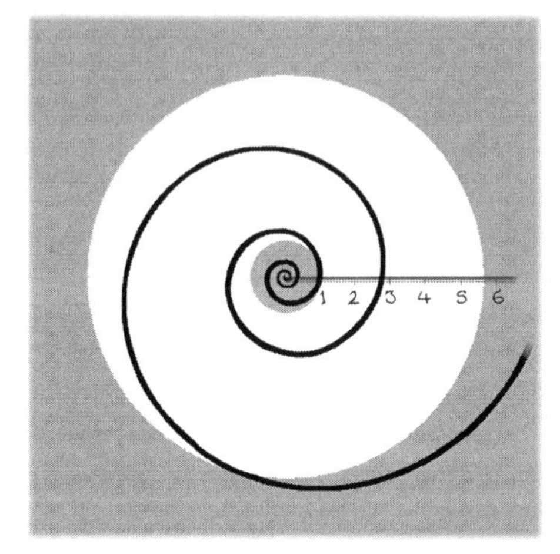

Note that the shaded-in circle with radius 1 is too small to see in the left-hand image.

...add the result to the logarithm of 250 (that gives 7.230...), subtract 1 (which gives 6.230...) and, finally, multiply the result by 250, to give 1557.525... Clearly, this is a much improved approximation (recall that the 250th prime is 1583). The percentage error is now just 1.61%.

The first approximation predicts that the ten millionth prime number is about 161 180 957 and the second predicts that it is about 178 980 382. In actuality, it's 179 424 673. In this case, the first method gives 10.17% error and the second gives 0.25% error.

VISUALISING THE CONNECTION BETWEEN SPIRALS AND PRIMES

We'll now look at a couple of visualisations which show just how directly the prime count can be linked to equiangular (logarithmic) spirals. The first involves stretching a rope out along the number line (it's fixed to the zero point). At any number you reach, to get an approximate number of primes up to and including that number, simply pull the rope taut, and walk in a circle until you reach the "base-*e*" spiral, as shown:

Having chosen the number 44 and walked to it on the number line, the rope is pulled taut.

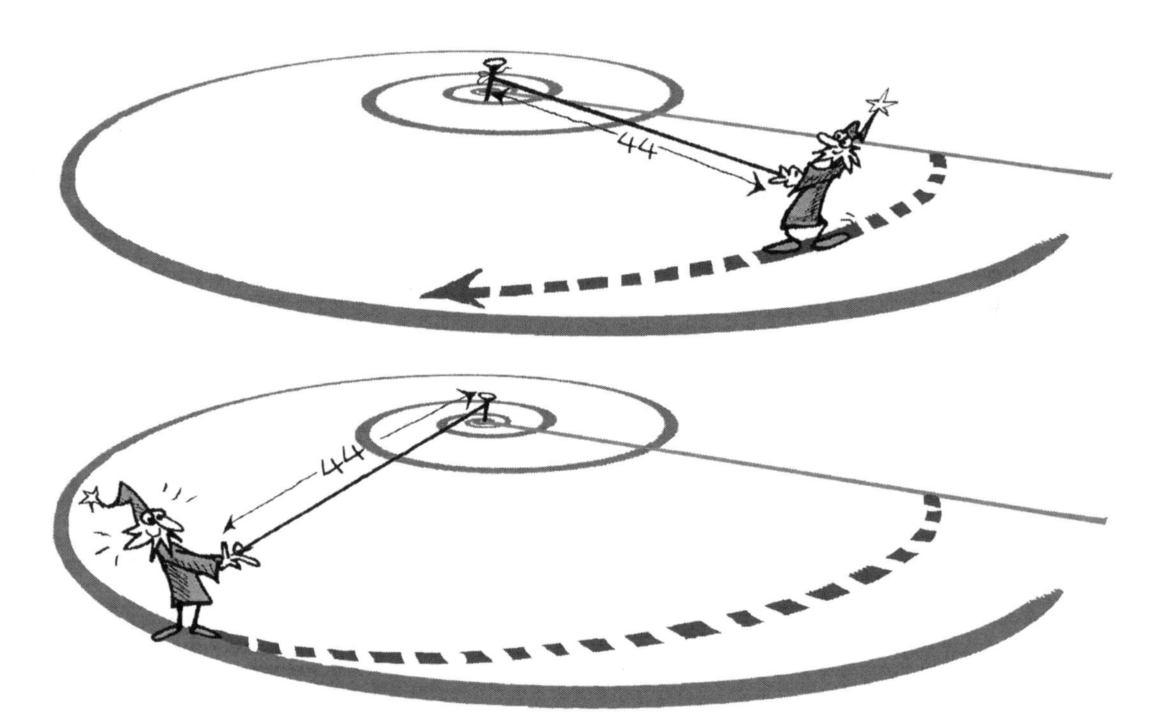

Keeping the rope taut and walking clockwise, the unique point on the spiral which is at a distance of 44 units from the centre is eventually reached.

Now, drop the rope and follow the spiral back to the "1" point on the number line, counting the number of coils you walk (that won't generally be a counting number, of course), then divide your original number by that. This is a direct application of the Prime Number Theorem. We have used the (exact) geometry of the logarithmic spiral to learn about the (approximate) distribution of primes.

Interestingly, a "reversed" version of this visualisation allows us to use the (exact) distribution of primes to construct an (approximate) logarithmic spiral. The fact that this works can also be deduced directly from the PNT with some effort.

Again, we start with a rope fixed to zero on the number line. We walk out along the line, stopping at each counting number to make a note of how many primes we have encountered thus far. We multiply the counting number by itself, divide by the number of primes found and then (keeping the rope taut) walk that many steps along a circular path in the anticlockwise direction. We then stop and light a candle.

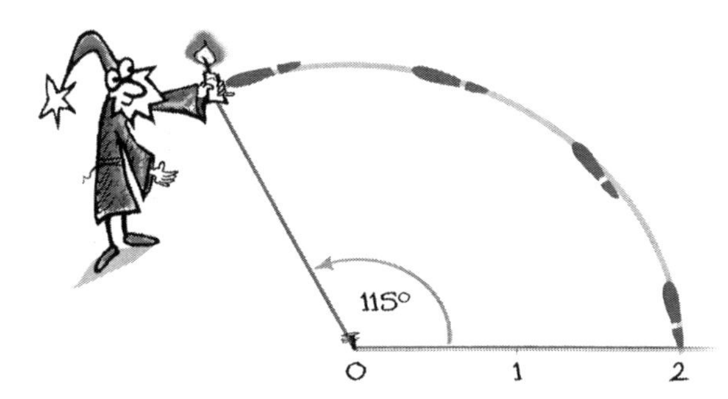

At 2, having found just one prime (2 itself), we divide 2 × 2 by 1 to get 4 steps.

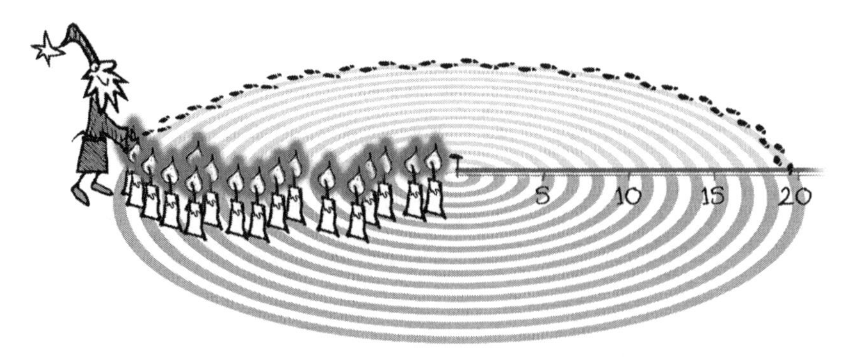

At 20, we've found 8 primes, so we divide 20 × 20 (that's 400) by 8 to get 50 steps.

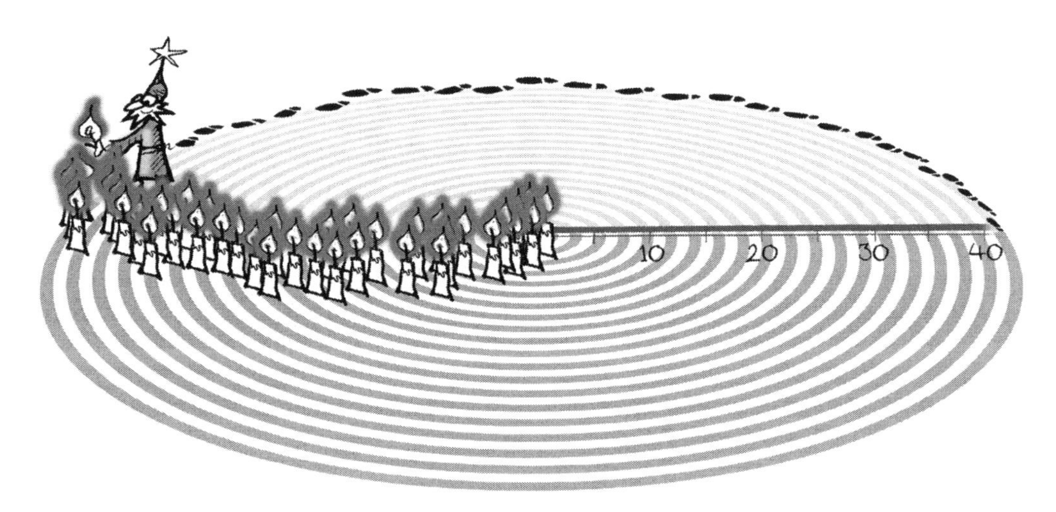

The footprints shown here are only symbolic. The real number would be 40 × 40 divided by 12 (the number of primes less than 40), which is a bit more than 133.

Now return to the number line and continue to the next counting number. Follow the same instructions. Keep doing this forever…

If you think about it, you'll realise that what you get depends on the size of your steps when walking anticlockwise on those circular paths. As long as you keep your steps all the same size, you'll get an approximation of a logarithmic spiral – the bigger your steps, the tighter the winding of the spiral. You can measure the size of your steps in units, where the space between each counting number and its successor on the number line corresponds to one unit of length. If your steps are about 6.28 (that's 2 times π) of these units, which is the distance around the circle with a radius equal to one unit…

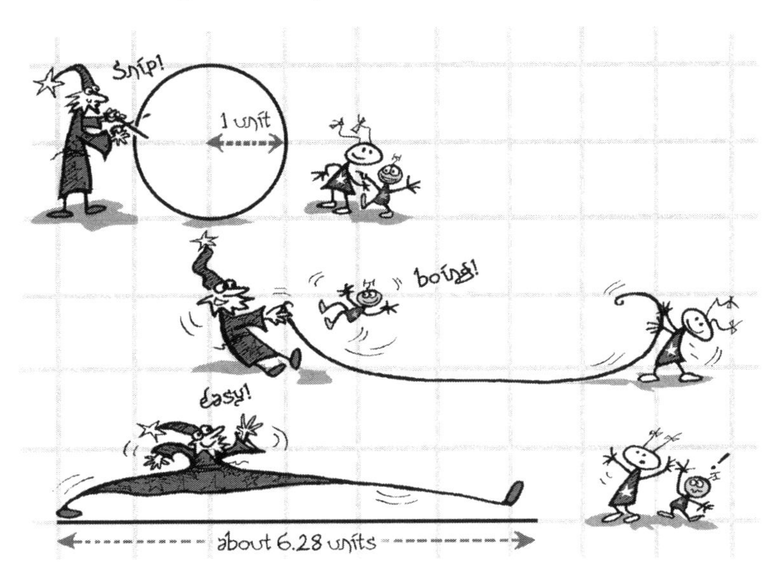

...then you'll get an approximation of the "base-*e* spiral" which we've mostly been working with. Using another step size will produce an approximation of an equiangular spiral with a different base. Here we see what's produced at various distances along the number line using the 6.28-unit steps:

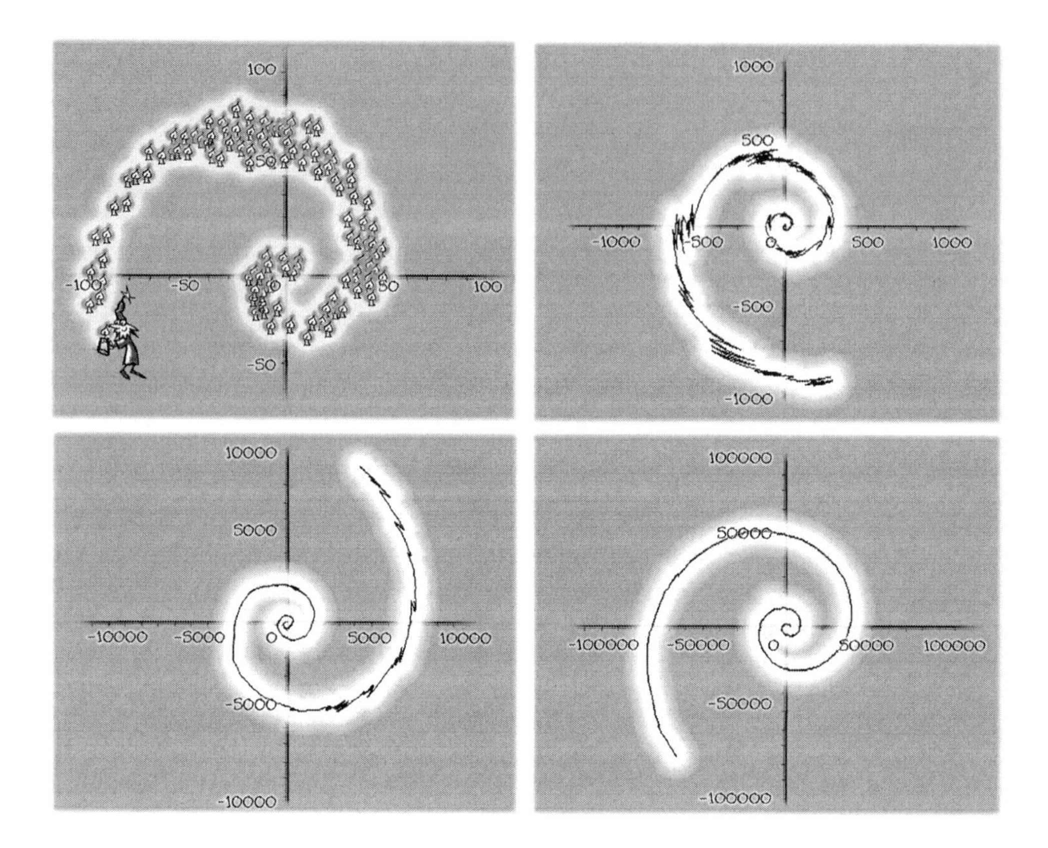

You needn't worry too much about why this works. The main point is that we have been able to produce something very close to the image of an equiangular spiral *without* using any knowledge of spiral geometry. We're just counting primes, multiplying, dividing and walking certain numbers of steps around circles. Knowledge of the prime numbers is enough to produce these approximate spirals.

This reverses the idea of the PNT, where we can use logarithmic spirals to obtain approximations for the amounts of primes in various chunks of the number line.

These two visualisations should have communicated to you the essence of how the distribution of prime numbers can be related to the "uncoiling of a snailshell". It's worth pointing out that ideas involving logarithms (like the distribution of primes) are *not* usually explained using spirals. In textbooks and lecture halls, it's always formulas and functions. We've just seen how easily this stuff can be explained in terms of spirals, but for some reason, no one ever seems to do this.

A COUPLE OF HELPFUL ANALOGIES

Returning to our "astronomical" analogy for looking into the number system, there's a very beautiful metaphor which someone once suggested to me – a spiral galaxy. Many galaxies have a geometric form involving a logarithmic spiral. Astronomers have a pretty good explanation for why certain galaxies can end up like this (in terms of gravitational physics), but we don't need to worry about that. The point here is that the logarithmic spiral suggested to us very clearly by what we see through our telescope *is only a very rough approximation*. Galaxies consist of huge numbers of stars, so when you get close enough, you see only a "splatter" of these, arranged seemingly at random. That is, you see "individual eccentricities". From a distance, though, you see the "social statistics" of the stars, and these are undeniably related to spiral geometry.

This is highly reminiscent of how the prime numbers are distributed. Viewed "up close", you see a random-looking splatter, but viewed from afar, you see very clearly a regularity, linked to spiral geometry. But, unlike the stars, which are spread out across three-dimensional space, the prime numbers are distributed along a one-dimensional number line, so you can't exactly *see* the spiral (you need visualisations, like the one with the candles). Still, the PNT tells us that it is indisputably *there.*

Reconsidering the "fish" analogy for primes, just as some modern fishermen use electronic sonar devices to detect where schools of fish are located, in our infinitely deep "number pond", we can use the Prime Number Theorem to get a sense of where the primes are located. The point of this analogy is that the fishermen are not trying to pinpoint the location of an *individual* fish they seek to catch but rather to get a general, "statistical" idea of where the fish are most numerous, in this way increasing their likelihood of catching something. Although it doesn't particularly help in the ongoing quest to find ever larger prime numbers, the PNT can be used to predict the likelihood of finding primes in any given stretch of the number line.

SURPRISE AND ASTONISHMENT CAUSED BY THE PNT

> "*There are two facts about the distribution of prime numbers of which I hope to convince you so overwhelmingly that they will be permanently engraved in your hearts. The first is that, despite their simple definition and role as the building blocks of the natural numbers, the prime numbers ... grow like weeds among the natural numbers, seeming to obey no other law than that of chance, and nobody can predict where the next one will sprout. The second fact is even more astonishing, for it states just the opposite: that the prime numbers exhibit stunning regularity, that there are laws governing their behaviour, and that they obey these laws with almost military precision.*" (Don Zagier [3])

Notice that Zagier says the second fact is "even more astonishing", which implies that he finds the first fact astonishing. The first fact is another way of expressing something I quoted back in Chapter 8 from Apostolos Doxiadis' novel about the apparent lack of "divine geometry" in the primes. So, he's astonished that the primes don't seem to have an obvious pattern – "seeming to obey no other law than that of chance" – and he's *even more astonished* that they *do* follow a pattern, or at least that they "exhibit stunning regularity". The regularity he was talking about is what we see when we zoom out from our prime count staircase:

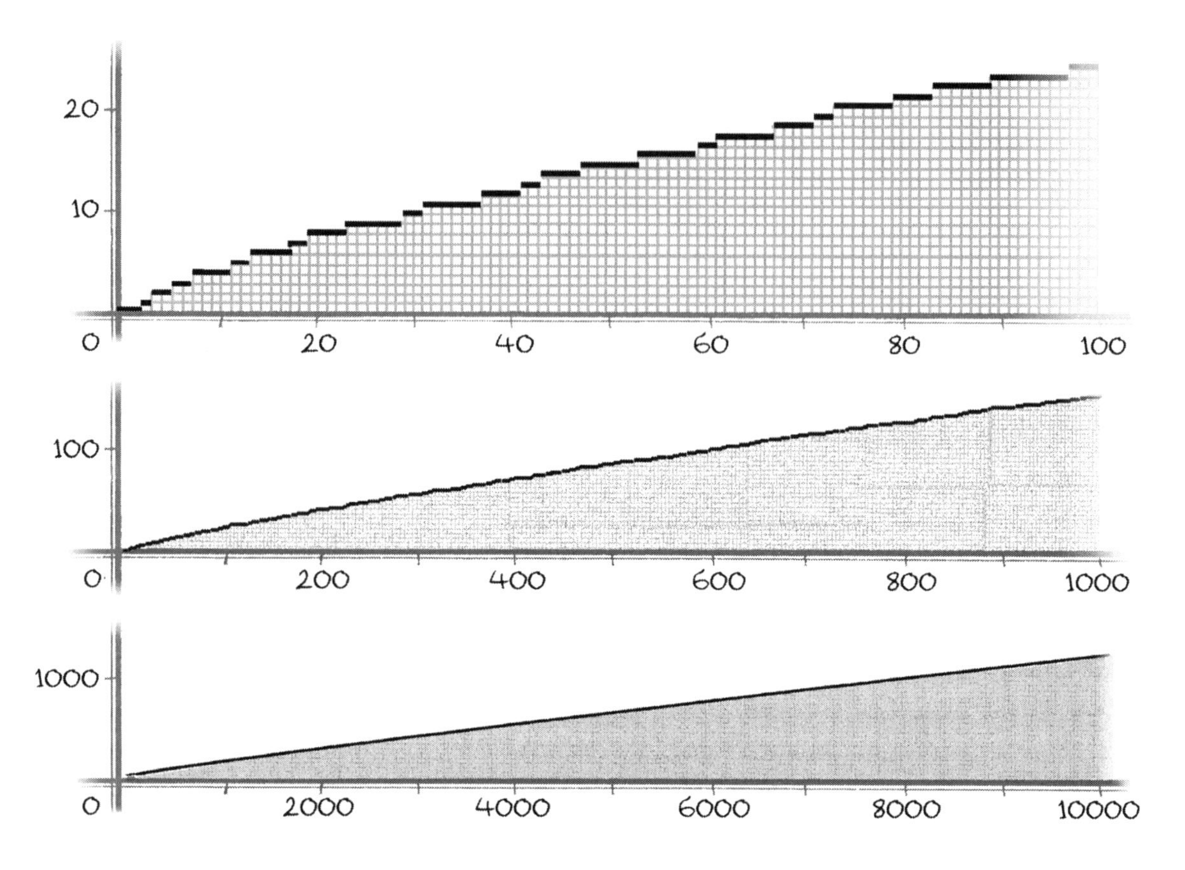

The same staircase graph seen at three different "zooms": 0-100, 0-1000 and 0-10000.

> "*For me, the smoothness with which this curve climbs is one of the most astonishing facts in mathematics.*" (Don Zagier[4])

As we know, this is not a "curve", nor has it any "smoothness", for if we examine it closely, we see that it's made entirely from bits of horizontal and vertical line. So what he really means is:

> *For me, the way this "staircase" graph appears, at these scales, to be a smooth curve is one of the most astonishing facts in mathematics.*

"What did he expect?" asked my friend Tim (a Buddhist poet with little mathematical knowledge) when shown this quotation. "Why don't you ask him what he expected it to do?" Although Tim's view may seem naive, he does have a point. The fact in question is mathematically straightforward, and as "natural" and indisputable as anything can be, so why is a leading mathematician openly claiming to be *astonished* by it?

Don Zagier is not an isolated weirdo mystic. An article based on the lecture containing these remarks of his was chosen as the first article in the first ever edition of *The Mathematical Intelligencer*, an academic mathematics journal published between 1979 and 2006, characterised by its "*conversational but scholarly tone*"[5]. Much of the mathematics community would admit to some sort of empathy with his comments. Interestingly, at this point, we're no longer talking about mathematics. We're back to *feelings* – in particular, *astonishment* – and feelings, as I've already pointed out, don't come into mathematics as such[6]. It's perfectly valid and reasonable to discuss such reactions to, and feelings about, mathematical truths (although as of 2010 there's almost no literature whatsoever on this topic). But we must keep in mind that these feelings are separate from the actual mathematics which they're about.

Many students of mathematics (and even some textbooks) express a similar astonishment, or at least *surprise*, that logarithms should have a role to play in describing the "behaviour" of the prime numbers. Logarithms are not generally associated with the simple properties of counting numbers, having been taught to generations of reluctant youth for entirely different reasons (mostly their applications in engineering and in the calculation of compound interest).

The pattern which Gauss and Legendre detected in the primes is now a proven reality and is very well understood, but the fact that prime numbers obey a rule like this only "statistically" (that is, it's a pattern of the "imperfect" variety) also seems to be a source of surprise for some people, I've found. It's been mentioned already that the PNT can be used to determine probabilities for finding primes in certain regions of the number line. It just *doesn't seem right* to some people that things like *probabilities* and *statistical thinking* should be so directly applicable to something as eternal and absolute as the distribution of primes within the sequence of counting numbers.

Further instances of *surprise* and *bafflement* associated with the prime numbers were discussed in connection with Ian Stewart's remark back in Chapter 4.

Professor Zagier ended his lecture with these words:

> "*I hope that…I have communicated a certain impression of the immense beauty of the prime numbers and the endless surprises which they have in store for us.*"[7]

Notice the use of "endless". This "surprise" theme will be revisited in Volume 3.

WEEDS: ATTITUDES TOWARDS NUMBER AND NATURE

Both of the following quoted passages about the distribution of prime numbers involve the image of *weeds*.

> "*The first* [*fact*] *is that, despite their simple definition and role as the building blocks of the natural numbers, the prime numbers… grow like* ***weeds*** *among the natural numbers, seeming to obey no other law than that of chance, and nobody can predict where the next one will sprout. The second fact is even more astonishing, for it states just the opposite: that the prime numbers exhibit stunning regularity, that there are laws governing their behaviour, and that they obey these laws with almost military precision.*" (Don Zagier[8])

> "*When you look at a list of them stretching off to infinity, they look chaotic, like* ***weeds*** *growing through an expanse of grass representing all numbers. For centuries mathematicians have striven to find rhyme and reason amongst this jumble.*" (Marcus du Sautoy[9])

The psychology underlying this choice of imagery is interesting, as it may provide clues about Western civilisation's interrelated attitudes towards Nature[10] and number concepts.

A weed is nothing more than a plant which some people have decided is "undesirable". There's no biological characterisation of a "weed". The emergence of the "weed" concept is very much related to the gradual loss of contact between Western civilisation and the plant world. Each plant in a human-inhabited ecosystem used to be known to the local people for a variety of practical, medicinal and/or culinary uses. With the emergence of large-scale agriculture, particularly modern "monoculture", the vast majority of plants (those not being grown as crops or considered sufficiently beautiful) came to be dismissed as annoying, irrelevant "weeds". The fact that they are no longer of any interest to humans (and that they disrupt our agricultural or aesthetic schemes) has blurred into our applying a *value judgement* to them, almost as if there's something intrinsically "wrong" with these plants.

Even plants which are appreciated for their beauty or used decoratively *go in and out of fashion* as time passes, which is an equally silly projection of ephemeral and arbitrary human values onto remarkable organisms which have been quietly growing, dropping seeds and gradually evolving for millions of years. I imagine it could be quite difficult to explain this Western phenomenon to a "primitive" tribe who had an encyclopaedic knowledge of all the plant species in their immediate environment, passed down by oral tradition through countless generations.

The immediate feeling of many modern Western humans when first presented with the basic facts about the distribution of prime numbers is that the number system *shouldn't behave like that*. There's a moment of denial, a desperate grasping for pattern. Mathematicians (both professional and amateur, in their very different ways) pursue patterns in the prime numbers driven by some belief that the number system *must* ultimately be orderly, *must* obey laws.

Look at those two "weeds" quotes again. The Western establishment has traditionally associated the following words with virtue: "regularity", "laws", "governing", "obey" and "military". These words it has traditionally frowned upon: "chaotic", "weeds", and "jumble". We're supposed to respect our military and those who govern us and to obey laws in order to preserve the "regularity" of society. We're not supposed to want chaos, weeds or jumble (although there has always been a small margin of "difficult" people who do).

The "culture over Nature" mindset which prefers straight lines to curves and fields of uniform crops to wildflower meadows has arisen in relatively recent human history, although how far back its roots go is hard to say. My feeling is that it's related to the tendency discussed in Chapter 1 to marginalise anything *that can't be quantified*. Back there, I suggested that this might be linked to our relationship with the number system – that we've "fallen under the spell of quantity" and that this has unbalanced us at the level of our collective "cultural psyche".

This association of weeds with prime numbers illustrates certain relationships between humans, number concepts and Nature. The extreme quantity-obsessed human behaviour we currently see so much of ties in with the deeper and deeper links being forged between humanity and number, as well as with the simultaneous weakening of the bonds between humanity and Nature.

There's a conflict built into our relationship with the primes. This is evident in the many references to "surprise" and "astonishment" which accompany descriptions of the distribution of prime numbers and surrounding mathematical issues. *Our expectations or assumptions are conflicting with the reality* (it's that kind of conflict which underlies any kind of "surprise").

I have come to look at the prime numbers as possessing a certain *wildness*, like a magnificent, ancient gnarled oak tree. Western poets can still perceive and express the magnificence of the gnarled oak, but its form and structure are entirely at odds with the modern world they must inhabit – just compare it to a shopping centre.

> "*...the prolific and immensely influential master mathematician Leonhard Euler... expressed in 1751 his bafflement about the impenetrability of the primeland thicket...*
>
> *Since primes are the basic building blocks of the number universe from which all the other natural numbers are composed, each in its own unique combination, the perceived lack of order among them looked like a perplexing discrepancy in the otherwise so rigorously organized structure of the mathematical world.*" (H.P. Aleff [11])

The word "thicket" again suggests Nature in a wild, unmanaged state, having reached a point of becoming (like weeds) somehow problematic.

Looked at in this way, the prime numbers are like weeds that are ruining Western civilisation's beautifully trimmed conceptual lawn. Naturally, Western civilisation has *studied* the prime-weeds. It can't eradicate them or force them into a more obviously regular pattern, so it attempts to "explain" them, to capture them within the net of rational thought.

There's something which I must confess before we get started on the next chapter. Despite the impression that I may have created, as well as the rough, statistical pattern which Legendre and Gauss detected, there is *also*, in a very deep and subtle sense, a "perfect" pattern in the primes. Unfortunately, though, to be able to see *that* pattern, we first need to become familiar with some mathematical concepts which are normally considered beyond the grasp of people without extensive mathematical education. You'll be gently guided through these concepts eventually, but first we must look a bit more closely at what we've already found.

chapter 12

the deviation

Since being proved in 1896, the Prime Number Theorem has provided a guaranteed "fairly good" approximation for the amount of primes up to any given counting number. The approximation becomes more and more accurate as we choose ever larger numbers. Mathematicians, though, are the last people on Earth to be satisfied with a "fairly good" approximation – they're on a continual quest for greater precision. So, once the truth of the PNT was established, number theorists were immediately seeking *better approximations*.

It's perhaps not obvious to you what I mean by "better approximations". As with the PNT, any such approximation must possess the property of eventually remaining as close to 100% accuracy (or 0% error) as you insist, if you're prepared to look far enough down the number line. But it must also achieve *higher accuracy sooner*. That is, if you demand, say, 99% accuracy, a better approximation than the PNT would have to guarantee that such accuracy persisted after a smaller number than the PNT's "anything with more than 65 digits".

In the last chapter, we saw how the PNT provides us with a smooth curve which follows the "prime count staircase" reasonably well (although there's certainly room for improvement). A better approximation would produce a different smooth curve which would more closely hug the staircase, that is, where the gap between curve and staircase grows much more slowly as we proceed along the number line.

A better approximation which can be easily described is given by taking your chosen number and dividing it not by its logarithm but rather by *its logarithm minus 1*. So, for example, if we were interested in the amount of prime numbers less than 100 000, the PNT tells us that we can divide 100 000 by its logarithm, which is about 11.512. That gives 8685.889..., or approximately 8686 primes. But this improved approximation gives 100 000 divided by 11.512 – 1, that is, 100 000 divided by 10.512 or approximately 9512 primes. The actual number is 9593. So, at the 100 000 mark, the PNT approximation is 90.5% accurate, whereas the other approximation is considerably more impressive, achieving 99.2% accuracy.

One way of visualising this improvement is to go back to the explanation of logarithms which involved counting the visible coils of a spiral between a pair of circles. If you increase the size of the inner circle like this...

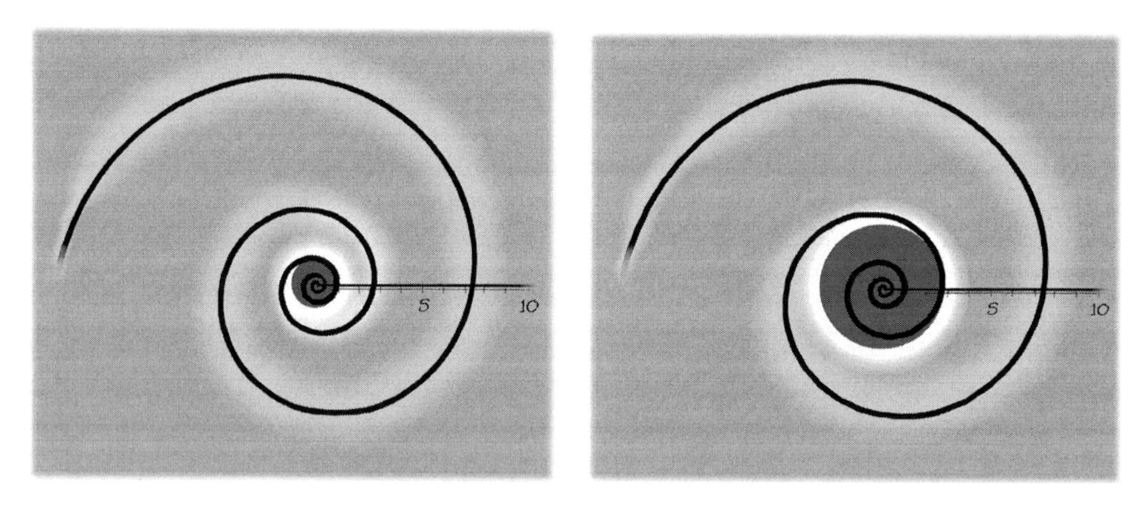

The circle on the left has radius 1. The circle on the right has radius 2.718... (that number "e"), which is where this spiral has its first crossing after 1.

...then you're just counting one less coil. The coil-count inside the new visible region is then going to give us the logarithm minus 1.

This improved approximation is what Legendre came up with[1], while Gauss' starting point was the simpler (but less accurate) approximation. The PNT is traditionally given in Gauss' form – a number divided by its logarithm – but the "minus 1" version gives a mathematically equivalent result. That means that before the PNT was proved in its familiar form, it was proved (quite easily) that if Gauss' version were true, then Legendre's version must also be true, and *vice versa*. Likewise, if Gauss' version were false, Legendre's version would be, and *vice versa*. The two propositions stand or fall together. They are logically and mathematically tied together. So, when Gauss' version was proved in 1896, Legendre's version was instantly known to be true. Basically, they're two variations on the same truth – it's just that Legendre's relates to a more efficient approximation. The situation can be compared to receiving two different sets of directions to get somewhere, both turning out to be accurate (so that they're in some sense "equivalent") but one being considerably quicker and more direct than the other.

Here's a table showing the relative accuracy of the approximations:

number	primes up to and including the number	by Gauss's approximation	accuracy	by Legendre's "minus 1" approximation	accuracy
100	25	21.7	86.9%	27.7	89.0%
1000	168	144.8	86.2%	169.3	99.2%
10000	1229	1085.7	88.3%	1217.9	99.1%
100000	78498	72382.4	92.2%	78030.4	99.4%
100000000	5761455	5428681.0	94.2%	5740303.8	99.6%
10000000000	455052511	434294481.9	95.4%	454011971.3	99.8%

Here, we see the graphs of the two approximations and how closely they follow the prime count staircase:

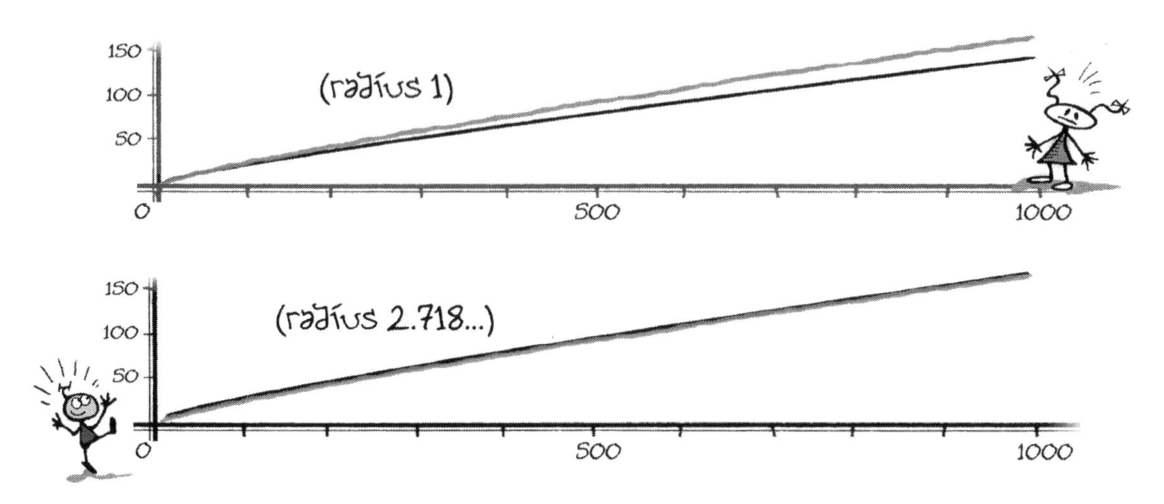

The top function graphed in black approximates the prime count staircase (grey) by counting coils inside the visible region with inner circle radius 1 (we divide by logarithms). The bottom function graphed in black uses inner circle radius 2.718... (we divide by "logarithm − 1").

Even the more accurate of these was not *entirely* satisfying to those mathematicians seeking to understand the primes. And so a couple of much better improvements were eventually discovered – their graphs "hug" the prime staircase even more closely. They're still based on logarithms but are considerably more difficult to describe. If you're feeling mathematically confident and want to know more, have a look at Appendix 8.

ANOTHER WAY TO COUNT PRIMES

Because what we've just been looking at are *approximations*, there will always be a certain amount of error involved. Sometimes they'll predict more primes than there actually are, sometimes less. The aim of trying to improve the PNT is to make the error *as small as possible, as quickly as possible*. But this seeking of improvements to the original PNT turns out to lead us down a dead end of mathematical exploration. Instead, the way forward involves *changing the way we count primes*. So, rather than delving further into the improvement of the basic PNT approximation, we'll look at this other approach to counting primes which, as we'll come to see, turns out to be much more "natural" (even if it might not seem so at first).

There are two differences:

First difference: We don't just count the primes, we also count all of the *powers* of each prime. A "power" of a number is what you get when you multiply that number by itself any amount of times. So we don't just count 2, we count 2, 2×2 (4), $2 \times 2 \times 2$ (8), $2 \times 2 \times 2 \times 2$ (16),... We don't just count 3, we count 3, 3×3 (9), $3 \times 3 \times 3$ (27), $3 \times 3 \times 3 \times 3$ (81),... Recall the image of counting numbers as "prime clusters" introduced in Chapter 6. We simply look at the factorisation of each counting number, and if it has only one prime number involved (however many times), it gets counted. If it has a "mixed factorisation", it doesn't. In this way, we count *primes and their powers*. Because every number can be thought of as a power of itself (the "first power"), we could think of this method as simply counting *powers of primes*.

Second difference: Normally, when you count something, whether sheep or prime numbers, you add 1 to your count for each sheep, prime number (or whatever) that you encounter. In this approach, though, we don't just add 1 to the count when we find a power of a prime. What we add depends on the number itself. More specifically, we add *the logarithm of the prime involved*. So, for each of 2, 4, 8, 16,..., we add the logarithm of 2 (which is about 0.693). Similarly, for each of 3, 9, 27, 81,..., we add the logarithm of 3 (which is about 1.098) and for each of 5, 25, 125, 625,..., we add the logarithm of 5 (which is about 1.609).

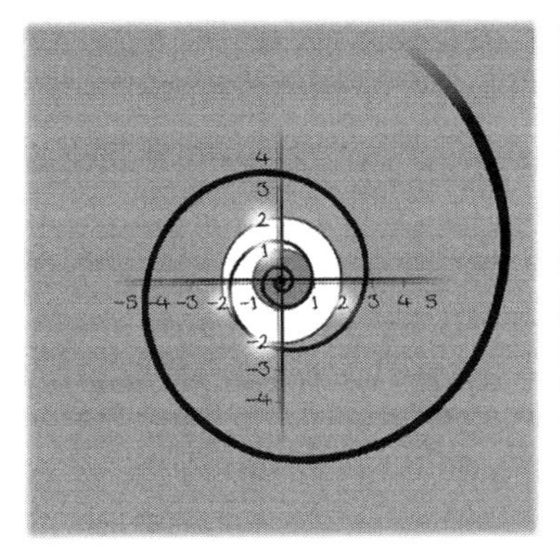

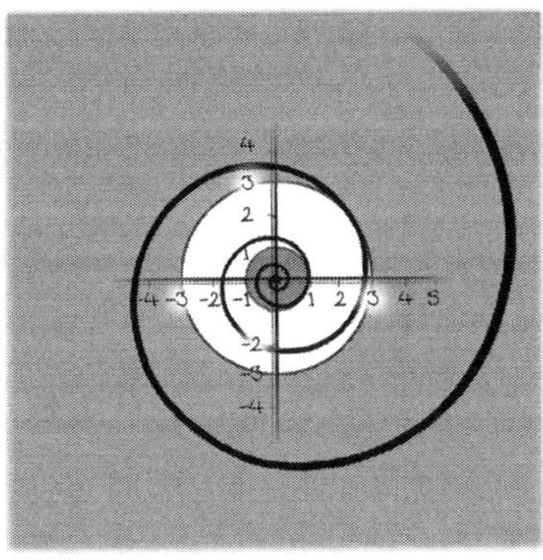

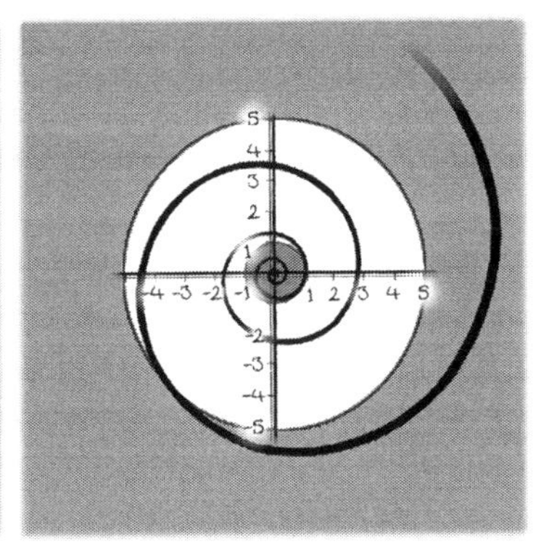

The logarithms of 2, 3 and 5 shown as numbers of coils (in the white regions).

While the usual method which we've seen just counts the primes up to a given number (adding 1 to the count for each prime), this new method can be thought of as counting (or perhaps "measuring") something slightly different. Although it's not standard mathematical terminology, I like to think of what's being counted (or "measured") as the *amount of "primeness"* up to a given number. Larger primes (and their powers) contribute more to this total amount of "primeness" than smaller primes (and their powers), and all the powers of a single prime contribute equally.

The usual method of counting gives us the amount of primes up to a given number, and this, according to the PNT, is approximately that number divided by its own logarithm.

This new method of counting – which, importantly, *involves logarithms* – gives an "amount of primeness" up to a given number *which is approximately equal to that number*. It says, for example, that the "amount of primeness" (the sum of all those logarithms for all the primes and powers of primes) up to 12 947 978 approximately equals 12 947 978. In actuality, it's 12 949 044.496... That's 99.99% accuracy (and this accuracy just keeps increasing, getting as close to 100% as you want if you're willing to consider big enough numbers. It might be helpful if you visualise it like this: if you were to walk along the number line counting "primeness", the amount you'll have counted at any stage will approximately equal the number describing where you are.

The fact that this kind of prime counting gives the approximation just described is mathematically *equivalent* to the fact of the PNT, in the same way that the "Gauss" and "Legendre" versions of the PNT approximation which we looked at are equivalent to each other. So, before the PNT was proved in 1896, it had already been shown that if the PNT *were* true, then this other proposition involving the new way of counting primes (or "primeness") would also be true, and *vice versa*. Proving the PNT could then be achieved by proving that the amount of "primeness" (as defined above) up to a given number is approximately equal to that number, in the sense that the error in this approximation comes as close to zero as you like if you're prepared to look at big enough numbers.

This is how a lot of mathematical research is done. Certain mathematical propositions which are *believed to be true* but not yet proved are proved to be equivalent to each other. Then, *if any one of them* is proved, all the rest are immediately known to be true. Quite often, you can prove something by finding an equivalent proposition which is easier to prove and then proving that.

Recall that the PNT was *believed* to be true for a whole century before it was actually proved. Although a century of calculation had produced exact prime counts which were entirely in keeping with the PNT, it was fully understood by all involved that *no* amount of numerical evidence would ever prove it, because it concerns something which goes on *forever*, something which is truly *infinite*. Any evidence which might be accumulated can say nothing which applies with any certainty beyond the tiniest sliver at the beginning of the sequence of counting numbers.

The relative simplicity of the new approximation suggests that we have been asking the wrong question. Rather than asking "How many primes are there up to a given number?", we should be asking "How much primeness is there up to a given number?", where "primeness" is measured in the way which I've described[2].

Despite the existence of the improvements to the PNT which were mentioned earlier, we're going to put those aside and stick with this more subtle way of counting because it better reveals the true nature of what's going on with the number system. "Logarithmic" prime counting is somehow *more natural*, as we'll go on to see. So we can now forget about the usual method of prime counting, even if it seemed a bit easier, and adopt this new (and improved, for most purposes) method, along with the associated (and equivalent) version of the PNT.

We've seen that the prime numbers thin out in a way related to "how a snailshell uncoils", so that if you count them in the usual way, you find a "statistical" pattern involving logarithms. Alternatively, if the primes are counted in this new way which *directly involves* logarithms, with bigger primes (and their powers) contributing more heavily to the count, then you end up with an "amount of primeness" which doesn't thin out but instead remains approximately (and becomes increasingly) steady forever.

To summarise: you can either count primes "uniformly" and get a "logarithmic" pattern, or you can count them "logarithmically" and get this much simpler "uniform" pattern. We're adopting the latter approach.

ANOTHER STAIRCASE

Just as with the original method of prime counting, we're able to build a "prime count staircase". As you'd expect, this one looks a bit different, with stones of different sizes involved – think of it as the "primeness count staircase":

Here, a new stone is added to the height of the previous column only when the factorisation of the number in question is "pure" (involves just one kind of prime factor). The height of this stone is the logarithm of that prime. Because 6 has mixed factors (2 and 3), no new stone is added to its column. 37, like 2, 3, 4 and 5, has a "pure" factorisation, so a stone of height 3.610... (the logarithm of 37) is added.

If we zoom out from this staircase, we begin to see something which is almost indistinguishable from a 45° slope (that is, the diagonal of a square):

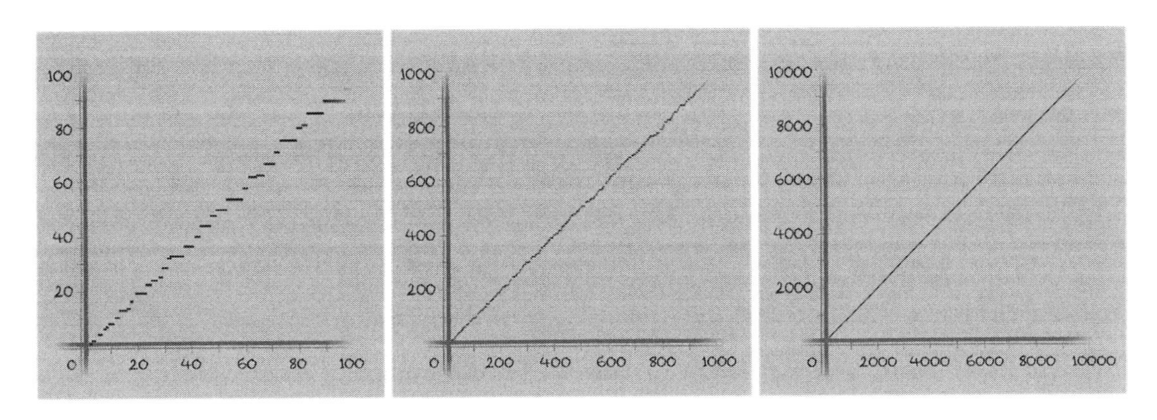

This helps us to see why the logarithmic approach can be described as the "most natural" way to count primes. If you stood at any point on the number line and looked up, the "height" of a 45° sloping line above you would be exactly the same as the number describing your position. The steadily sloping tendency of the staircase we've built means that at any given number, the "amount of primeness" (rather than the number of primes) we'll have found is approximately the same as the number itself.

This logarithmic counting idea is due to Pafnuty Chebyshev[3], who first published it in the early 1850s[4].

LANDSCAPING THE NUMBER LINE

Most of us would be content with such a good approximation, but as I've already suggested, mathematicians wouldn't. They're compelled to look "further down" or "deeper in", so they seek patterns in the *deviation*, that fluctuating gap between the actual prime (or primeness) count and its approximation. *Any* approximation will produce a deviation, and this can then be studied in detail. An important reason for adopting Chebyshev's logarithmic counting approximation is that what you find hidden in its deviation is *much* simpler and more beautiful than what you find in the deviation associated with any other approximation.

We're going to imagine walking along the number line again, but rather than building another staircase, we're going to build something which goes up *and down*. We're going to "landscape" the number line. Imagine we're counting both primes and their powers, as I've described. At each point along the line, you can compare your actual "primeness" count to the estimated one, that is, to the number which describes where you're standing. If the amount of primeness exceeds the estimate, then we'll build upwards by the amount by which it's too high. If it's *less than* the estimate, then we'll dig down by the amount by which it's too low:

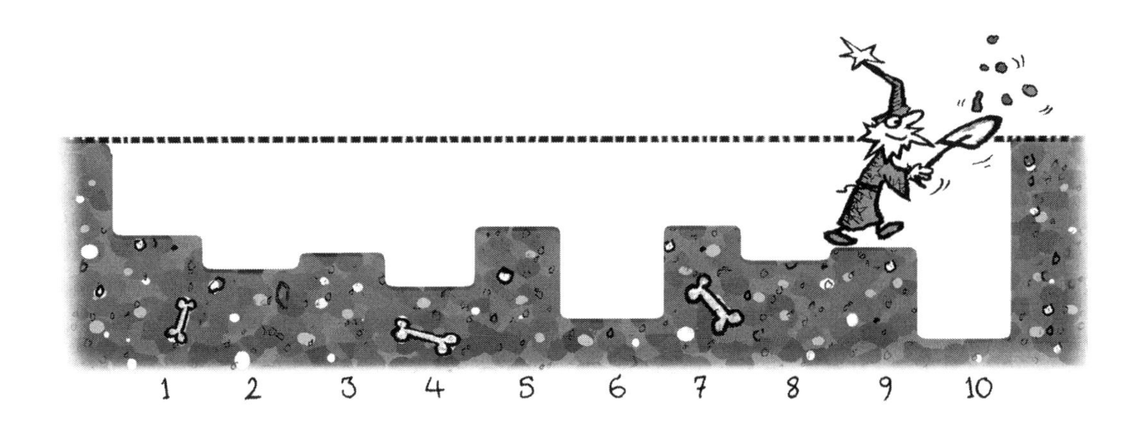

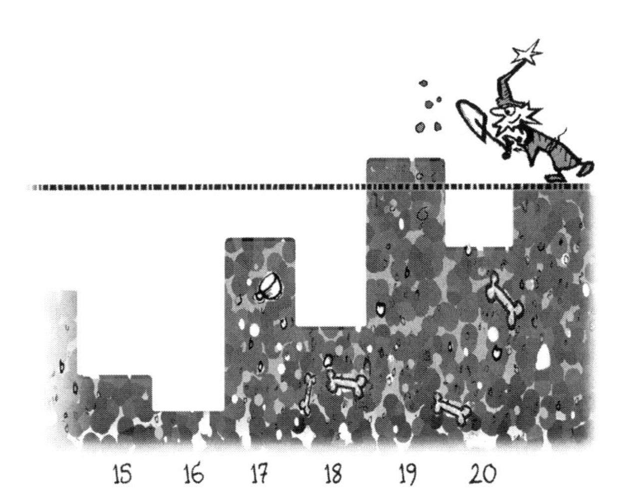

At 10, the primeness count is about 7.83, which is 2.17 less than the estimate (10), so we dig. At 19, the count is 19.27, so we build up slightly. At 20, it's still only 19.27, so we dig, as that's less than the estimate (20). At 1240, the count is 1243.001, so we build up by just over 3 units.

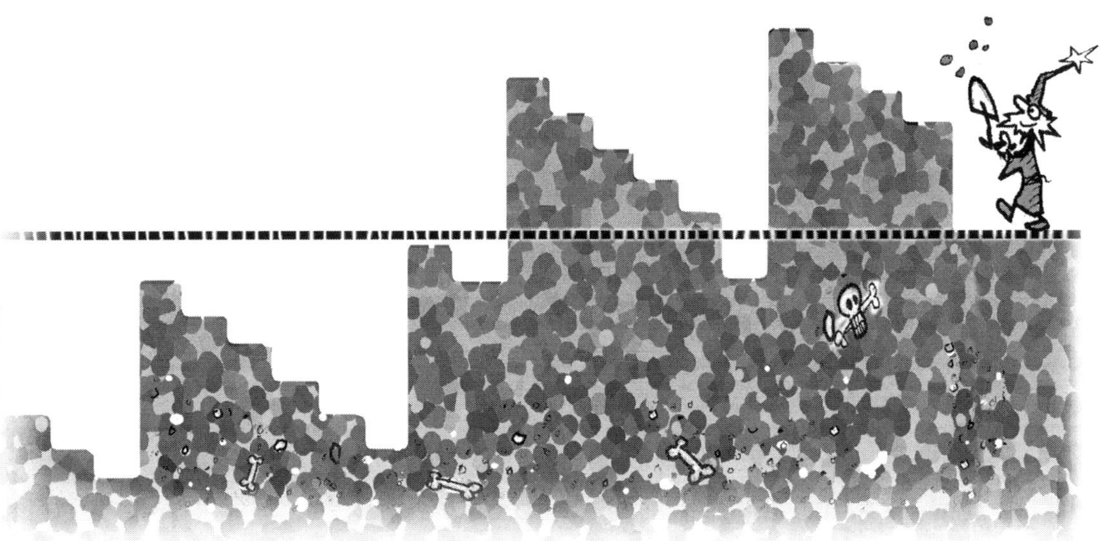

If we continue doing this far enough along the number line, we produce a rather striking "landscape" which looks like this (the horizontal number line corresponds to the level of the ground which was "landscaped"):

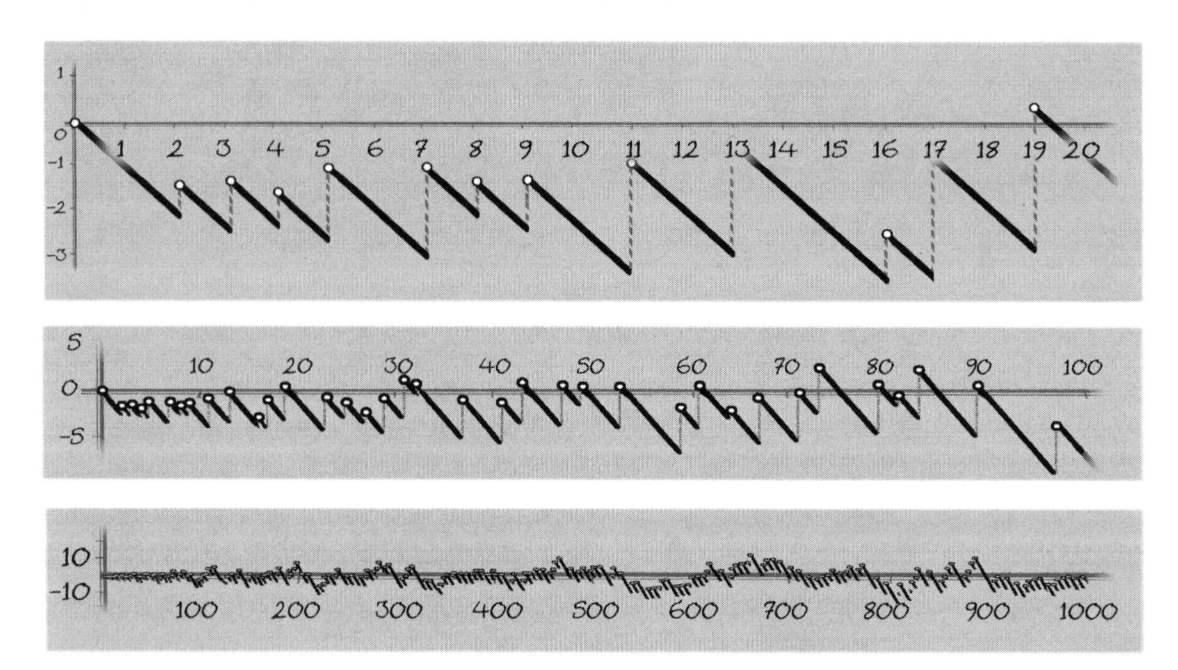

Notice that these graphs are made of diagonal segments rather than the blocks seen in the previous three images. There's a very good reason for this, which will be explained next.

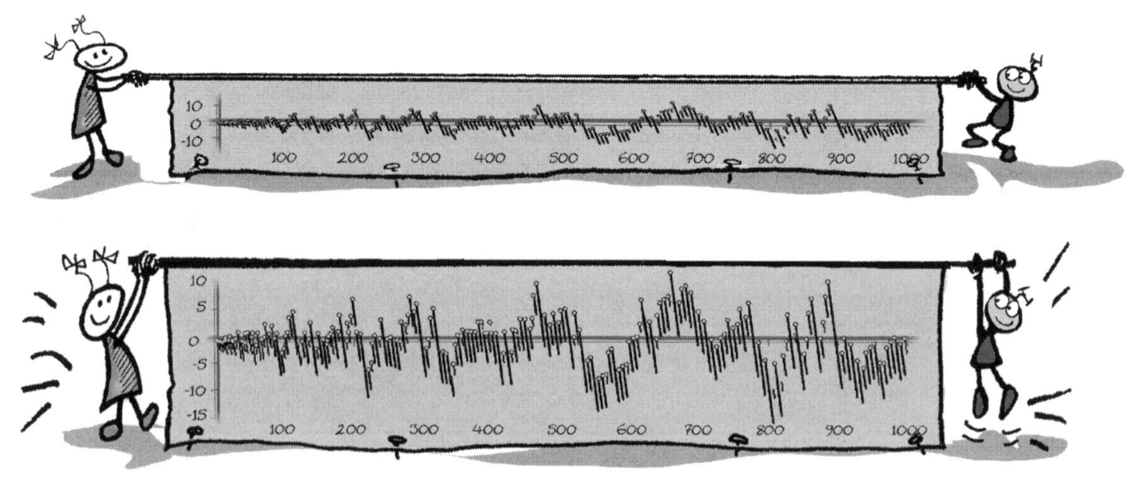

In the 0–1000 range, the deviation remains rather small, so to get a better sense of what it's doing, we can stretch the graph vertically, as shown. This is known as "rescaling" a graph.

What we're looking at here is the graph of a function – we'll call it the *deviation function*. It measures the difference between how much "primeness" there actually *is* less than a given number and how much there "should be" according to our simple estimate. In regions of the number line where prime numbers (and their powers) are unusually dense, there's "more primeness than there should be". This is reflected by the fact that in such regions, the graph we've just seen sticks up above the horizontal number line. In regions where primes and their powers are unusually sparse, there's "less primeness than there should be", reflected by the fact that the graph dips down below the number line.

In the "landscaping" images on page 241, we were only considering this "deviation" measured at the position of each counting number along the number line. But we could just as well make the comparison between the amount of primeness found and our location on the number line at other positions, between counting numbers. To see the result of such comparison, we'll go back to the staircase and the diagonal line:

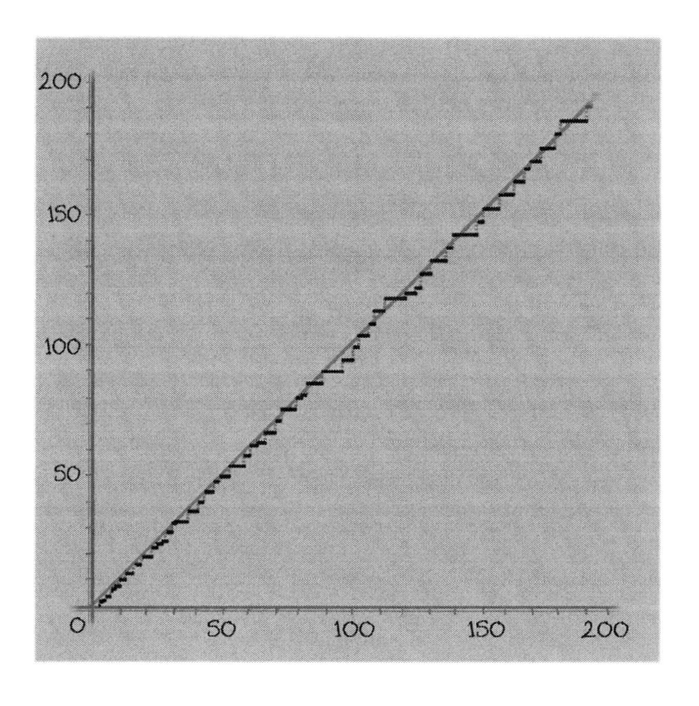

Wherever the staircase goes up, that indicates the location of a prime or a power of a prime. We'll draw a vertical line through each of these locations, like this:

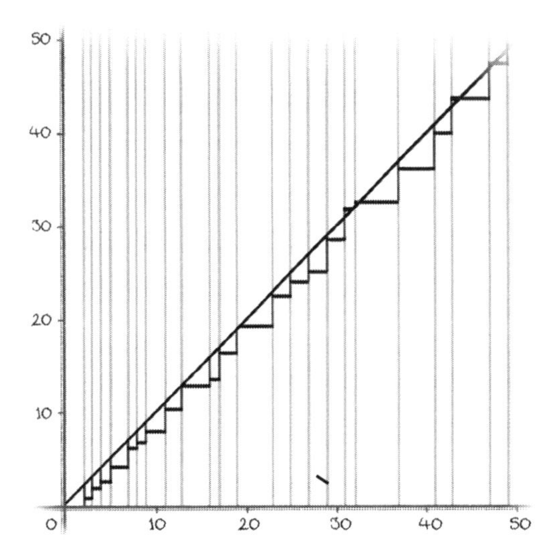

Consider each vertical strip separately, noticing how each one contains a piece of horizontal line (corresponding to an actual amount of primeness) and a piece of diagonal line (corresponding to the approximation). The amount of space between them, and which one is above the other, will vary, and it's this variation which the graph of the deviation function shows. So, let's slide each of the vertical slices so that the piece of horizontal line ends up meeting the number line, like this:

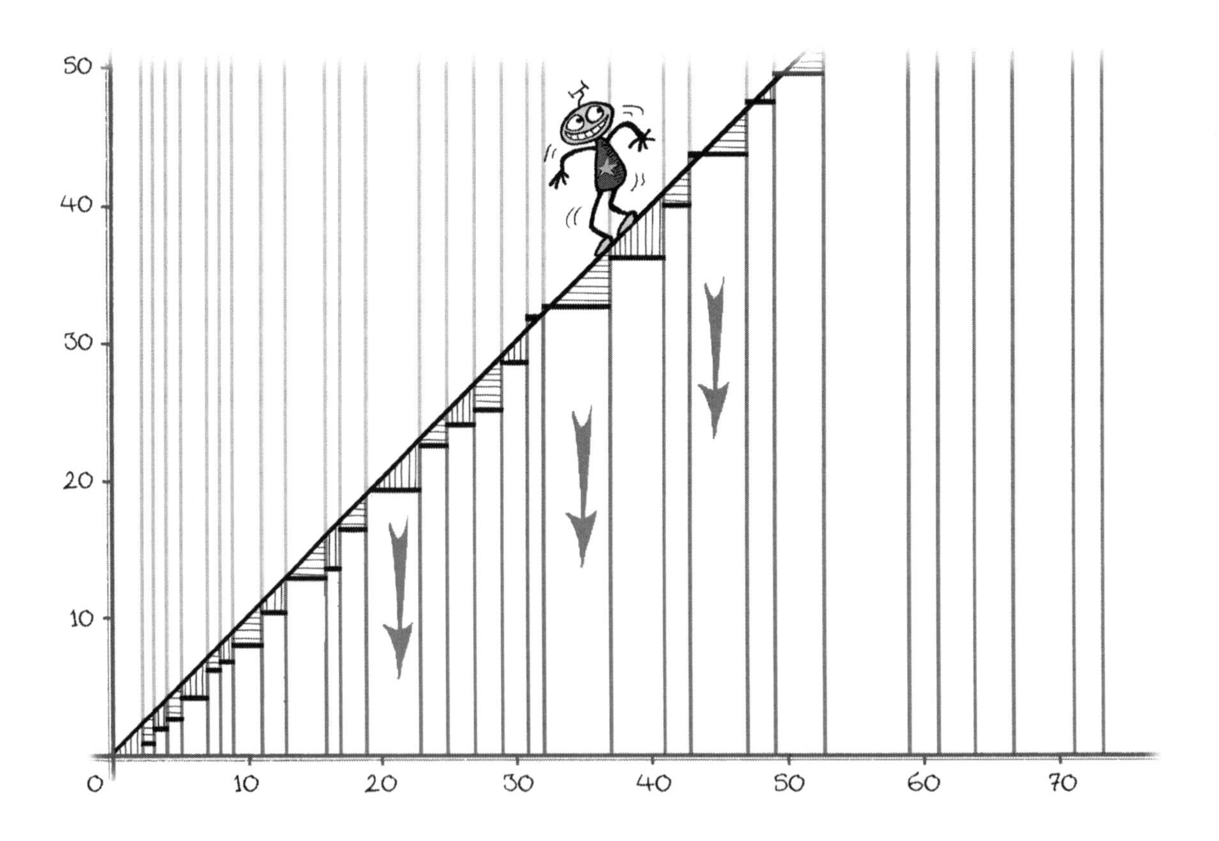

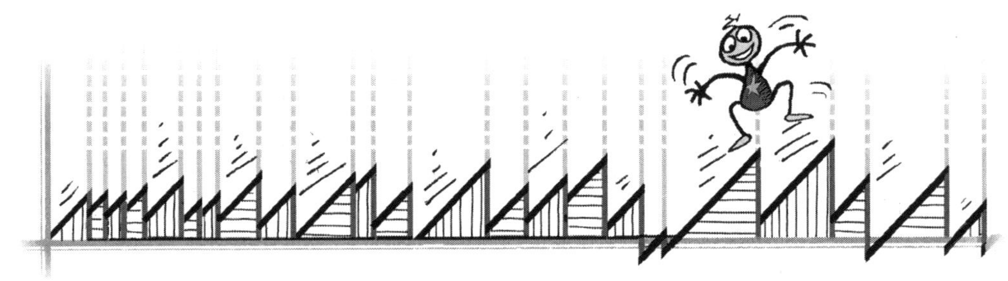

All of the horizontal segments from the last picture now lie on the number line. In this way, the (few) shaded pieces sticking up above the diagonal in that picture are now *below* the number line, and the rest, which were hanging below the diagonal, are now above it.

If we do this for that section of the number line from 0 to about 500, we get this...

...which, when flipped over, is just the graph of the deviation function:

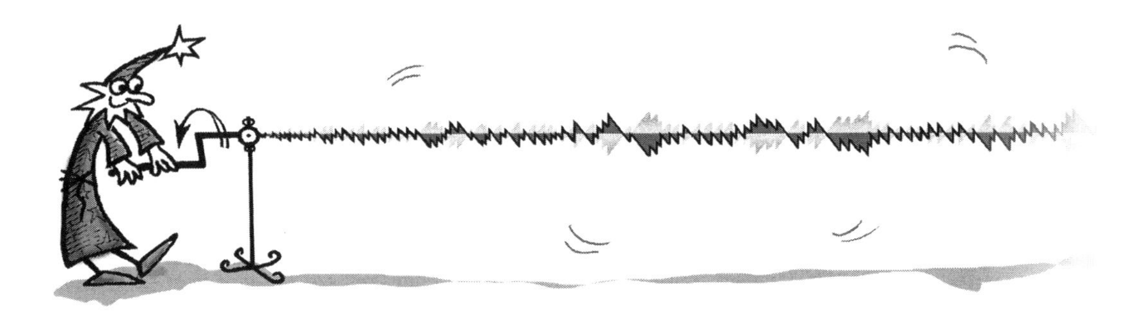

This graph shows us the fluctuating error associated with the approximation for the amount of primeness less than a given number (that approximation being the number itself). In other words, the graph sticking up above or dipping below the horizontal number line indicates that there is more or less primeness than the estimate suggests there "should" be up to that point.

The fact that this jagged graph appears to "centre itself" on a horizontal line lends credence to the validity of the approximation. It sometimes strays too high, sometimes too low, but it always hangs around this "centre"[5]. If there was an overall tendency for the graph to stray upwards or downwards from the line, that would indicate a tendency to consistently overestimate or underestimate primes, not the sign of a good approximation.

SUMMARY AND CONCLUDING THOUGHTS

To summarise:

- ☆ We know that the prime numbers continue forever.
- ☆ We know that they thin out.
- ☆ We know that they thin out, on average, according to a mathematical law related to the geometry of a spiral (this can be slightly improved upon and presented in various ways, but these always relate back to the logarithm, hence to the spiral).
- ☆ By counting in a different way which *directly involves logarithms*, where larger primes (and their powers) contribute more than smaller ones, we can measure an "amount of primeness" whose growth, on average, remains steady.
- ☆ Despite the steadiness of the new "primeness" count, the actual amount of primeness still deviates from the approximation.

Before proceeding, we'll make a few observations about this deviation, as represented by the jagged graph we saw on pages 242 and 245.

First, any such graph we might construct is only going to show the deviation up to some finite number – however "big" we might think that is – so we'll only ever see an infinitely small sliver of the whole picture (since the number line extends infinitely).

Second, this odd-looking jagged graph, however much or little of it we can actually see, is *there*, built into "our" system of counting numbers, whether we notice it or not. In some sense, then, it's embedded in the very structure of what we call "reality". No one (as far as we know) knew about this until the 19th century, but it's inescapably *there*. And it looks like *that*.

Third, when examined closely, the graph is found to be made of disconnected bits – little pieces of diagonal line, each a counting number of units wide (and tall).

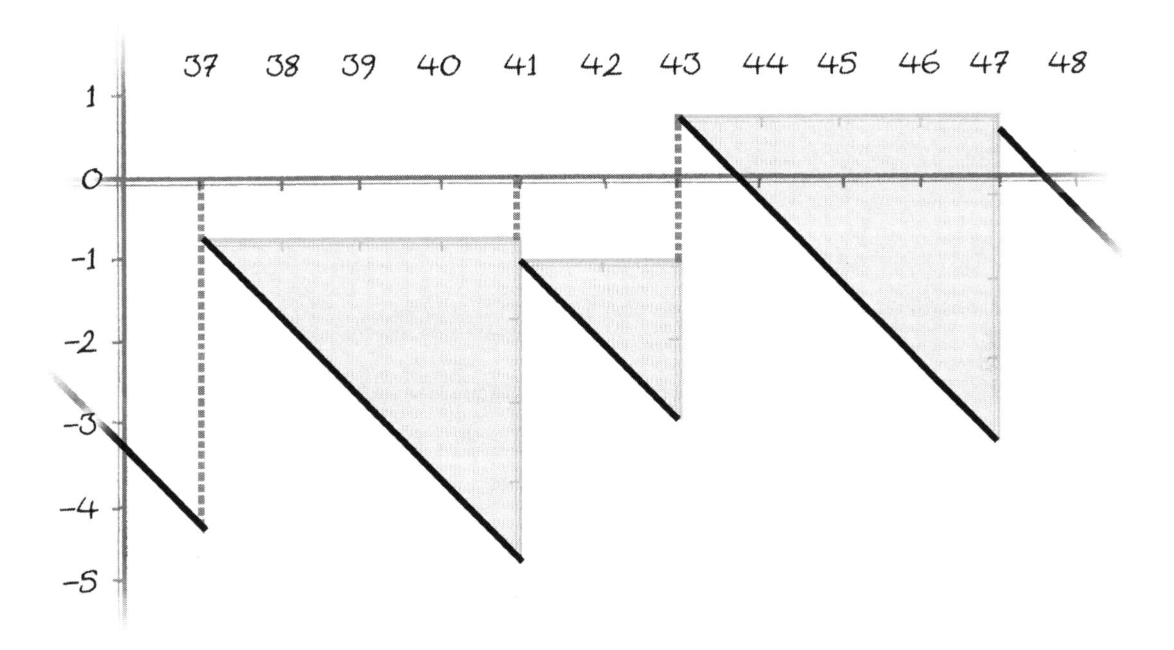

The "imperfect pattern" in the prime numbers which the PNT (or its more "natural" equivalent) tells us about is the only *order* we've found in the primes thus far in our exploration. By looking at how the primes *deviate* from this pattern, we're effectively stripping out this order and revealing what's left, showing how the behaviour of the primes deviates from, or fluctuates around, this pattern. This could be thought of as the "chaos"[6] or disorder inherent in the primes. However, this disorder is only *apparent*. Just because the graph looks like a jagged, irregular mess doesn't mean that it can't contain some kind of order or pattern once you know how to look at it correctly.

So, it's this deviation function that we'll focus on for the rest of this volume.

Chapter 13
Harmonic Decomposition

The deviation we've been looking at has been presented as a graph (the jagged looking landscape we've been considering), but that graph represents a *function*. As explained in Chapter 11, a graph acts as a guide to the workings of the function that it depicts. A function, remember, can be thought of as something which takes points on the number line and, according to a fixed rule, transports them to somewhere else on the number line. A function's graph allows us to work out where any given point will get transported to (that is, what the function "does" to any given number).

In this case, the *deviation function* takes a point on the number line, considers how much "primeness" (the sum of logarithms which we've seen) has occurred up to that number, subtracts the number itself, and transports the original point to the result of this subtraction. If the amount of primeness is bigger than the number itself, then subtracting will give a familiar positive number. If the amount of primeness is less than the number, then we're subtracting a bigger number from a smaller one (for example, at 55, we'd have $53.455 - 55$), which gives a negative number. Dipping below the horizontal number line as they did, the first two graphs on page 242 showed that the deviation function has a strong tendency to produce negative numbers (less primeness than there "should be") in the initial stretch of the number line. As we get out into the 0–1000 range, though, a balance between negative and positive numbers starts to emerge.

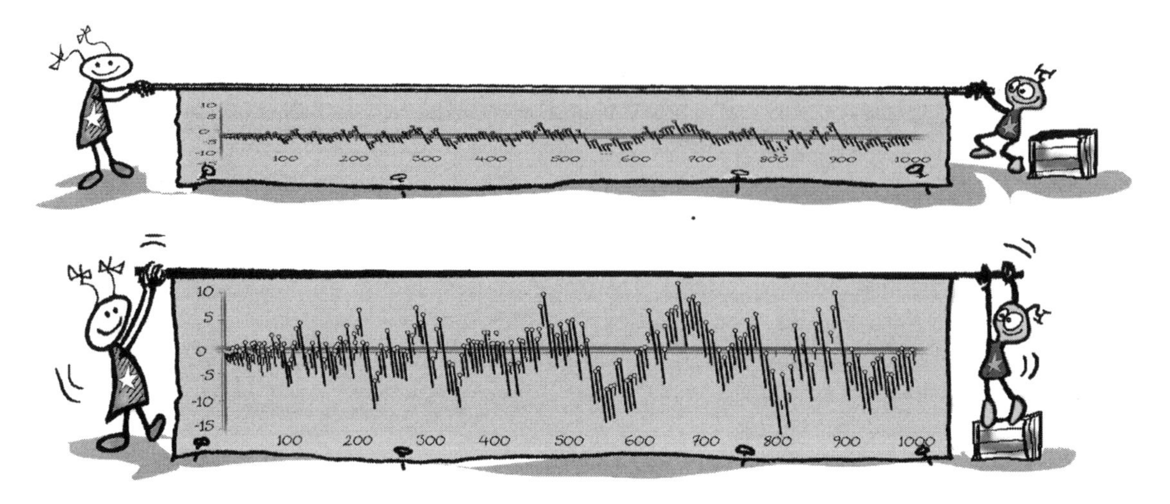

Stretching the graph vertically, we can see that despite an early bias for it to tend towards the negative, things start to balance out fairly soon.

The graph of the deviation function visually represents what the function does, as just described. So, given a point on the number line, the location it's going to get transported to by the function is shown by how high (or low) the graph is above (or below) this point on the number line. The graph being *above* the number line at a point indicates that the function transports that point to a *positive* number (the distance above telling you which one). The graph being *below* the number line indicates transport to a *negative* number (the distance below telling you which one).

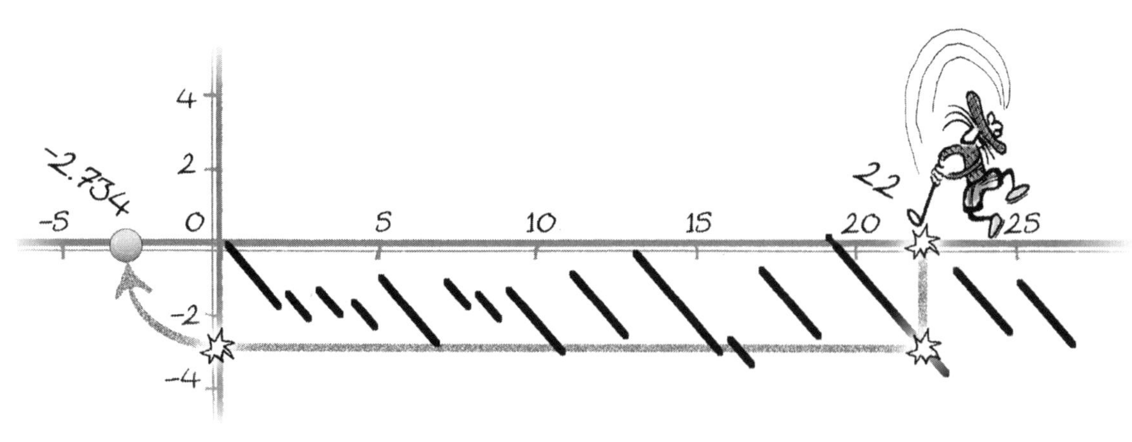

At 22, the graph is about 2.734 units below the number line. This tells us that the "deviation function" transports the number 22 to, approximately, the (negative) number −2.734.

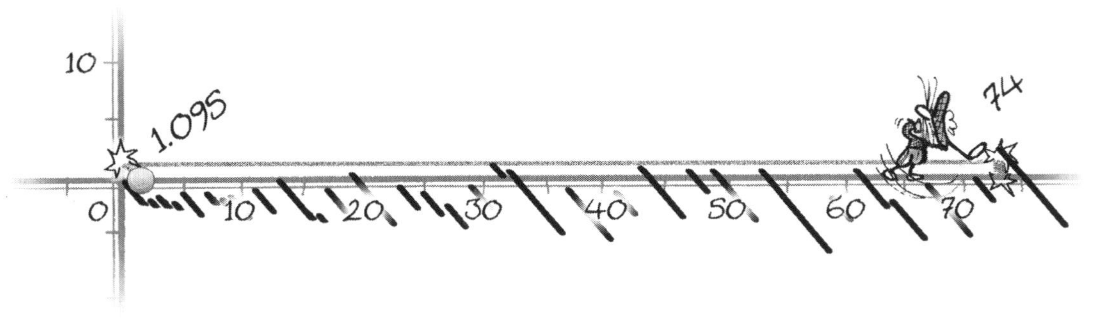

At 74, the graph is 1.095 units above the number line. This tells us that the function transports the number 74 to the number 1.095. Here, the quarter circle that the ball follows after hitting the vertical line is almost too small to see.

THE SOUND OF THE PRIMES

If you've ever done anything involving sound recordings on a computer, or watched someone else doing so, you'll probably recognise images like these:

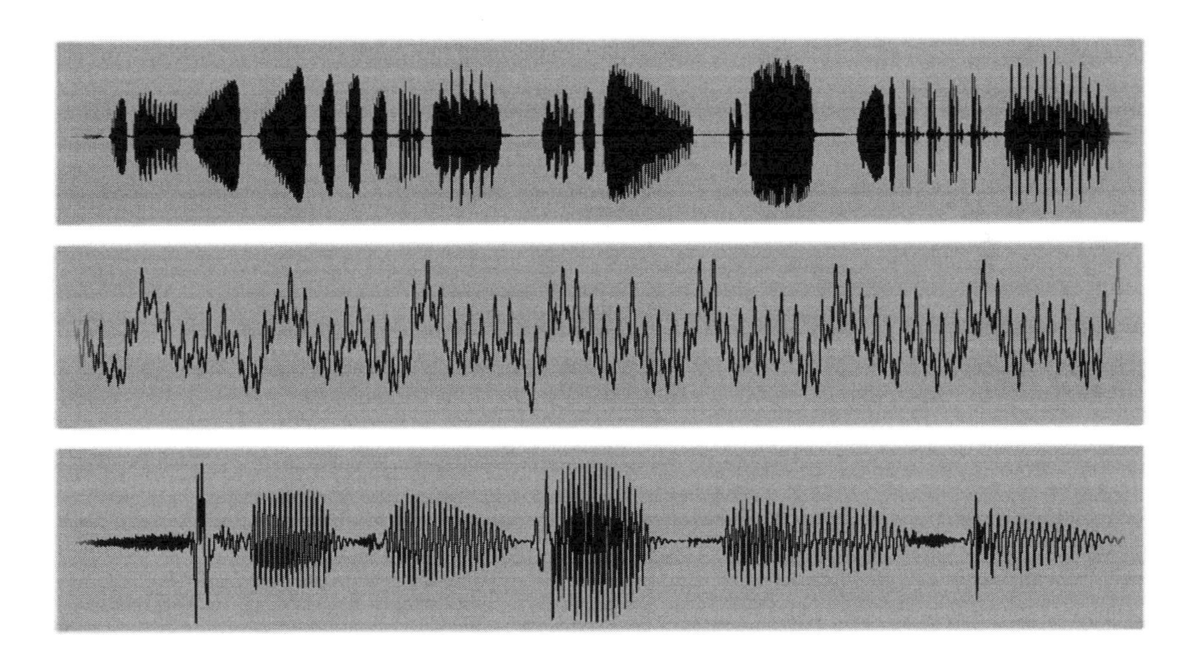

These are "waveforms" – visual representations of sound. The first is the sound of a skylark singing, the second is the sound of Eric Dolphy playing a low C on his bass clarinet and the third is the sound of me saying the word "Swedenborgianism". The length of the waveform (horizontally) corresponds to the length in time of the sound in question (in these cases 4.374, 0.09 and 1.332 seconds) and the fluctuating height relates to the volume, pitch and other qualities of the sound over that time.

To someone familiar with this way of graphically representing sound, it would be natural, on looking at the graph of the "primeness count deviation" seen on page 250, to wonder what it would *sound like*. And, indeed, using a computer, it's fairly straightforward to convert the graph into sound. There are a few problems with this, though:

- ☆ There's no obvious timescale to work with, so it's not clear how fast or slow the sound should be played[1].
- ☆ As already observed, no graphical representation we produce can ever show more than the tiniest piece of the entire deviation of the distribution of primes from its approximation (which continues on down the number line *forever*). Likewise, we'll only ever be able to hear an infinitely brief little introductory sound at the beginning of an infinitely long "piece".
- ☆ As far as we can tell, however much of it you play and at whatever timescale, it sounds like a strange noise, entirely unmusical to most ears[2].

It might seem like this idea of trying to "listen" to the primeness count deviation, possibly seeking to hear some kind of order, is a bit of a fanciful distraction. But, in fact, there's actually something in this. To understand what, we'll need to look at some of the basic ideas associated with these sonic waveforms.

You've almost certainly heard of *sound waves*, *radio waves* and *microwaves*. These are (roughly speaking) particular patterns of the movement of energy through space. There's now a highly developed body of mathematical theory by which physicists are able to model, produce and manipulate these waves. The same mathematics applies to waves and ripples moving through water, which is why the already-familiar word "wave" has been used for these invisible phenomena. In some ways, light also behaves like a wave[3] (optical theory also makes use of the same mathematics).

We can use this mathematical theory to design musical instruments, concert halls, radio and telecommunications technology, lenses, telescopes, microscopes, analogue synthesisers, lasers, *etc*. The mathematics can get very deep in places, but it's all ultimately built up from one simple idea – the *sine wave*:

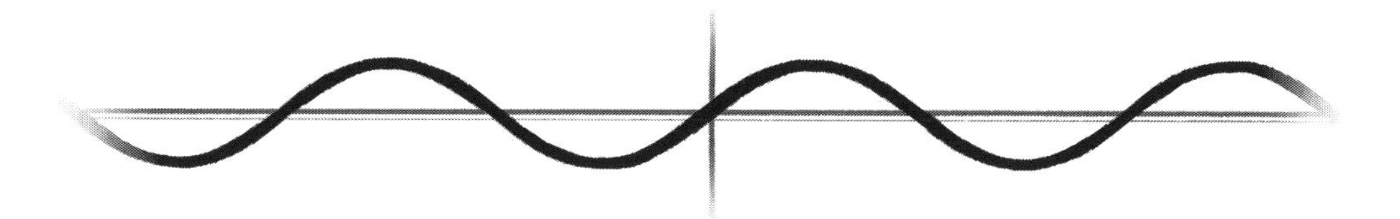

SINE WAVES

A sine wave is perhaps most easily understood in terms of motion around a circle:

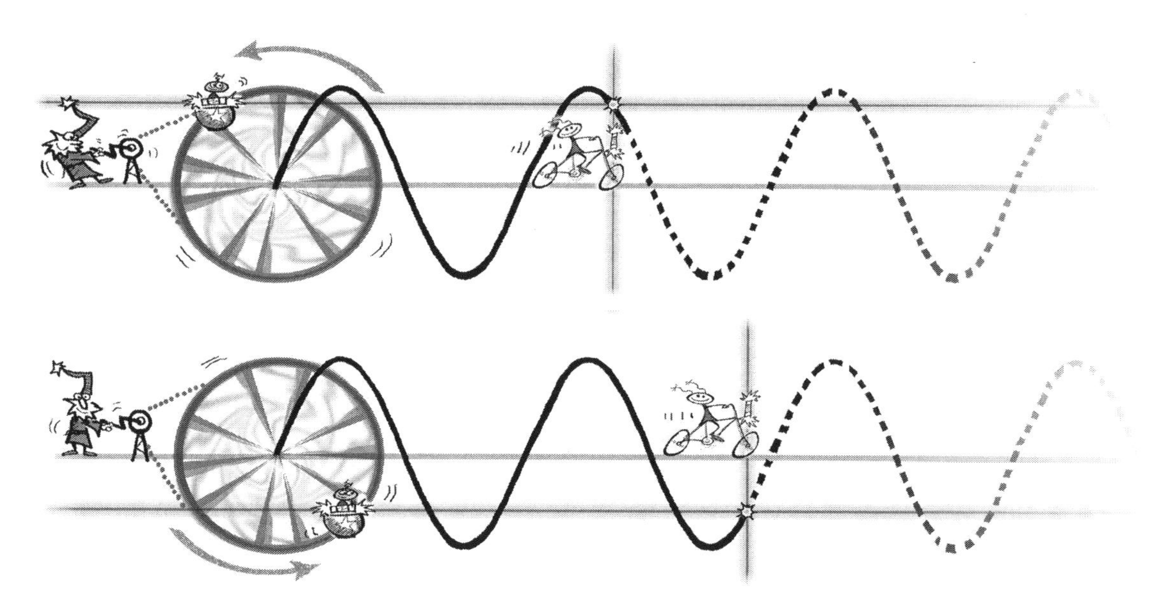

The wheel is cranked at a constant speed and the rider in the carriage shines a horizontal beam. The rider of the bicycle, moving at a fixed speed along the horizontal line through the centre of the wheel, shines a vertical beam. The point where the two beams cross traces a sine wave.

There are three things which we can adjust in order to affect the shape of the wave that's produced:

(1) The *size of the wheel* is directly linked to how high (or low) the highest (or lowest) points of the wave will be. This is called the *amplitude* of the wave.

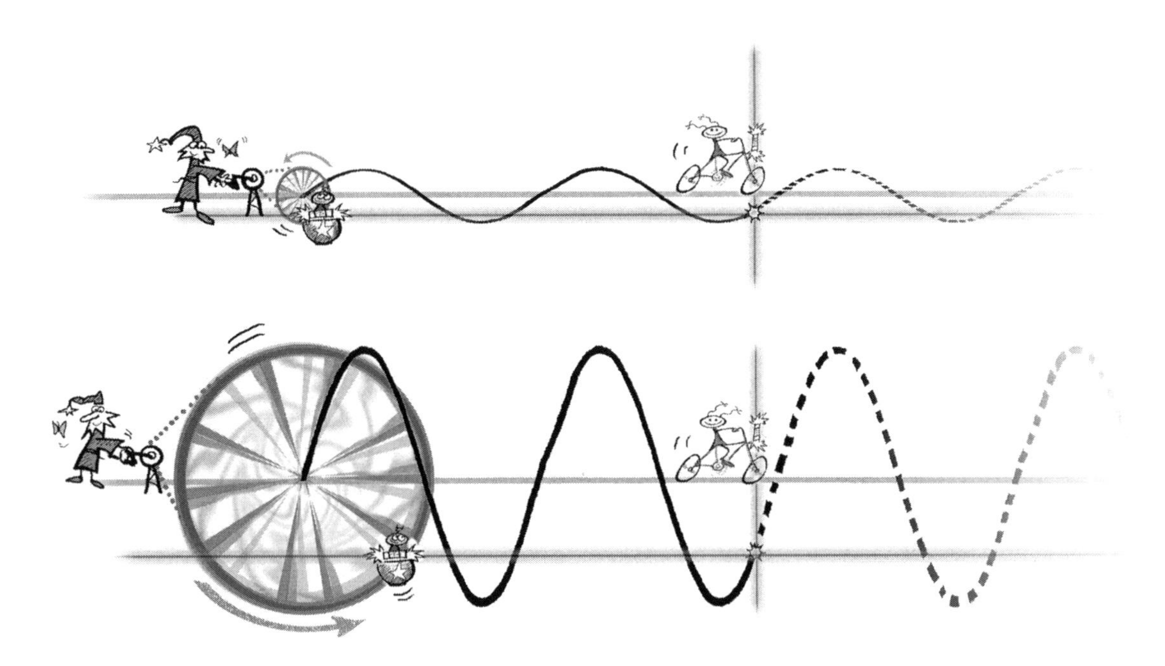

A smaller wheel will produce a wave with a smaller amplitude, while a larger wheel will produce a larger amplitude. The height of the peaks of the wave above the horizontal line (and the depth of the "troughs" below it) will always be the same as the radius of the wheel.

(2) Keeping in mind that the bicycle travels at a fixed speed, the *speed at which the wheel rotates* will affect how "bunched together" or "spread out" the wave will be – that is, the distance between the peaks of the wave. This is called the *wavelength*. Wavelength is directly linked to something called the *frequency* of the wave, which measures how often (how "frequently") the wave reaches a peak. Clearly, the larger the wavelength is, the smaller the frequency, and *vice versa*. The frequency of the wave is the same as the frequency with which the wheel returns to its original position (the number of times

per hour – frequencies are often measured in *cycles per second* (or *Hertz*), but whether we use seconds, minutes, hours or years makes no difference here, as long as we stick with our original choice and stay consistent).

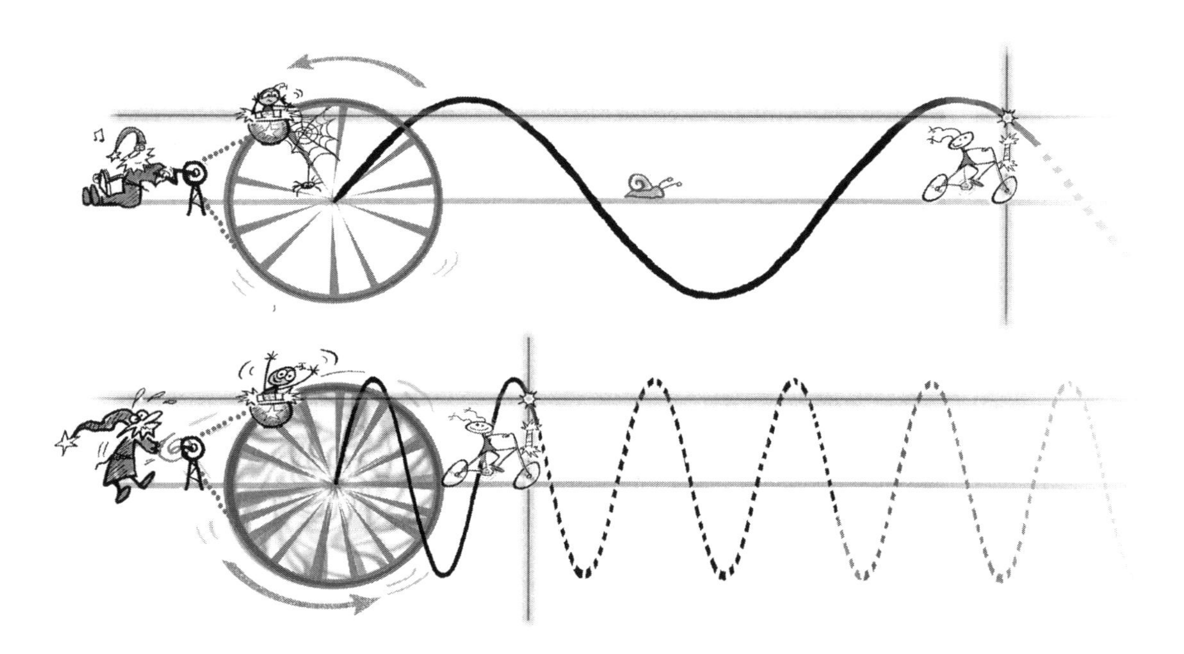

If the wheel is rotated slowly, peaks of the resulting wave will occur less frequently and so will be spaced farther apart (this means the wavelength is large, but the frequency of the wave is "low"). If the wheel is rotated more rapidly, the peaks will occur more frequently and so will be bunched closer together (so the wavelength is small and the frequency is "high").

(3) We'll treat the horizontal line as a number line and assume that the bicycle starts off at the "0" position (the hub of the wheel). The *starting position of the carriage on the wheel* doesn't affect the *shape* of the wave but rather how it's positioned on the number line. If the carriage on the wheel starts where the number line crosses the wheel (the "3 o'clock" or "9 o'clock" positions), then the wave will pass through 0. If the carriage starts somewhere else on the wheel, then you'll get the same wave, just shifted (mathematicians would say "displaced") along the number line. The amount of displacement is known as the *phase* of the wave.

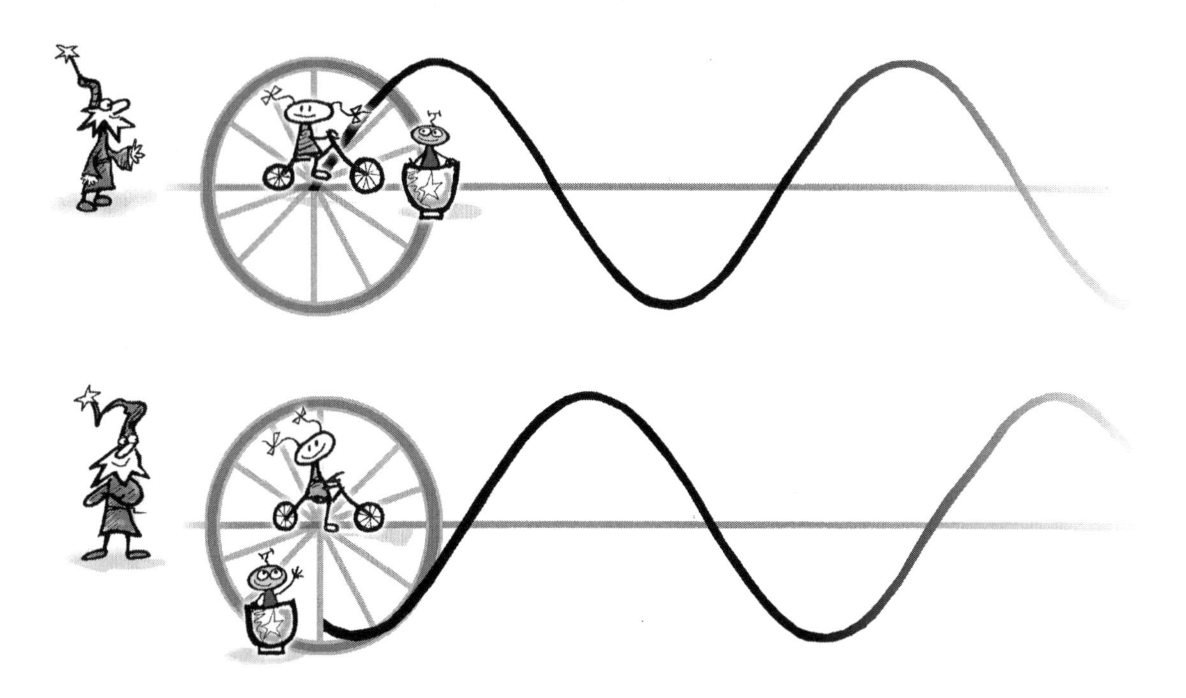

The two waves shown have the same amplitudes (because the wheels are the same size) and the same frequencies (because the wheels are rotated at the same speed), but they have different *phases* due to the different starting points of the carriage on the wheel.

By varying the amplitude, wavelength (or frequency) and phase we can produce a whole range of different sine waves, but they all share the same basic form. In terms of sound waves, a sine wave corresponds to a single, pure tone (not easily produced without electronics, although some well-tuned bells and bowls come close). Amplitude relates to the volume of the tone and wavelength/frequency relates to its *pitch* (how high or low the tone sounds). A larger amplitude would be a louder tone and a larger wavelength (or, equivalently, a smaller frequency) would be a lower-sounding tone. The phase of a sound wave is not something audible, although two waves which differ only in phase can produce noticeable *interference effects* when played together.

All sound engineering and acoustic science ultimately relates back to the study of sine waves and combinations of sine waves.

COMBINING SINE WAVES

Playing two pure tones simultaneously, we end up with a more complex sound wave, which looks something like this:

You may have seen an *oscilloscope*, an instrument which displays such waveforms on a (sometimes circular) screen. These waveforms could originate from a sound or from a pulsating electrical current:

What I'm calling "waveforms" would also be called *signals* in some contexts. There's a whole branch of physics called *signal processing*.

Here's a graphical explanation of how two waves can be "added together":

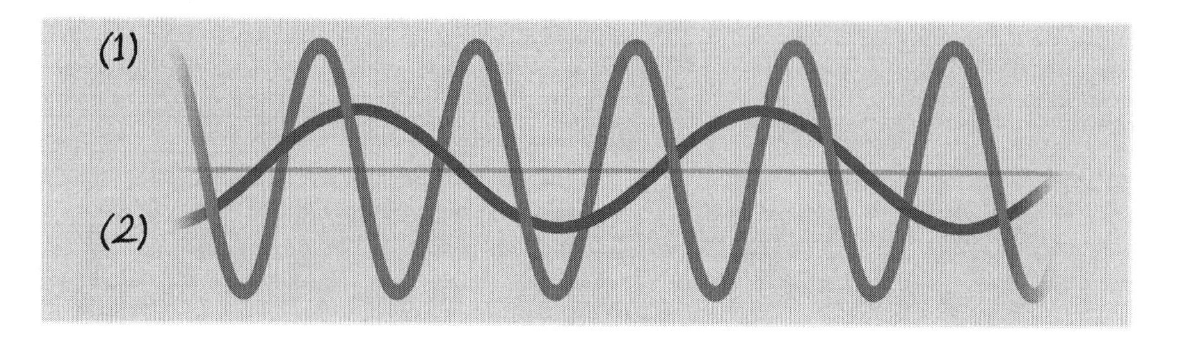

These waves have different amplitudes and frequencies but share the same horizontal axis.

At any chosen horizontal position, one apprentice will stand on the number line and the other on wave 1, holding a pole between them. The first apprentice is going to walk, at a right angle to the number line, until wave 2 is reached. The apprentices will keep hold of the pole in order to maintain a fixed distance between them, and will move in unison, always keeping the pole "vertical".

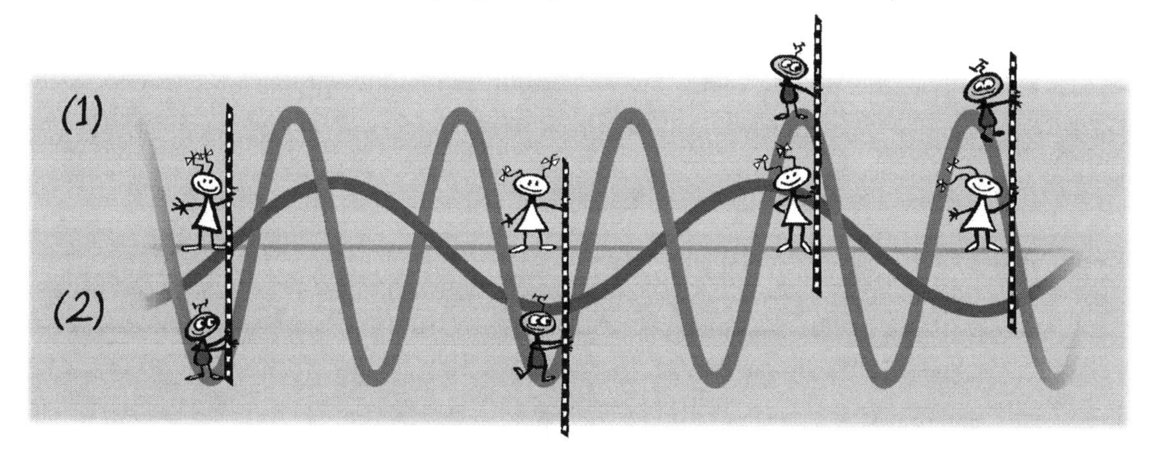

Four positions have been chosen to demonstrate the procedure. In each case, the apprentice on the number line is about to set off, "vertically" towards wave 2. The role of the apprentice on wave 1 is to preserve the distance between them.

Once the apprentice who started on the number line reaches wave 2, they both stop, and the other apprentice (who started on wave 1) marks its location. It's as if wave 1 is being pushed up or down at each point, depending on what wave 2 does there.

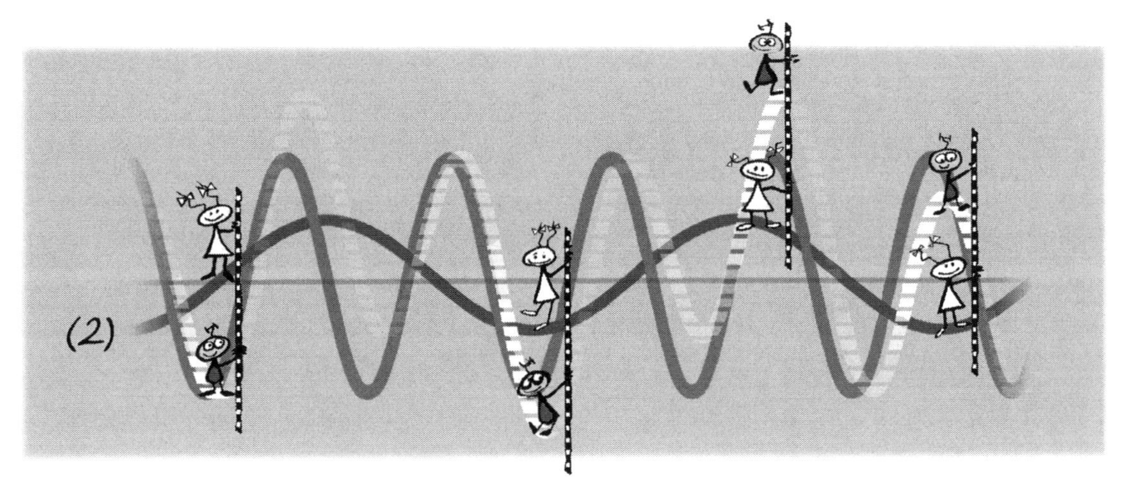

The dotted white wave indicates where the marks will be made. In these images, it's the positions of the apprentices' feet which should be understood as their "locations".

Here's a better view of what that dotted white wave looks like:

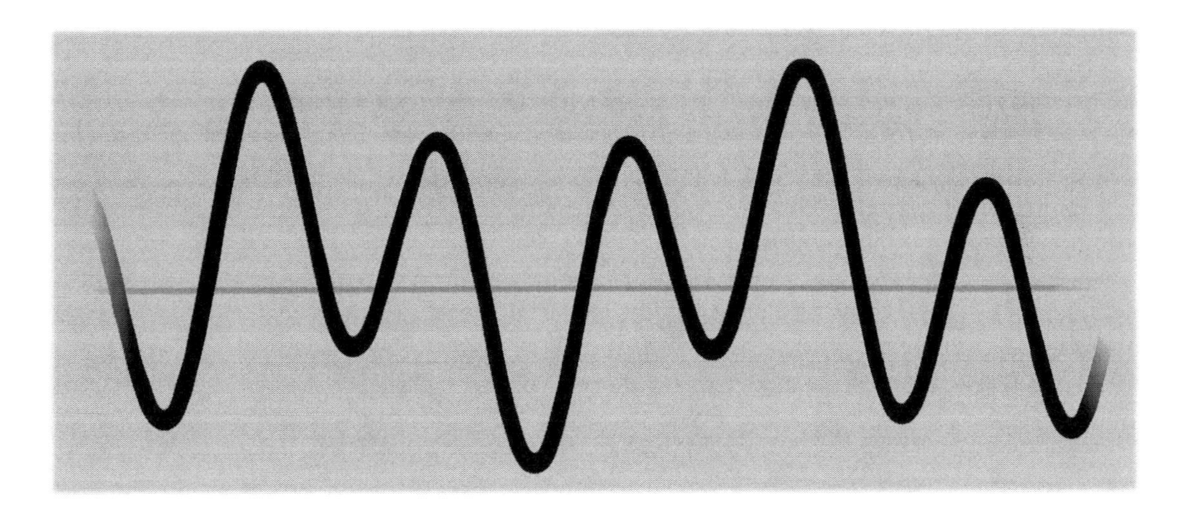

That's the graph of the waveform you get when you add wave 1 and wave 2.

The same idea could be applied to the graphs of any two functions, not just sine waves. In fact, the same procedure could be used to add *any number* of waves, waveforms, or other types of functions together:

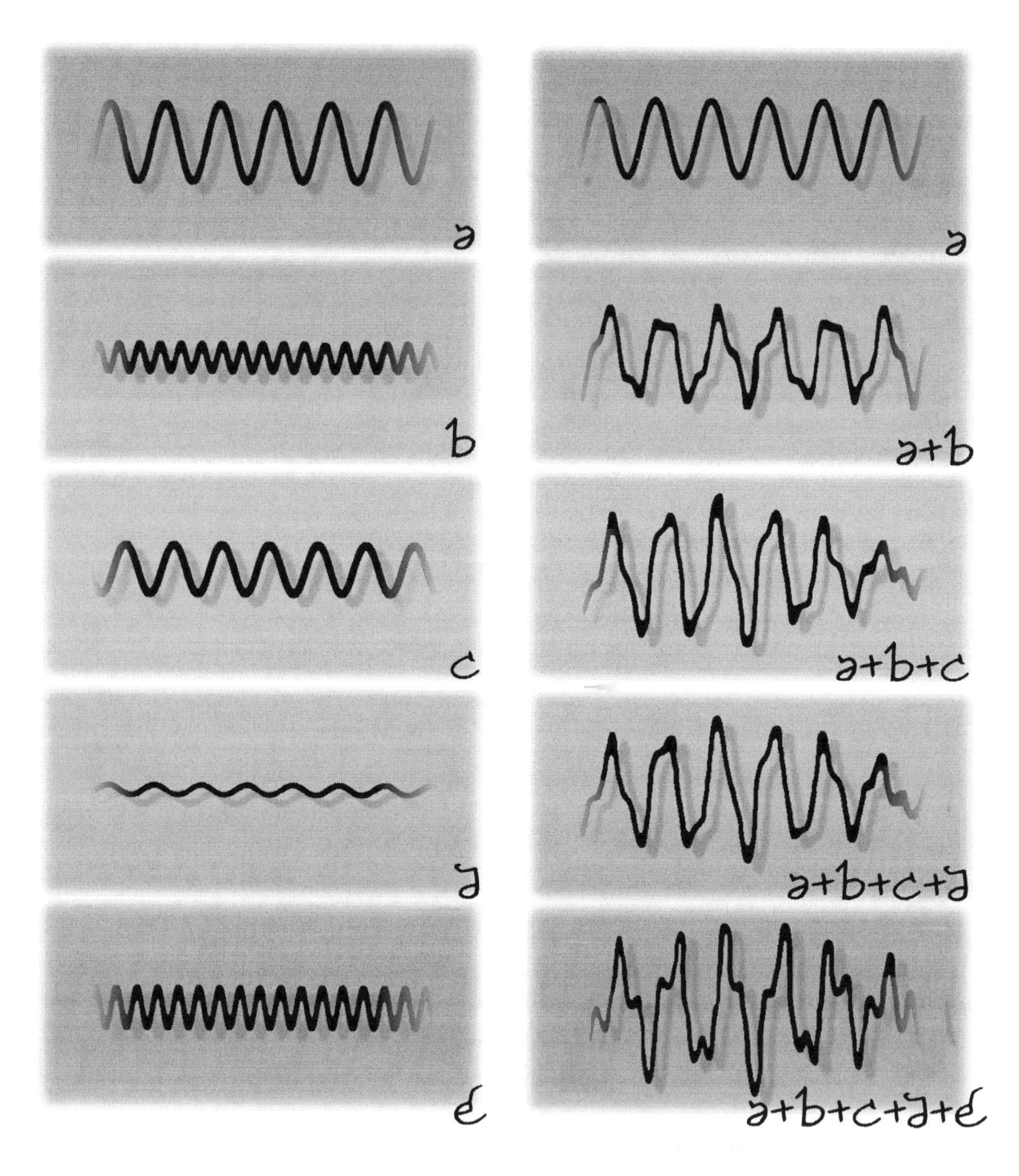

If you visually scan the right-hand column (the "running total" of waveforms which have been added together) without much thought or analysis, you'll probably get an intuitive sense of how this wave addition works. Notice that once you've added (a) + (b), you're no longer adding pure sine waves. You add (c) to the "(a) + (b)" *waveform*, then

(d) to the "(a) + (b) + (c)" *waveform, etc.,* but the basic adding process is the same.

BREAKING WAVEFORMS DOWN INTO SINE WAVES

This process can be reversed. That is, given a waveform like this...

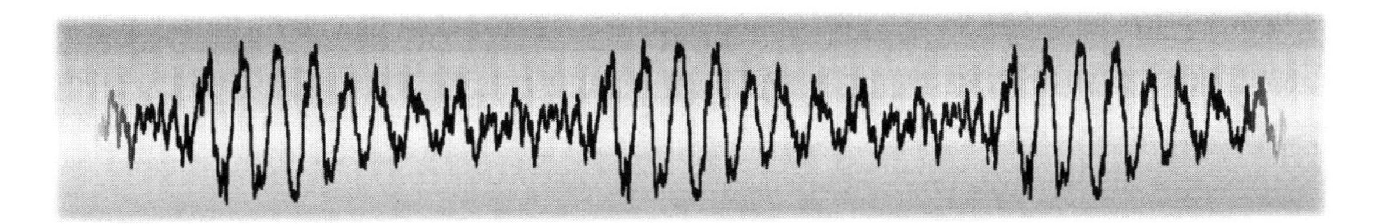

...we can attempt to find a set of sine waves which, when added together, produce it. There's a branch of mathematics called *Fourier analysis* which provides powerful techniques for doing this kind of thing. If you adjust the bass, treble or mid-range of a sound recording using a *graphic equaliser*, the technology is exploiting the mathematics of Fourier analysis in order to isolate waves with a certain range of frequencies from the overall sound so that they can be boosted or diminished.

In other contexts completely unrelated to sound, Fourier analysis can be used to analyse large amounts of random-looking data and detect tendencies suggesting cycles or "periodic" behaviour. This could involve tides, the migrations of animals, heartbeats, electrochemical fluctuations in the brain, or just about anything where you might expect to find some sort of *regularity of repetition*.

Sound and electrical engineers work with certain distinctive waveforms like these...

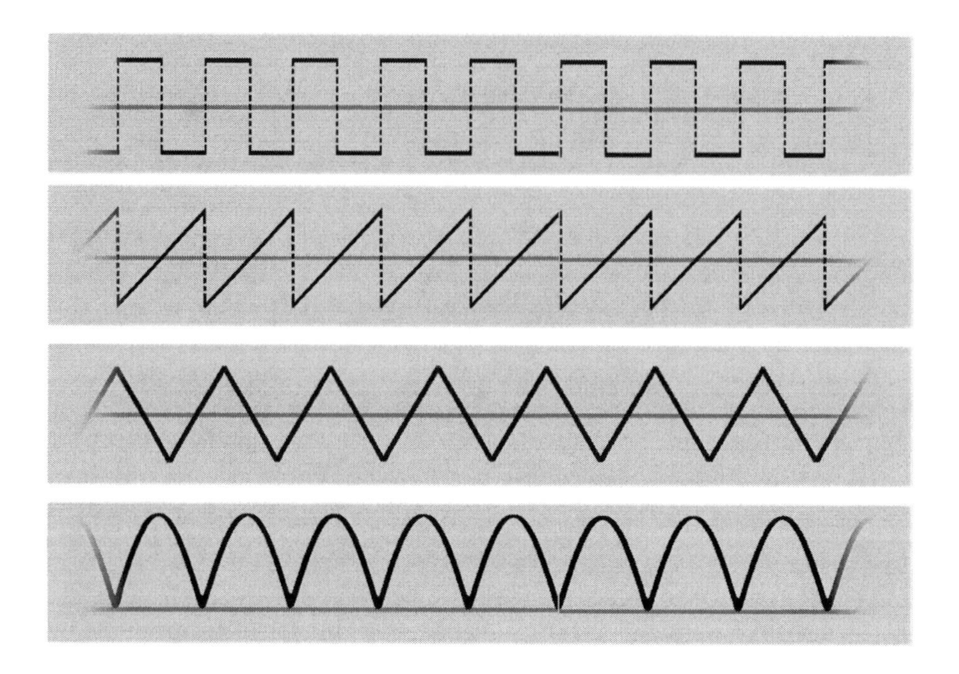

...which can be broken down, using Fourier analysis, like this:

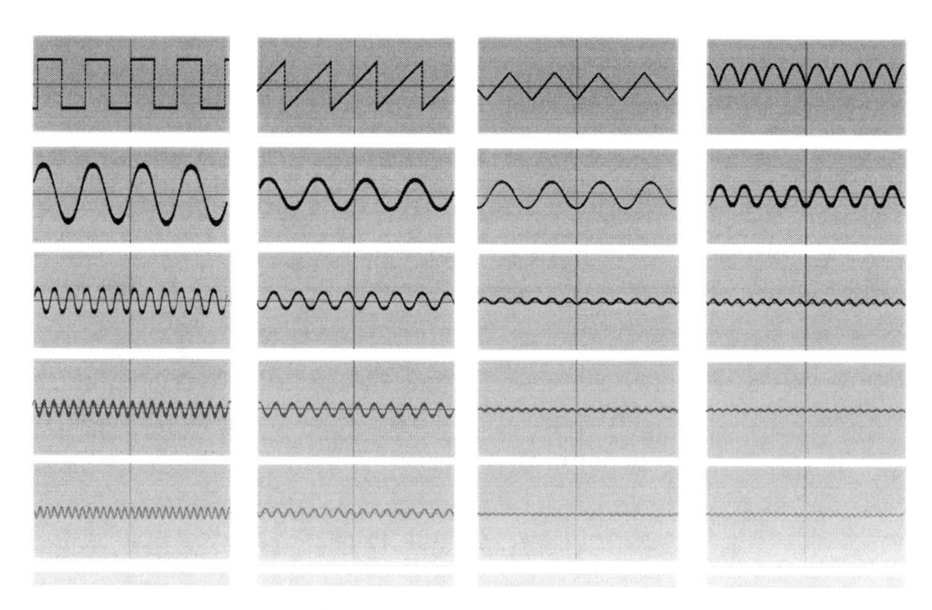

Each of the four special waveforms (known as "square", "sawtooth", "triangular" and "rectified") can be expressed as an infinite sum of sine waves, as suggested by the columns below them.

This process is known as *Fourier decomposition*. Sometimes it's called *harmonic decomposition*. The various pure sine waves involved in the breakdown are called the *harmonics* of the original waveform. As you'd expect, this terminology has its origins in the theory of music.

DECOMPOSING THE DEVIATION

Everything's now in place to reveal a most extraordinary fact about the distribution of prime numbers, a fact that came as a total surprise when it was first discovered.

It is that the deviation function which we recently met...

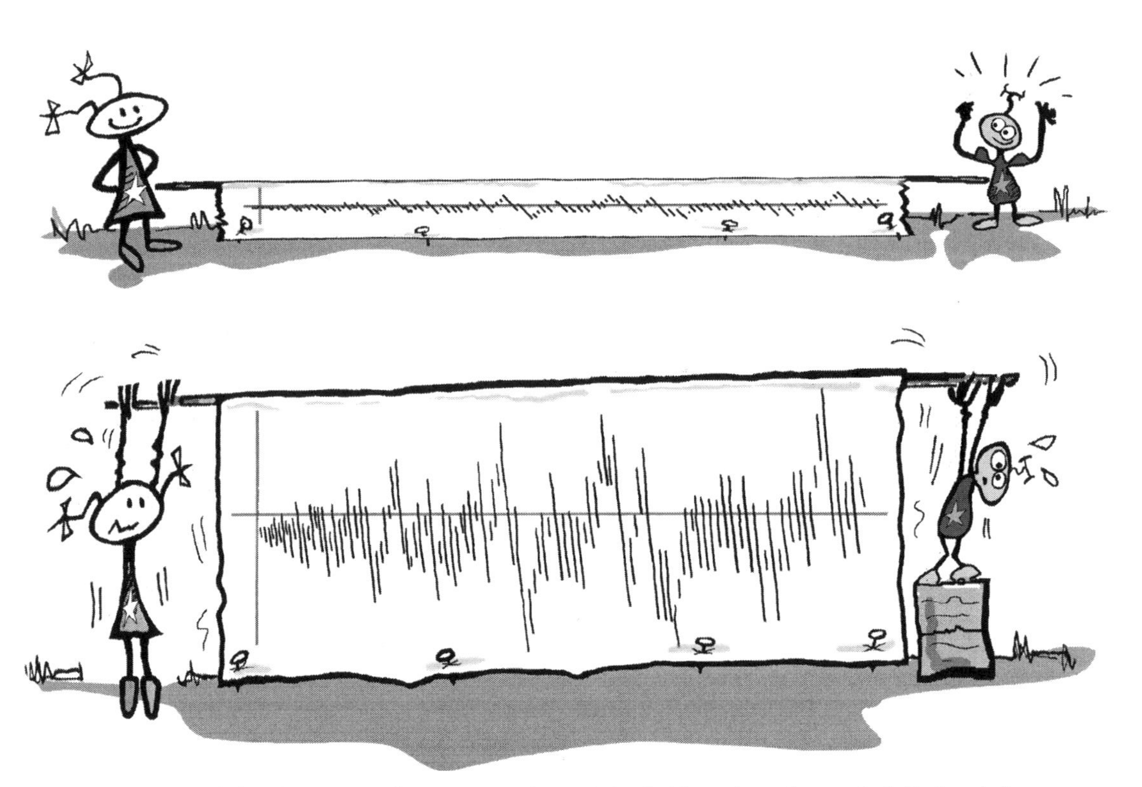

the graph of the deviation function in its original (above) and rescaled (below) forms

...(which tells us, roughly speaking, how much more or less "primeness" there is than there "should be" up to any given number) *can be harmonically decomposed*.

That is, it has a hidden "internal structure" made of simple waveforms (which are *very* closely related to sine waves). If this doesn't seem that remarkable to you, then to give you some idea of its significance, here's what Enrico Bombieri, a senior mathematician at the Institute for Advanced Studies (Princeton) once said about it:

> "*To me, that the distribution of prime numbers can be so accurately represented in a harmonic analysis is absolutely amazing and incredibly beautiful. It tells of an arcane music and a secret harmony composed by the prime numbers.*" [4]

Mathematicians do not easily use such awestruck language. I must stress that "noisy" looking waveforms like the primeness count deviation do *not* usually separate out nicely into a clear sequence of sine-wave-like components. In general, *there's no reason why they should*. But this one *does*, much to the collective amazement of those mathematicians who are aware of this (almost 150-year-old) fact.

Remember that the thing we've been calling the "primeness count deviation", as represented by the graph just shown, is an example of a mathematical *function*. As we've seen, there's an endless variety of possible functions, each representable by its own graph. These functions can all be submitted to a kind of harmonic decomposition, but *almost none of them* will decompose in this spectacularly "clean" way.

The tone of Bombieri's remark, as well as some other similarly awestruck comments on this matter which we'll see later, suggest that (in some strange sense) the seeming "jumble" of primes has somehow *conspired* to do this. For it turns out that if the arrangement of the primes were in even the tiniest way different, then the small difference this would make in the primeness count deviation would *completely* destroy the "incredibly beautiful" harmonic structure. Every prime number must be located precisely where it is for the whole thing to work (and, don't forget, there are *infinitely many of them*, of which we'll only ever see an infinitely tiny proportion). This runs completely contrary to the initial image of the primes as weeds, clutter or an untidy thicket. It's the image of something impossibly exquisite, infinitely perfect.

Despite having "been there" for at least as long as the earliest historical multiplication or division of counting numbers (some might say "since the beginning of time", or even "before the universe existed"), this harmonic structure was only discovered by *Homo sapiens* in the late 1850s. Those people who have absorbed enough higher mathematics to be able to learn about it very often express surprise and/or a sense of wonder or mystery when they first encounter it, but the fact that Bombieri is poetically enthusing about is still unknown to a huge number (possibly even the majority) of professional mathematicians, and just about everyone else alive. Strange, isn't it?

We will next see exactly *how* the deviation of the prime number distribution from its approximate average behaviour can be harmonically decomposed (broken down) into a sum of waveforms.

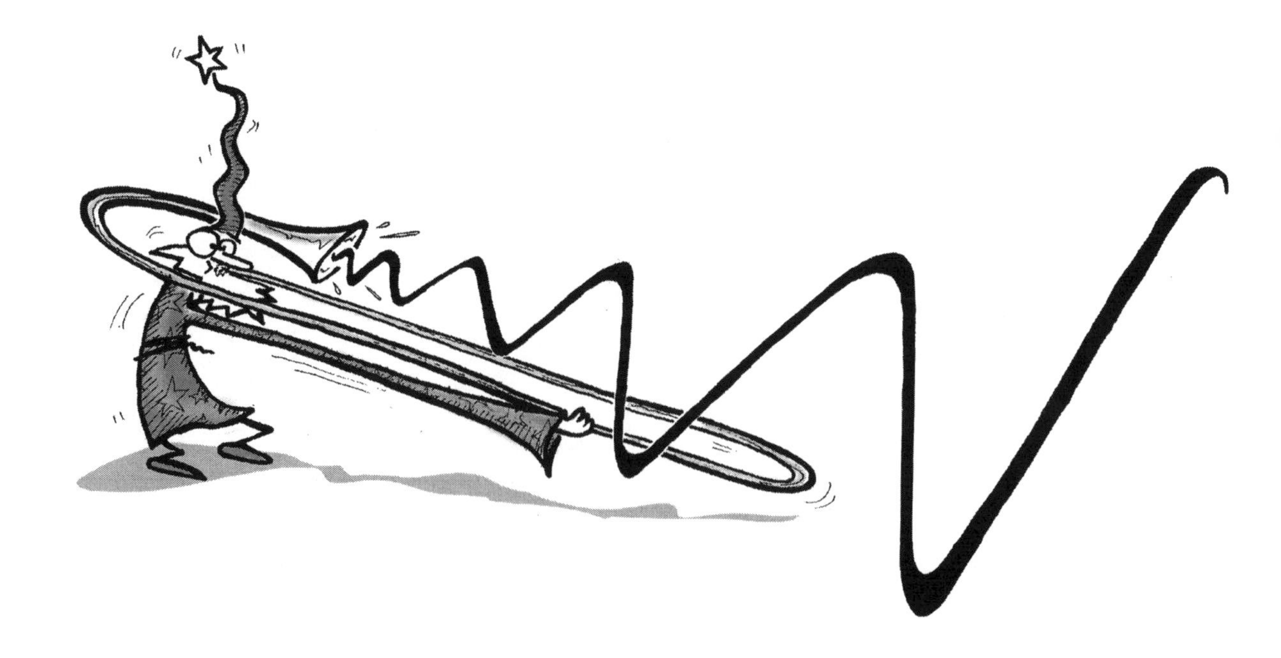

Chapter 14
spiral waves

You may have noticed that rather than claiming the primeness count deviation can be decomposed into sine waves, I stated that it can be decomposed into waveforms "which are *very* closely related to sine waves". That's because the deviation is not built out of these…

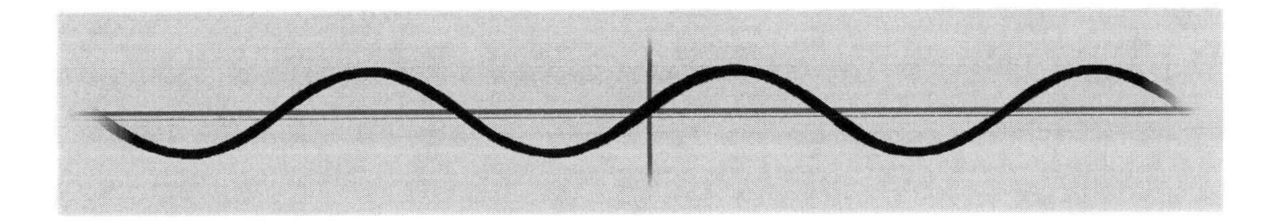

…but, rather, out of *these*:

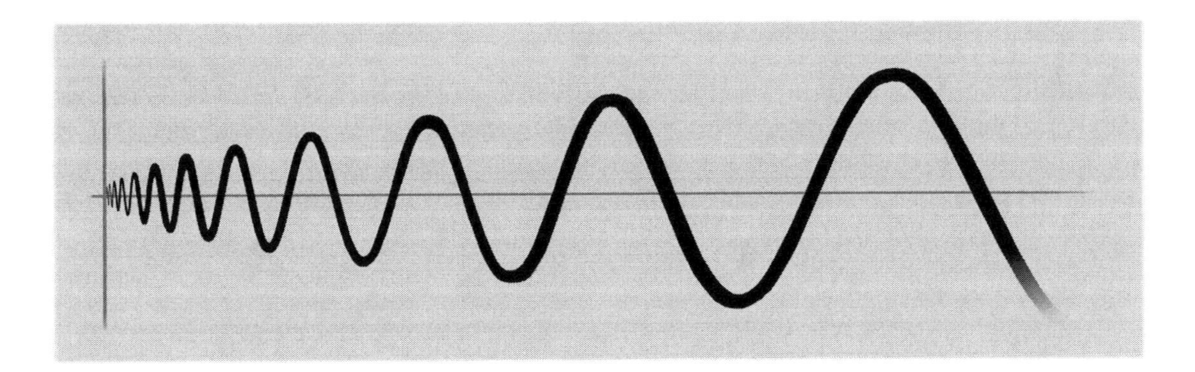

It should be clear that these latter waveforms are somehow related to sine waves, but there are some significant differences. The amplitude grows as we proceed along the number line, and the wavelength stretches too. We'll be calling these waveforms *spiral waves* because they turn out to be related to equiangular spirals in a very similar sort of way to how sine waves are related to circles. You won't find this term "spiral wave" in any mathematics literature, though – they're not common enough in other areas of maths and physics even to *have* a recognised name, so they'd usually be described as something like "logarithmically rescaled sinusoids".

It's possible to create these spiral waves with the same bicycle-and-Ferris-wheel scenario as was used to construct sine waves, except this time we're going to need a *spiral* Ferris wheel!

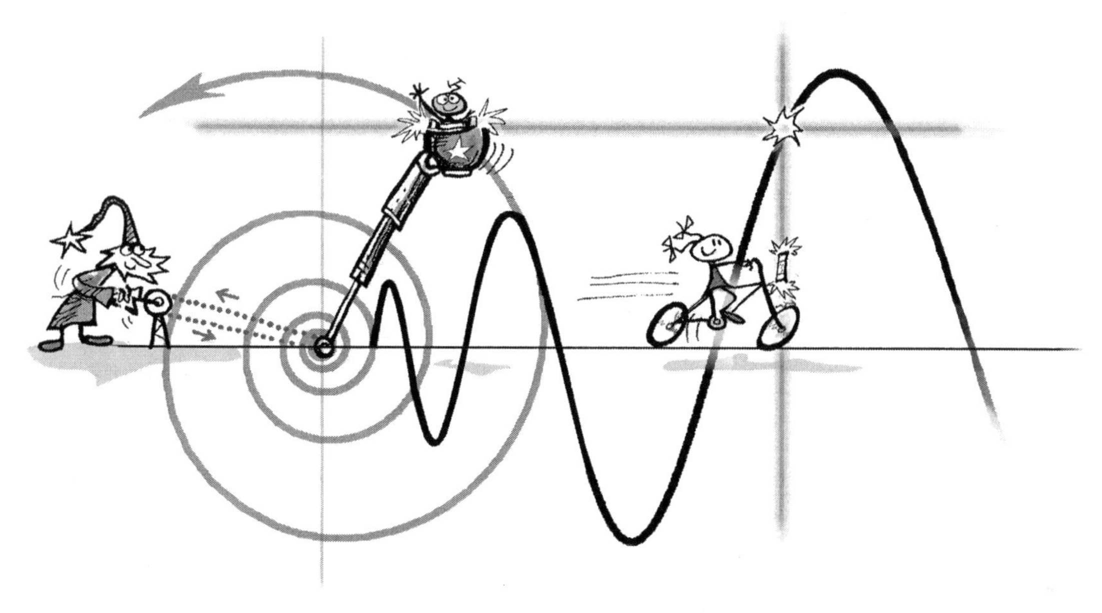

As with the sine wave construction, the wizard is cranking at a fixed rate so that the carriage does a fixed number of rotations per hour (that means it's speeding up considerably, as each coil is a lot longer than the previous coil). The bicycle is not moving at a fixed rate this time but is accelerating exactly like the ladybird we met in Chapter 9 – that is, *wherever it is on the number line, that's how fast it's moving.* The horizontal beam from the carriage intersects the vertical beam from the bicycle, and this point of intersection traces the spiral wave, as shown.

Starting the bicycle at 1, we'll only get that part of a spiral wave to the right of 1. To get the remaining bit between 0 and 1 is easy, though. We start both apprentices in the same places but turn the bicycle to face in the opposite direction and crank the carriage in a clockwise direction this time. That means the bicycle *slows down* as it travels from 1 to 0 (by the time it's reached 0.998, it's going at a speed of 0.998 "units per hour", by the time it's at 0.5, it's going at 0.5 u.p.h. (half its original speed) and at 0.003, it's only going 0.003 u.p.h. In this way, it turns out that the bicycle never reaches 0[1]. It just gets infinitely close (and infinitely slow). The carriage on the wheel travels *inwardly* this time, also continually slowing down in order to maintain a steady number of coils per hour (since each coil is much smaller than the previous coil). The intersecting beams this time create the little piece of spiral wave between 0 and 1[2]. One major difference between sine waves and spiral waves is that sine waves "live" on the entire number line, whereas spiral waves only "live" on that half of the number line to the right of 0 (the positive numbers, that is).

We've seen that there are three things we can adjust with sine waves: amplitude, wavelength (or frequency) and phase. With spiral waves, there are *four* things.

(1) We can change the *base* of the spiral. Thinking about it, you should be able to see that a loosely wound spiral (*below right*) will create a wave whose amplitude grows faster than a tightly wound spiral (*below left*).

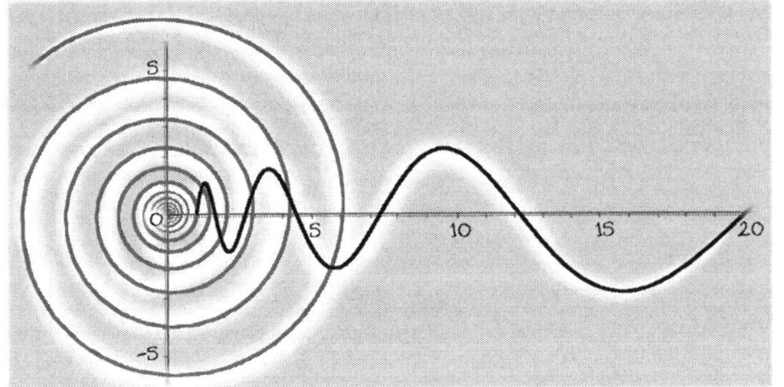

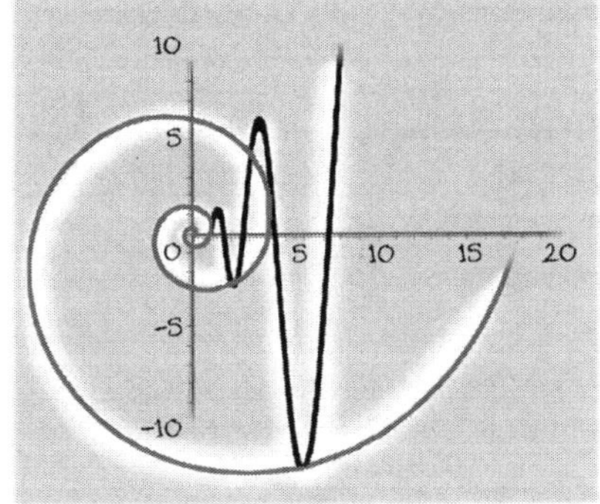

Notice how the peaks of each wave correspond to the highest points of the coils of the corresponding spiral, and the "troughs" to the lowest points. The fact that the spiral on the right opens out much more quickly means that higher peaks and lower troughs are achieved sooner.

(2) We can change the *speed of rotation* (coils per hour) of the arm which propels the carriage around the spiral track. This will also clearly affect the shape of the wave, seemingly having something to do with the (continually stretching) wavelength.

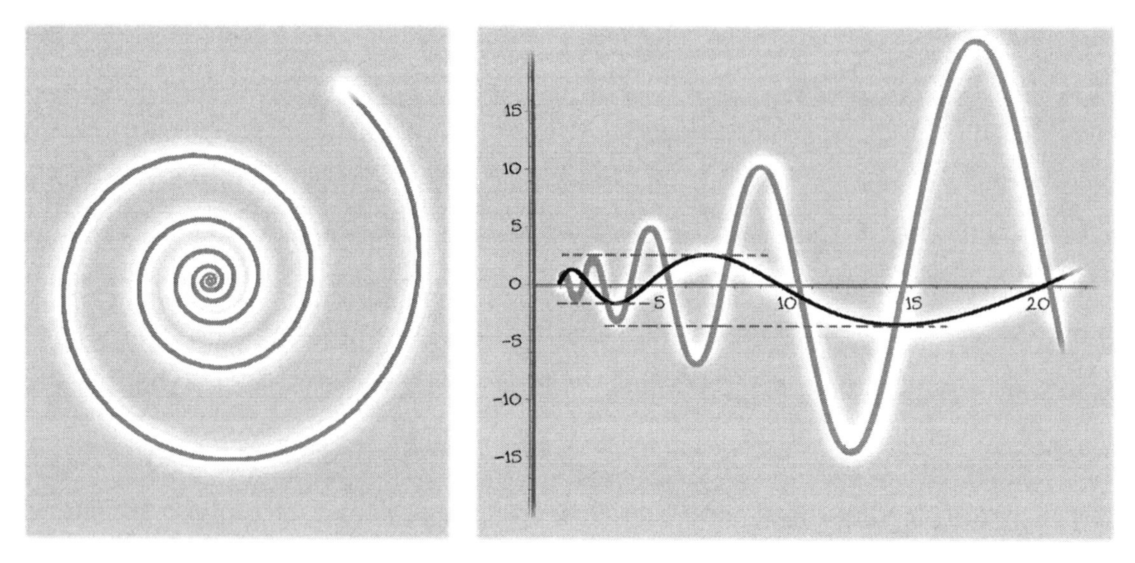

Here, the two spiral waves on the right are both generated by the spiral shown on the left, but the darker wave is the result of a slower rotation. The dotted lines have been introduced to demonstrate that both spiral waves have the same peaks and troughs – but the lighter wave achieves them sooner, due to the greater rotational speed which produces it.

(3) We can change the *distance from the centre* of the spiral where the carriage starts travelling outwards from – that's evidently something to do with the amplitude of the wave at the location 1 on the number line.

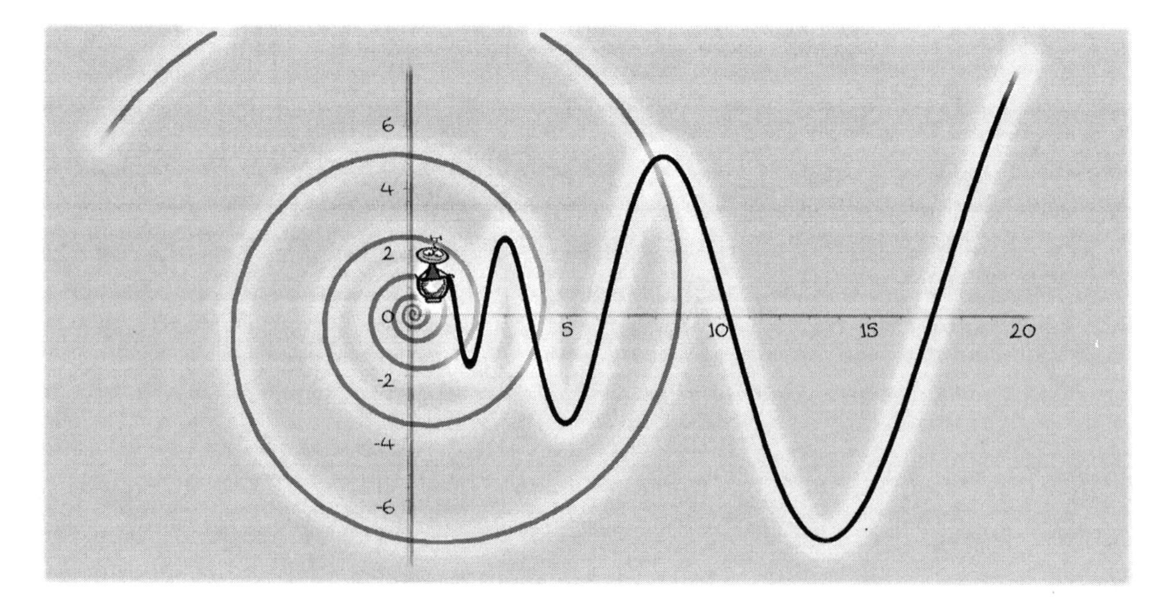

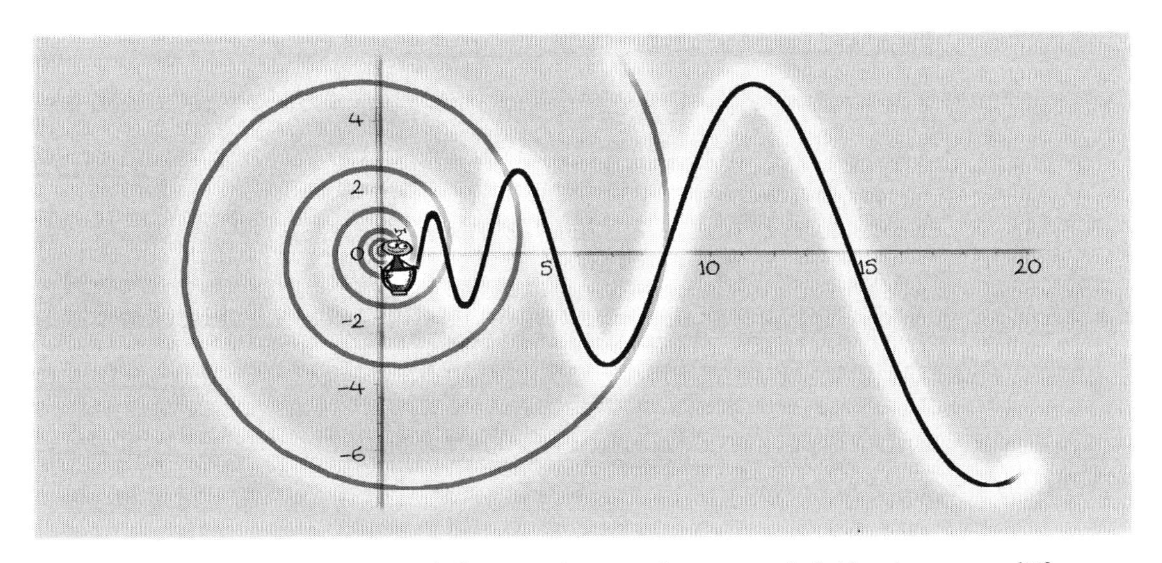

If we start the carriage at two different points on the same spiral (that is, at two different distances from the centre) then the "initial amplitude" (distance of the wave above or below the number line at 1) and hence the overall "scaling" of the spiral waves will differ.

(4) We can rotate the spiral around its centre. A given spiral may or may not pass through a given point on the number line. If not, we can always rotate it so that it does, but this will lead to vertical stretching of the spiral wave that's produced:

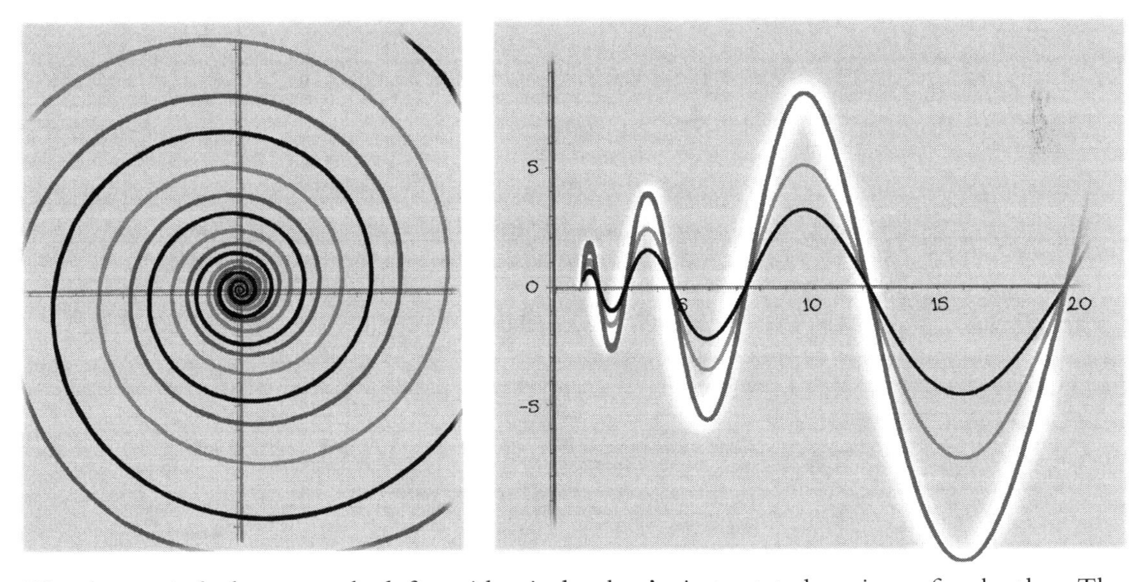

The three spirals shown on the left are identical – they're just rotated versions of each other. The spiral waves they produce all cross the number line in the same places, as shown on the right.

When we saw how to produce sine waves using the visualisation involving a circular Ferris wheel, the three things we could adjust were: (1) the size of the wheel; (2) the speed of rotation; (3) the initial position of the carriage. Although (1), (2) and (3) can be shown to correspond in some ways to the things (1)–(3) which can be adjusted in the "spiral Ferris wheel" visualisation, the correspondences aren't that straightforward – (1) for sine waves could also be related to (3) and (4) for spiral waves, for example.

Going back to our original construction of equiangular spirals, we're reminded of something interesting involving circles. You'll recall that we used a visualisation involving an elastic rope attached to a peg and a handle like this...

...where the pointer shows the direction to walk in (go back and have a look at page 169 if you can't remember how this worked).

By changing the angle of the arrow, we produce spirals with different bases (that is, with varying tightness/looseness of winding)...

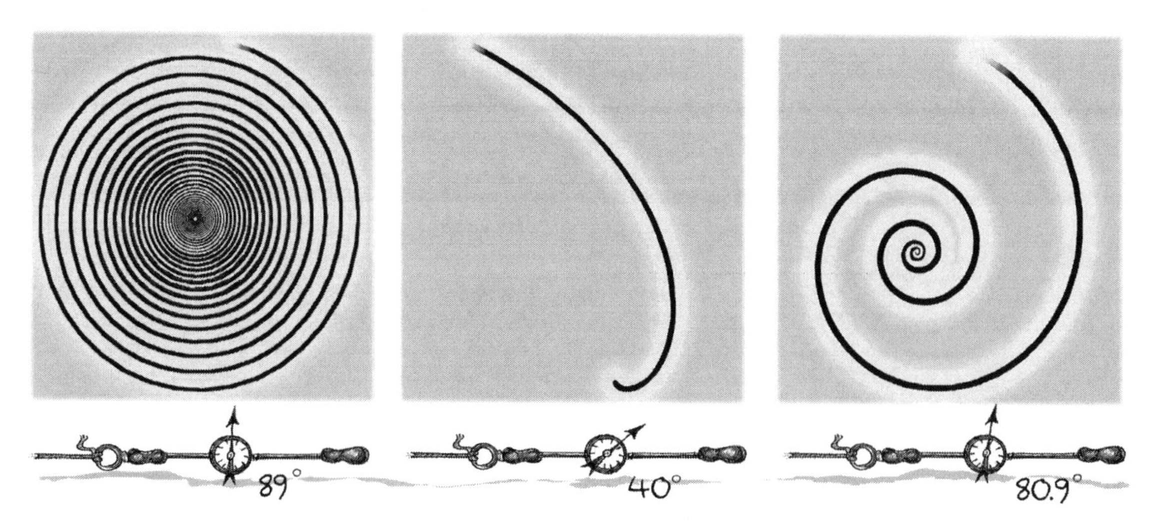

…and if we set the arrow at a right angle, *we end up walking in a circle*.

The rope is elastic, but because the pointer is at a right angle to it, no stretching will occur.

A circle, then, can be thought of as a kind of "extreme spiral" (mathematicians would say a "degenerate spiral") where the winding becomes so tight that it's no longer a spiral. Looking at it like this, we see that the circle is just one very special member of the family of equiangular spirals. In this way, the circular Ferris wheel we used to construct sine waves can be seen as a very special example of the kind of "spiral Ferris wheel" which we just used to construct spiral waves. So, in a sense, a sine wave can be thought of as a *particular type of spiral wave*. Mathematicians would say that spiral waves *generalise* sine waves (they're "more general" since they include sine waves as a special case)[3].

One key difference between sine waves and typical spiral waves is that "non-degenerate" spiral waves have *growing amplitudes* – their various "rates of amplitude growth" will turn out to be tremendously important in Volume 2.

Even though we haven't yet got a clear idea of what *wavelength* should correspond to in a spiral wave (it "stretches" as we proceed along the number line), it's easy to see that two different spiral waves can cross the horizontal axis (number line) at all the same points…

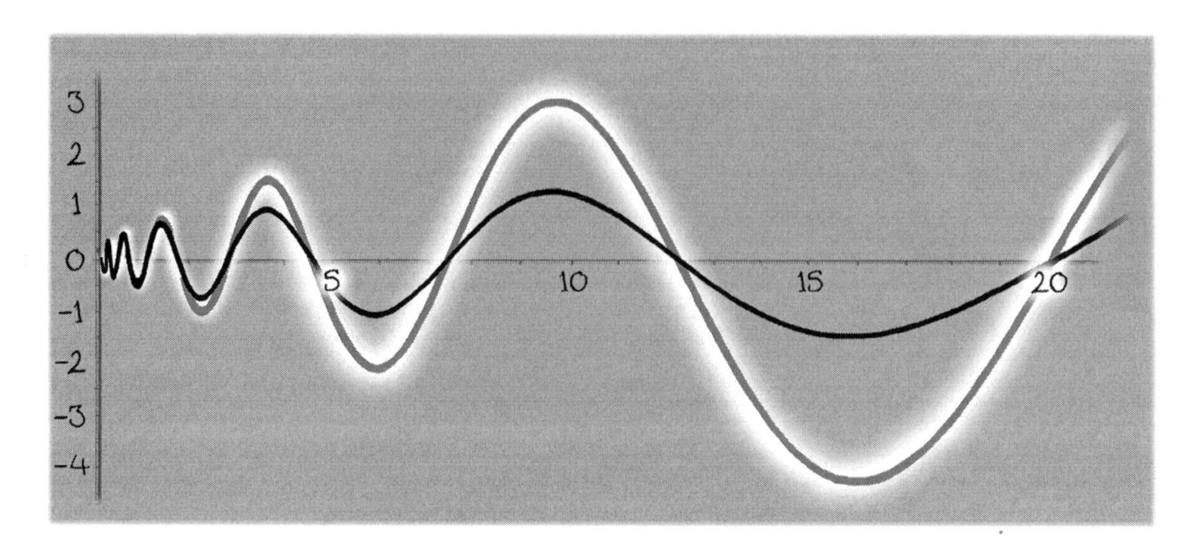

…so whatever "wavelength" (or "frequency") corresponds to in the context of spiral waves, it should be the same for these two. But they'd have different *rates of amplitude growth*: the peaks are getting higher faster in the lighter coloured wave.

DECOMPOSING THE DEVIATION FUNCTION INTO SPIRAL WAVES

It's into an infinite sequence of these spiral waves – rather than the sine waves familiar to sound engineers and electronic technicians – which the primeness count deviation separates. This is the wondrous "harmonic decomposition" which was referred to earlier. Although the deviation separates into an *infinite number* of spiral waves, importantly, the waves form a well-defined sequence – that is, the waves involved in the decomposition can be arranged into a list such that each one of them, eventually, will appear. This point may seem a bit subtle, but it's possible to have an infinite number of objects which *cannot* be listed in this way[4], and the fact that these waves *can* is highly significant.

Rather than taking it apart into spiral waves, we can work in reverse, *building* the primeness count deviation by adding together this infinite sequence of spiral waves. We've already seen how you can add any number of waves together. On the opposite page, we see five spiral waves (chosen at random) being added together sequentially:

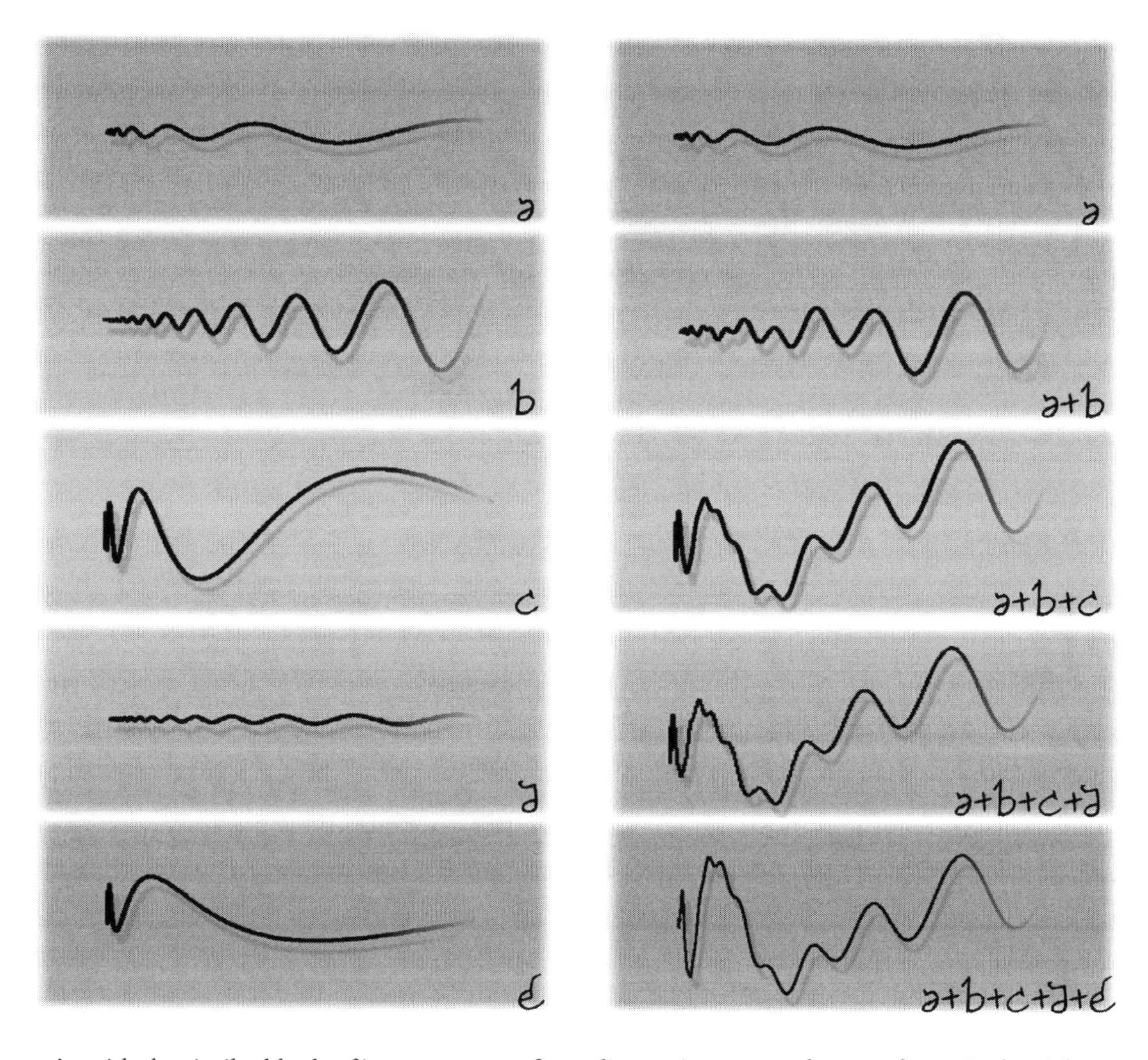

As with the similar block of images we saw for ordinary sine waves, the waveforms in the right-hand column can be thought of as a "running total" of the waves in the left-hand column.

If we were to work our way through the relevant sequence of spiral waves, adding them like this, we'd find that the waveform we had created would, at each stage, more and more closely resemble the primeness count deviation. It would take an infinite amount of time to add infinitely many waves, so we could never actually *do this*, of course. But we've seen that mathematicians are able to prove facts about situations involving infinitely many things (as with the FTA or Euclid's proof that there are infinitely many prime numbers). And it has been *proved* that this particular sum of infinitely many spiral waves produces the primeness count deviation exactly. Well, not *exactly*. The deviation is built from this infinite sequence of spiral waves *plus two other pieces*. They only affect the shape of the resulting graph very slightly, but if we're going to be mathematically precise, then they must be taken into account.

One piece (it's a function, to be precise, like the spiral waves are) looks like this:

That's its graph – just a horizontal line whose distance below the number line is the logarithm of the number you get when you multiply 2 times π (this "π" being "pi", which is about 3.14, you might recall). So, imagine taking this circle...

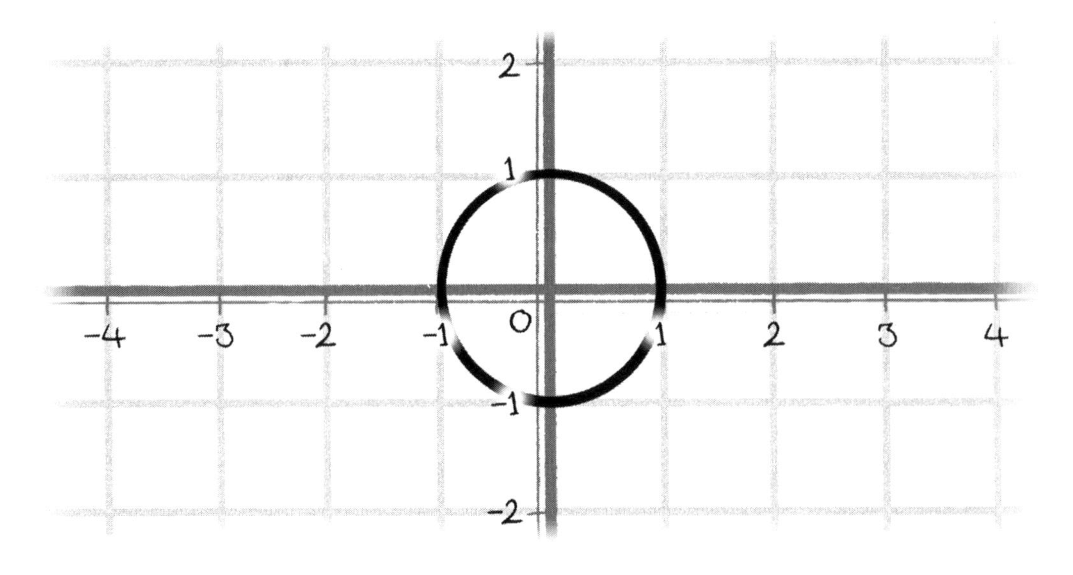

…and “unwrapping” it. We’d get a piece of line with length 2 times π, or about 6.28.

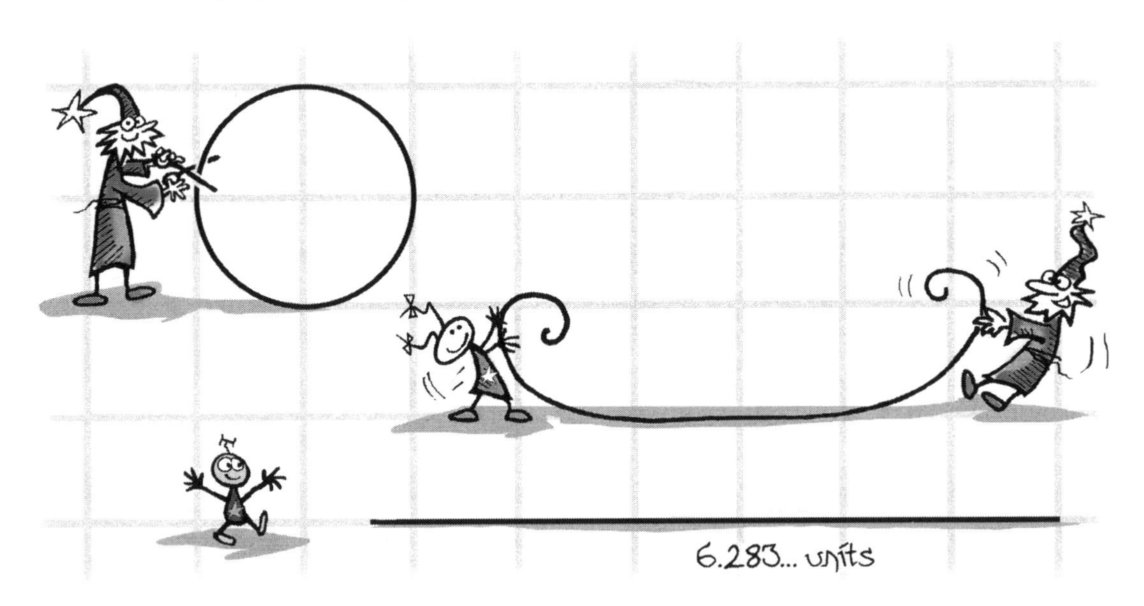

We could then make a circle from that (imagine pegging one end and walking)…

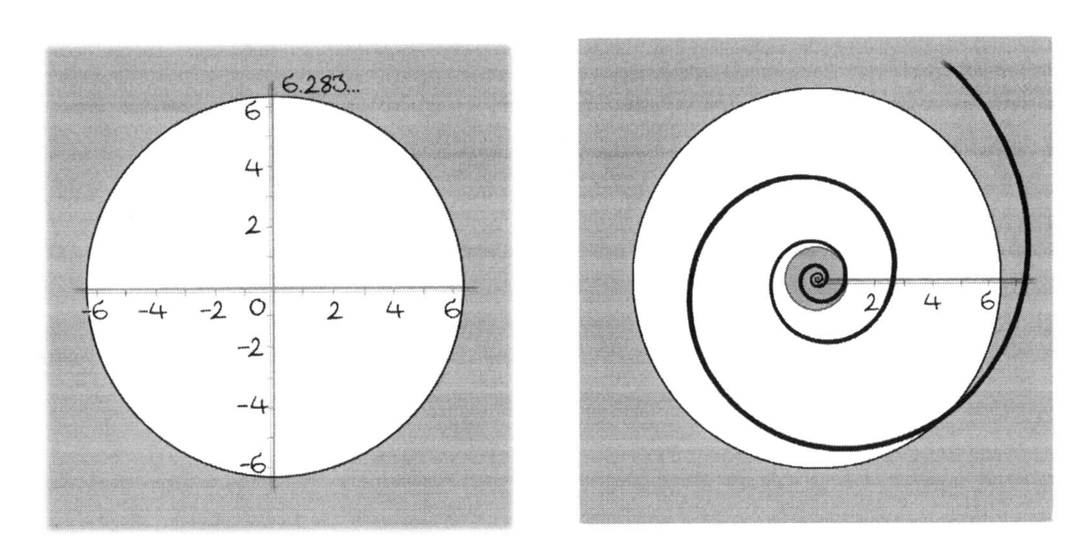

…and count coils to find that the logarithm of $2 \times \pi$ is approximately 1.838.

The other function which must be added to get *exactly* the primeness count deviation has a graph which looks like this:

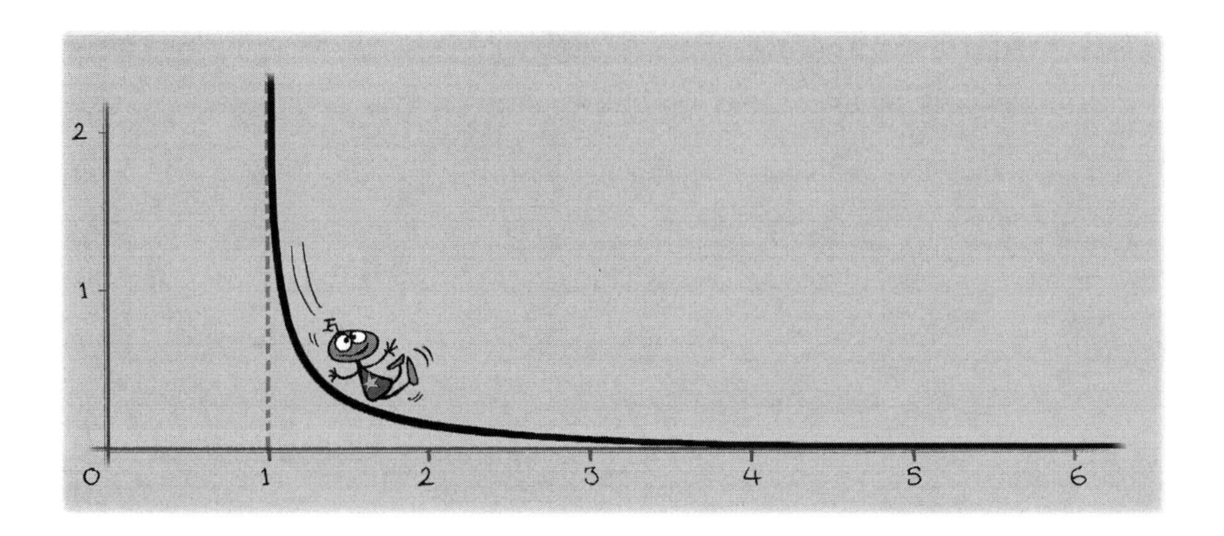

This one takes a bit more explaining, so I'll save it for Appendix 9. If you're not that interested in the details, rest assured that it can be explained entirely in terms of multiplication, division, subtraction and logarithms, so you'd be able to understand it by now if you *were* sufficiently interested.

In the next volume, we'll find out a bit more about these two extra functions which we need to add to the infinite sum of spiral waves in order to build the primeness count deviation. For now, the important thing to realise is that if we *can* construct the exact primeness count deviation by adding these various pieces together, then *exact knowledge of the various pieces will provide exact knowledge of the entire sequence of primes*. Why is this? Well, if we can construct an exact replica of the primeness count deviation...

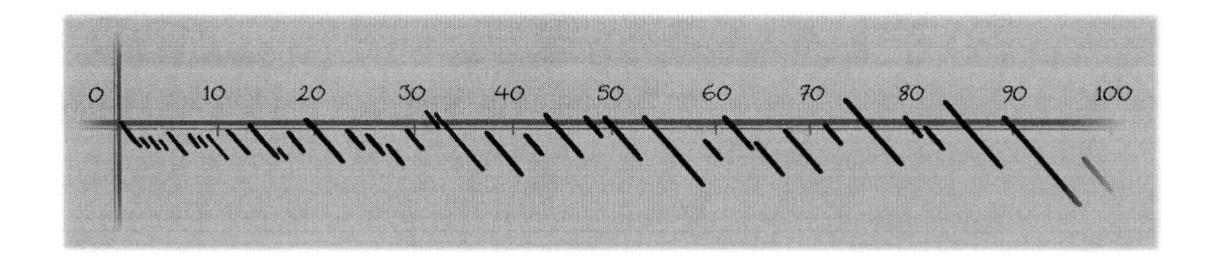

...and, treating it as a single waveform,

add it to this…

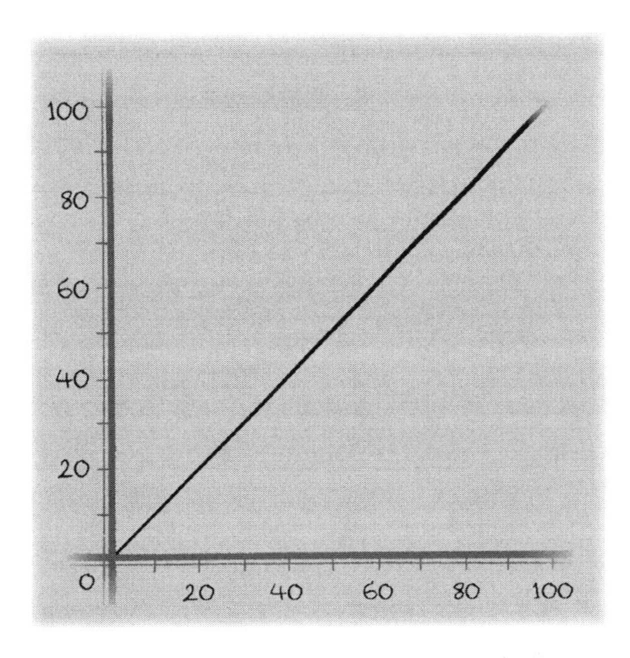

…then we get this…

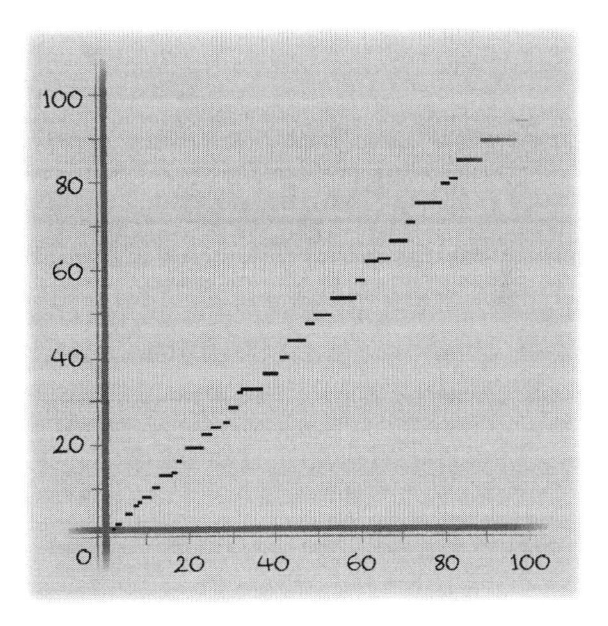

…which is the primeness count staircase (more "natural" than the first staircase we saw, based on counting prime numbers and their powers in the "logarithmic" way).

Why? Well, the deviation tells us exactly how the primeness count staircase *differs* (deviates) from the approximation of "how much primeness there should be" which the diagonal line graph represents. Where the graph of the primeness count deviation sticks up above the number line by some amount, it's telling us "there's actually this much more primeness than approximated". Where it dips down by some amount, it's saying "there's actually this much less primeness than approximated". So, by adding the deviation graph to the diagonal line (approximation) graph, we get a representation of *exactly how much "primeness" there actually is*. In other words, we get the primeness count staircase. (If you missed that, re-read this paragraph. You'll get it.)

Once we have the staircase, we can identify the points on the number line where it "jumps". Remember, this shows us where the primes *and their powers* are located.

We have jumps at 2, 3, 4, 5, 7, 8, 9, 11, 13, 16, 17, 19, 23, 25, 27, 29, 31, 32, 37,...

How do we separate the primes (2, 3, 5, 7, 11, 13, 17, 19, 23, 29, 31, 37,...) from the powers of primes ($4=2\times2$, $8=2\times2\times2$, $9=3\times3$, $16=2\times2\times2\times2$, $25=5\times5$, $27=3\times3\times3$, $32=2\times2\times2\times2\times2$,...)?

Looking at the staircase, we find that the jumps at 2, 4, 8, 16,... are all of the same height (being about 0.693, or the logarithm of 2). Similarly, the jumps at 3, 9, 27, 81,... are all of height approximately 1.099 (the logarithm of 3) and the jumps at 5, 25, 125, 625,... are all of height roughly 1.609 (the logarithm of 5). So, if we imagine climbing the staircase, whenever we find a "biggest jump yet", then we know we've found a prime number (as each new prime produces a jump bigger than the jumps at all of the earlier primes and their powers).

ANOTHER LAYER OF MYSTERY

At this stage of our exploration, the "mystery" of the distribution of prime numbers has effectively been replaced by a deeper mystery – that of the spiral waves. For we've seen that the seemingly jumbled nature of the primes, their seeming disorder, is only *apparent*. First, we found an "imperfect" or "statistical" pattern showing how the primes tend to thin out on average. We then looked at the deviation of the primes' *actual* behaviour from this average behaviour. At first, this appeared to be *totally* without order (in the form of a jagged, messy, "noisy" signal or graph), but we've since learnt that it can be broken down cleanly into a sum of spiral waves. Each of these waves can be described in terms of certain "specifications" (numbers a bit like the amplitude, frequency and phase which are used to specify particular sine waves – see pages 269–271). That information is now the key, for with it we can:

(1) construct each of the spiral waves involved;

(2) add them all together (plus the two extra bits we saw earlier);

(3) add the resulting waveform to the diagonal line graph;

(4) deduce the exact locations of *all* primes from the resulting staircase graph.

This "harmonic decomposition" of the number system (although in a slightly different form) was first discovered, then proved, by a desperately shy genius called Bernhard Riemann in 1859. At that point in history, recall, the PNT had been proposed and was believed to be true, but it hadn't yet been proved. Riemann was working on this problem and his discovery of the harmonic decomposition directly paved the way for others to prove the PNT some forty years later (although he didn't live to see it). Having discovered this breakdown of the primeness count deviation into spiral waves, he quickly established that the PNT was mathematically equivalent to a certain fact about the exact nature (the "specifications" mentioned above) of the spiral waves involved, so the problem of proving the PNT was reduced to the problem of proving this fact.

Regardless of the details of any human activity motivated by a desire to prove things, what all of this means is that *hidden within the overall "shape" of the prime number sequence*, if viewed in the right way ("through spiral lenses", if you like), *are endless layers of precise wave-like pulses or fluctuations*.

DECOMPOSING THE "OTHER DEVIATION"

We'll return just once more to the original, seemingly "obvious" way of counting prime numbers (where only primes, not their powers, get counted, and they all get counted as 1, no logarithms being involved). I've tried to convince you that this is not the most *natural* way to count primes. We'll now see another reason why.

This "obvious" way of counting primes produces a prime count staircase like this (steps appear directly above primes, and only primes, and they're all one unit high):

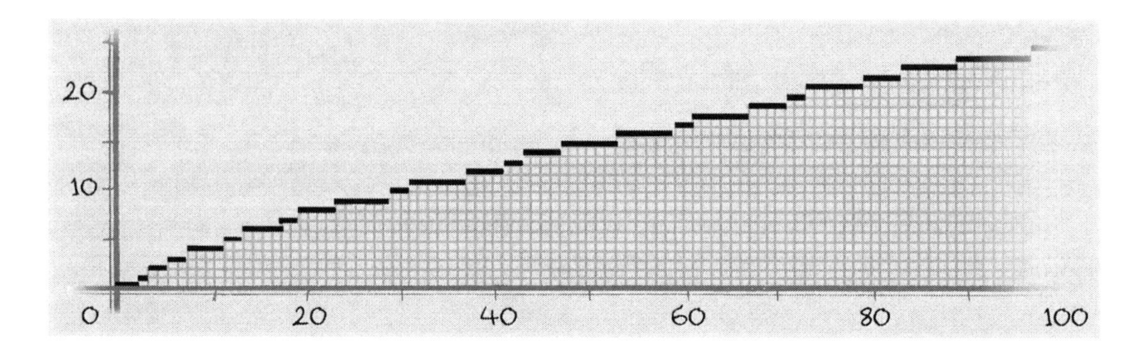

Like the other staircase, this one also has an "approximate behaviour" which we can describe with a graph. It's not a diagonal line, remember (we get that if we count primes and their powers "logarithmically"), but rather something like this...

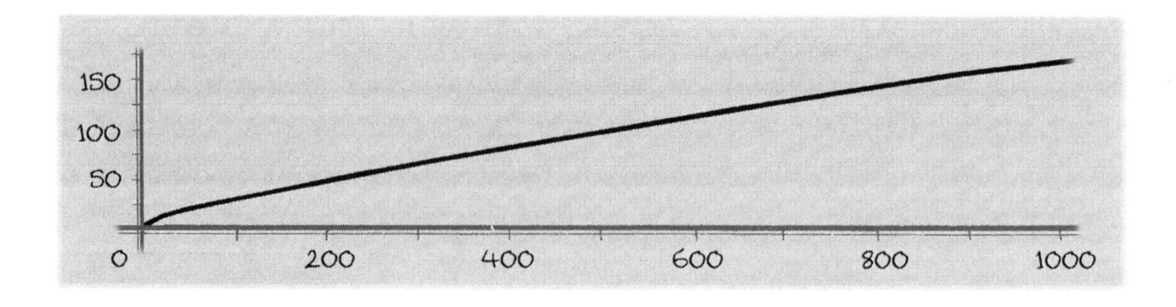

...which involves logarithms (it's the "li" function described in Appendix 8). Still, in the same way as before, we can look at the difference between the staircase and its approximation. Doing so, we find a deviation graph which looks like this:

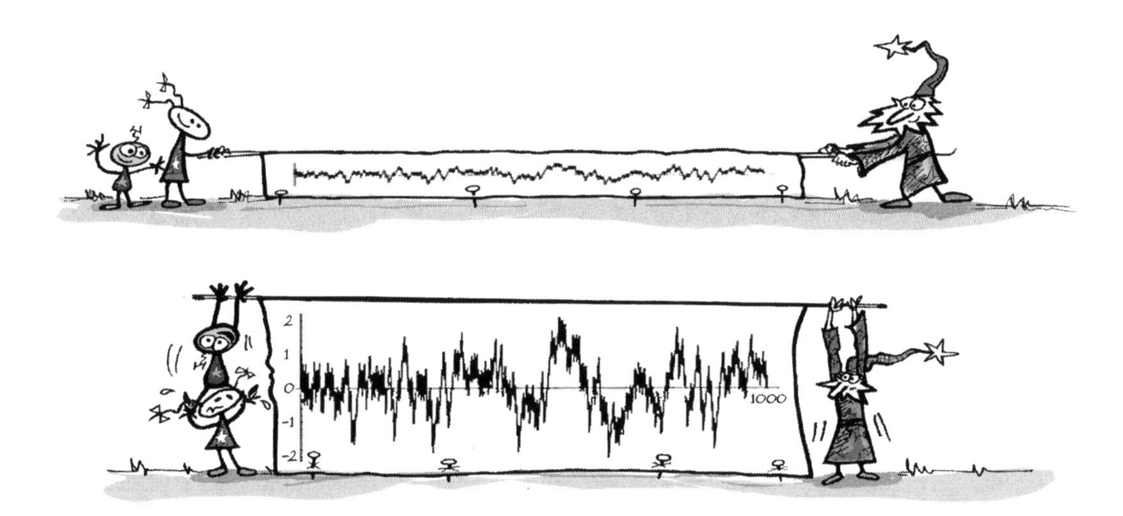

This graph certainly seems to have something of the same character as the original deviation graph, at least on first inspection. In fact, there's a sense in which it can be thought of as being just another version of the same thing. Like the original deviation, it can be decomposed into a sequence of waveforms. But they now look like this:

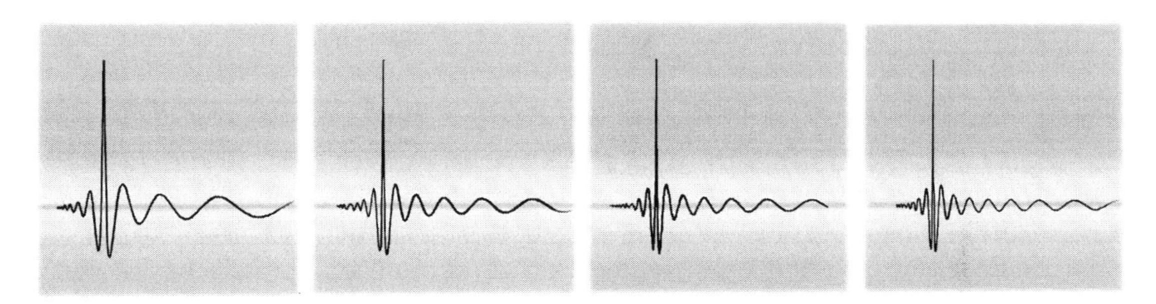

We're looking at the same harmonic decomposition, but now the waves have been distorted because we've counted primes in an "unnatural" way. I hope you'll agree that there's something more pleasing, beautiful or, at the very least, *simple* about these:

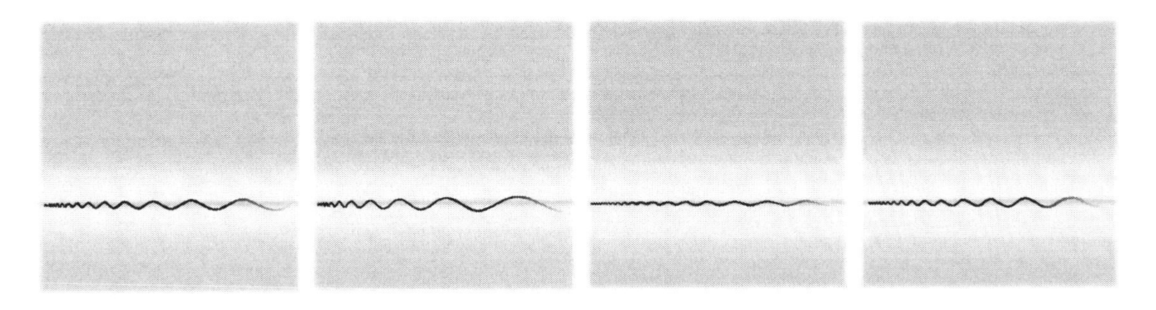

That agreed, we can now finally leave behind the original way of counting primes, which somehow distorts the beauty of the harmonic decomposition, and stick to the "logarithmic" one.

A SUMMARY

To summarise what we've found thus far:

☆ The prime numbers thin out along the number line in a way which can be related to equiangular (logarithmic) spirals.

☆ If we count the primes (and their powers) in the "logarithmic" way, we find that our count at any point on the number line is approximately the number of that point.

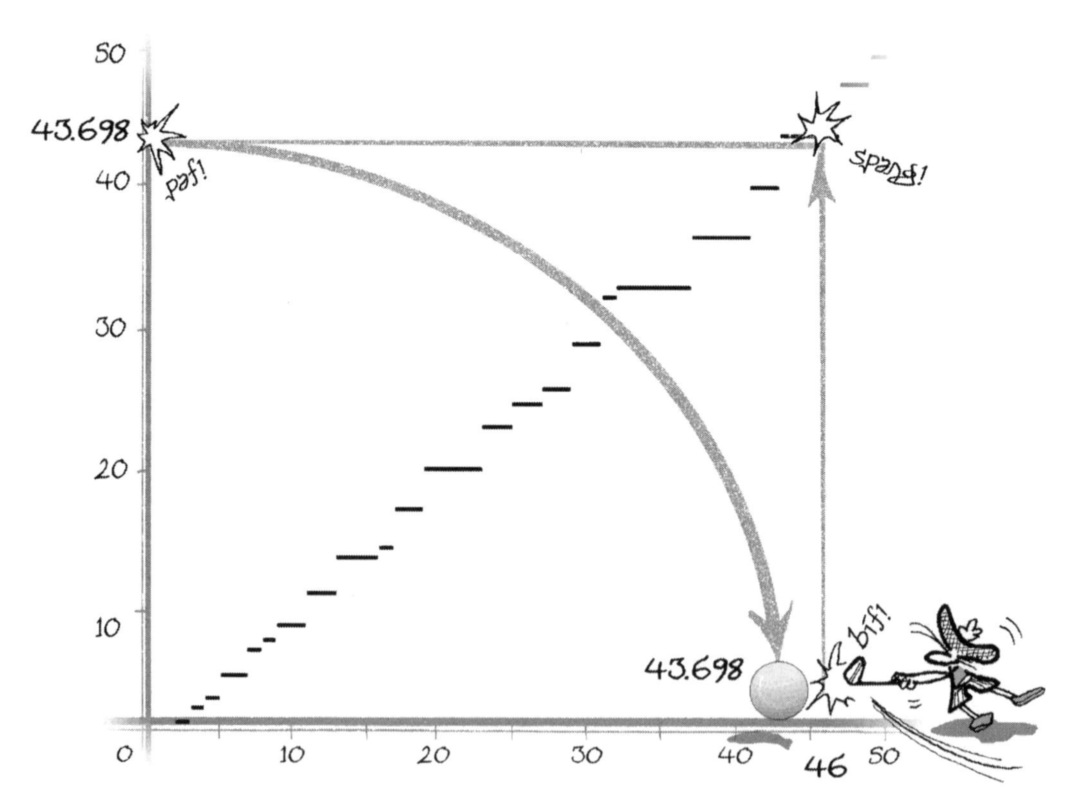

The staircase graph shows that the amount of "primeness" between 0 and 46 is 43.698…, as dynamically illustrated by the logarithmic prime counting function's golfing sprite here.

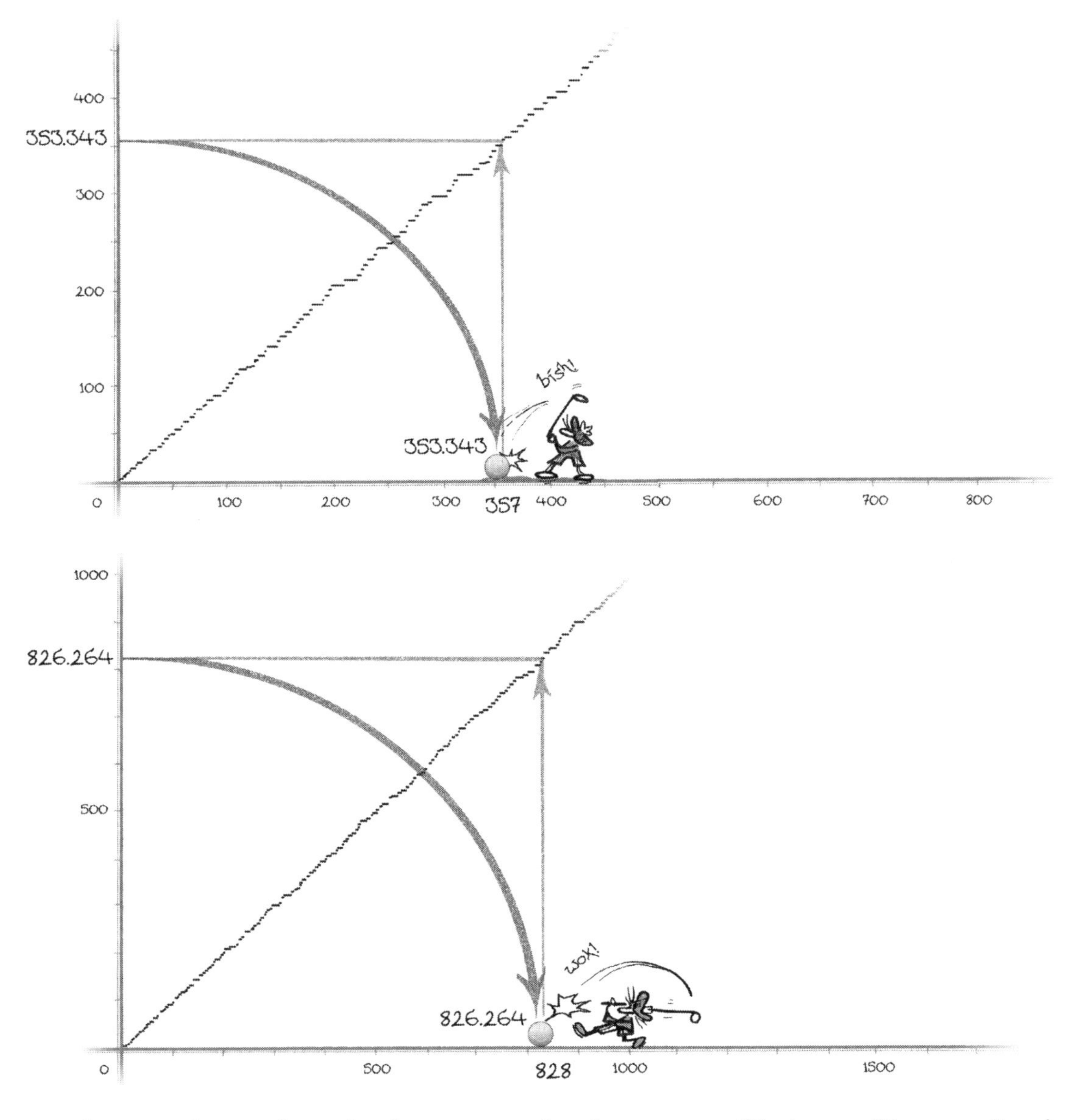

In the upper image, the sprite demonstrates that the amount of "primeness" between 0 and 357 is 353.343… In the lower one, we see that the amount between 0 and 828 is 826.264…

☆ If we consider (and draw the graph of) the deviation of the count from the approximation, we find a messy, "noisy" looking waveform or signal…

☆ …but this seemingly disorderly waveform turns out to peel away beautifully into a sequence of spiral waves (plus a couple of extra bits).

Because of the way they stretch out and grow, spiral waves appear not to have amplitudes and wavelengths (or frequencies) in the sense that ordinary sine waves do. However, we'll see that they do have similar "specifications", that is, numbers which can be used to describe them precisely.

At this stage of our journey into the heart of the number system, the origins of those particular numbers which describe the particular spiral waves into which the primeness count deviation decomposes are entirely mysterious.

Total knowledge of these numbers (which allow us to describe precisely each of the spiral waves involved) would be sufficient to reconstruct the entire sequence of primes, so this information could be said to *encode* the prime numbers.

Stop and think about it for a moment – all of this is present in your number system, continually there behind the scenes when you're counting out change in a shop or discussing the temperature or football results or anything else in which numbers play some part. And after millennia of intensive number usage, it was only discovered in 1859. Even today, only a *tiny* fraction of humanity has any awareness of it.

Chapter 15

Mysterious frequencies

Although I don't want to confuse matters by introducing any particular religious connotations, I do find the imagery below quite attractive and illustrative of what we're dealing with here (and I'll explain the reason for the two different angel sizes):

The overall sound made by the angels' trumpets can be thought of as a mighty cosmic chord containing an infinite number of notes, which, in some sense, *brings the prime numbers into being*. It is the sound of the primeness count deviation...

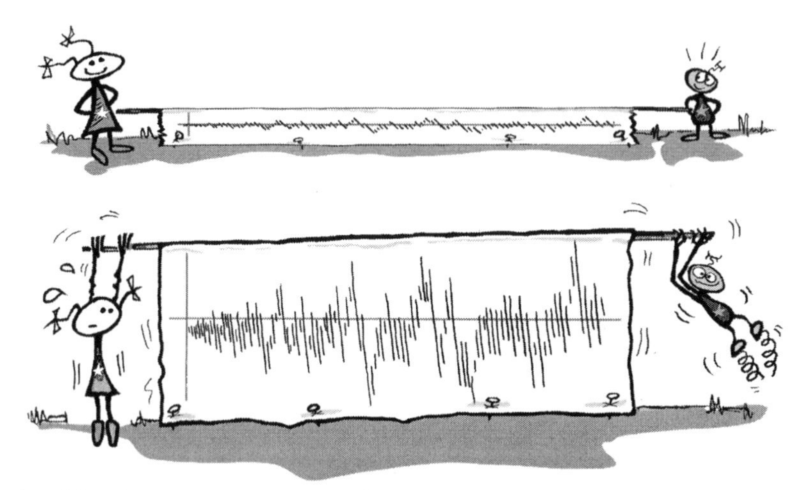

The graph of the deviation in its "natural" state, and then vertically stretched or *rescaled*.

...and when this "chord" is played over the top of a particular "background hum" made of three pieces...

These are the three functions we met in the last chapter – including the two "extra bits" which need to be added in with the spiral waves in order to reconstruct the "primeness count staircase".

…the “waveform” which is produced shows the distribution of primes in the form of the second staircase which we looked at, the one based on “logarithmic” counting.

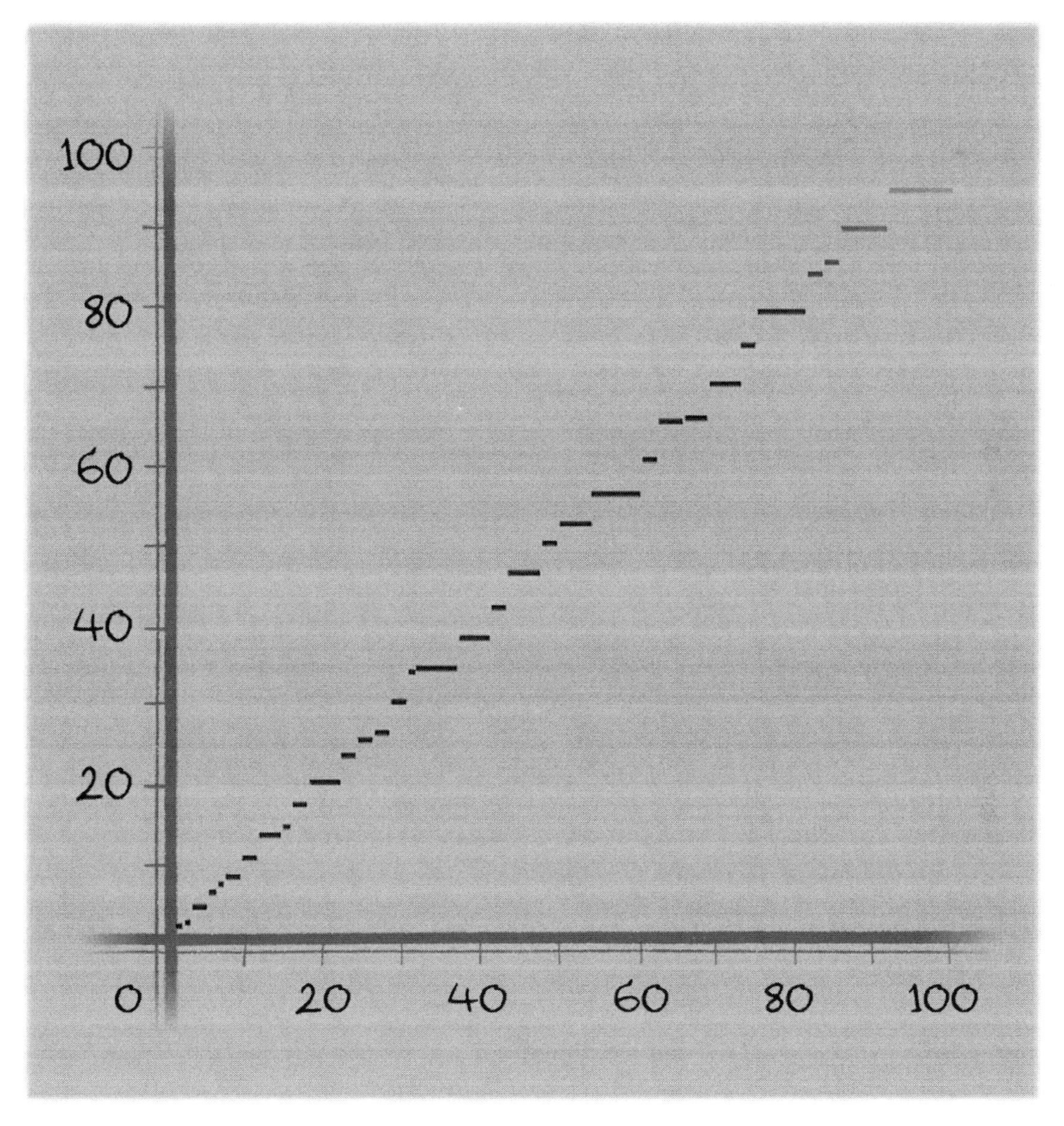

Remember, vertical jumps occur wherever there’s either a prime *or a power of a prime* on the number line below, and the size of the jump is equal to the logarithm of the prime in question.

Just to be *really* thorough about this, we’ll do it one waveform at a time:

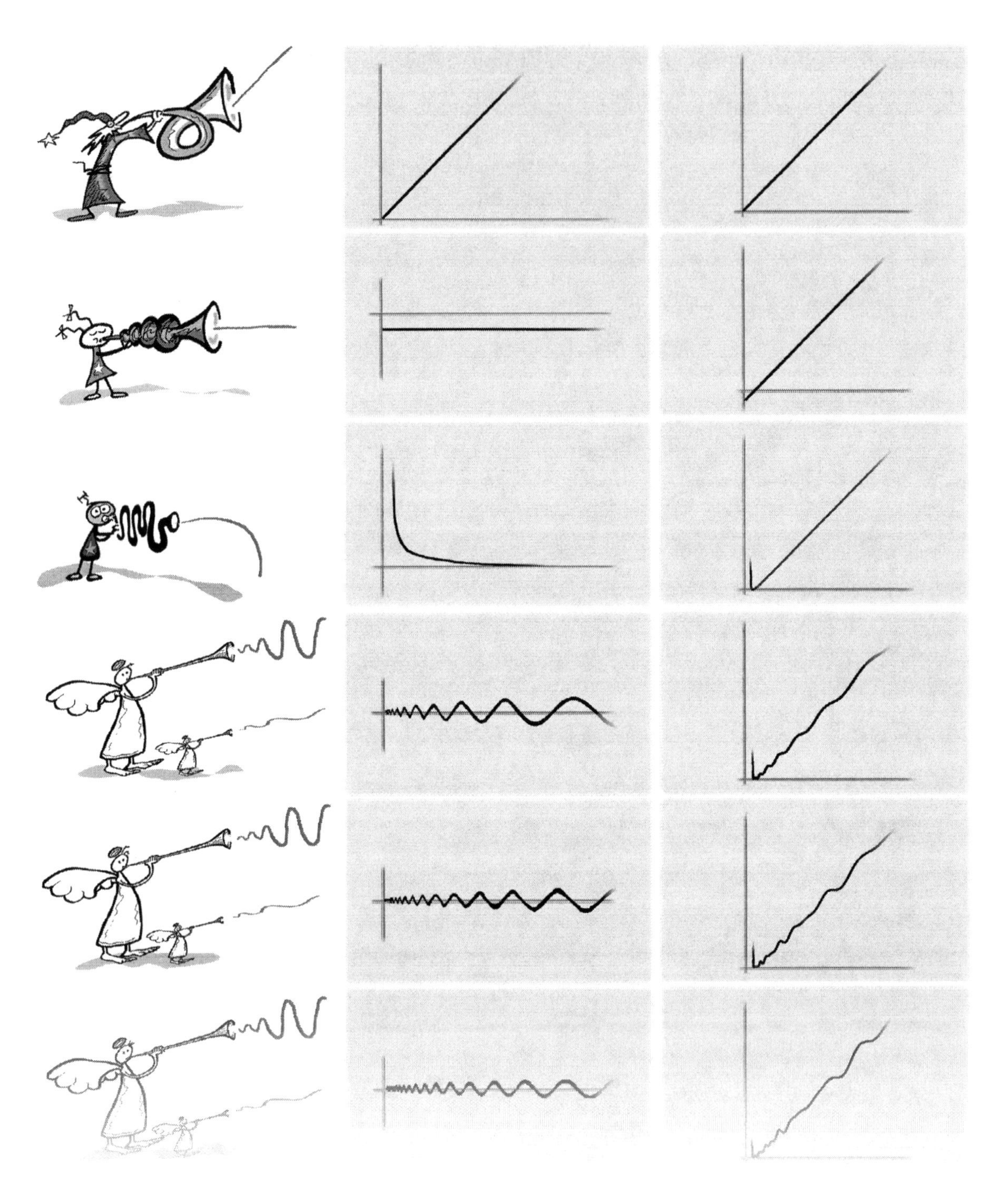

As with the similar tables in Chapters 13 and 14, the second column of graphs represents a “running total” of those in the first column. The second (horizontal)

graph, representing a fixed negative number (–1.837...), pushes the first graph down by that amount. The third one adds an infinite "hook" at 1[1] and then very slightly pushes the diagonal graph up. The spiral waves shown here are actually *pairs* of spiral waves added together – a "big" one and a "small" one. The small wave has no noticeable effect on the graph at this scale, so the sum looks like a familiar spiral wave (but both waves are needed to build the staircase). These seeming-spiral waves have been stretched vertically in the left column (very slightly) so that you can see them more clearly. All of the graphs show the number line from 0 up to 25.

Notice how, when you add in each successive spiral wave, it has an effect which could be casually described as "sloshing the primeness around". Wherever the spiral wave lies above the number line, it increases the amount of primeness represented in that part of the "running total" graph and wherever it dips below the number line, it decreases the amount of primeness represented there. Adding this spiral wave in...

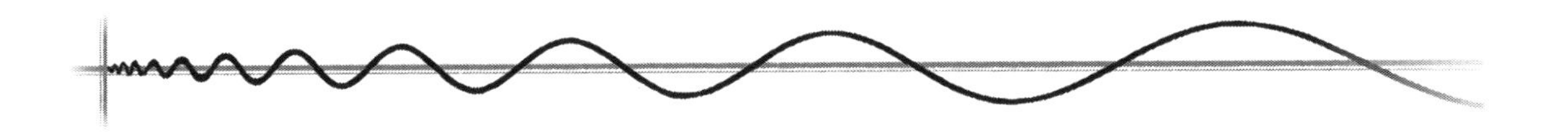

...you're basically saying this:

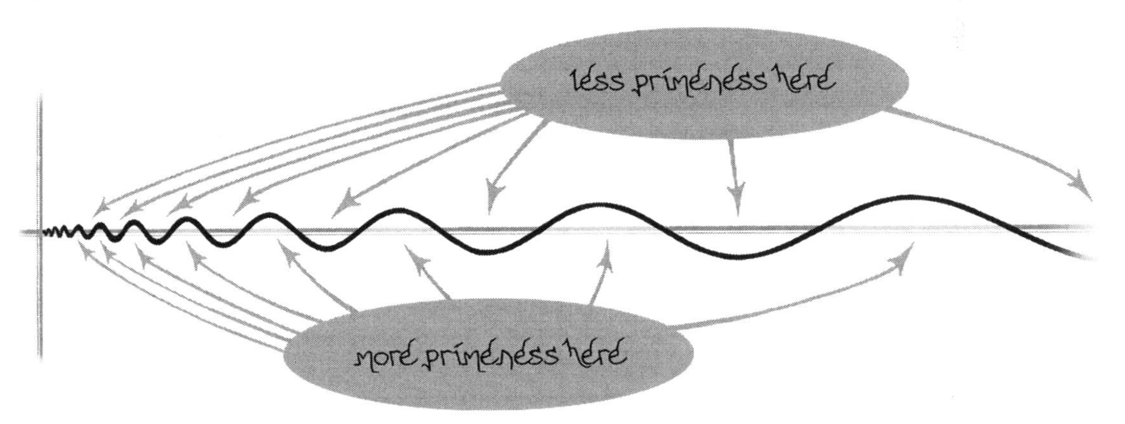

After an infinite amount of this (utterly precise) "sloshing"[2], the smooth, flowing waveform is sculpted into a staircase with perfectly straight edges and right-angled corners! (You can see this process beginning in the last couple of frames in the second column of graphs.) Not only that, it then contains the *entire* sequence of prime numbers as "encoded information". The general reaction from the mathematical community to these facts over the last 150 years has been something along the lines of "we had no reason to expect this to happen – it's totally extraordinary". Surprise!

What we've been looking at is a kind of "cosmic music" underlying the prime numbers, therefore underlying the whole number system and, arguably, *the whole of everything* [3]. Mathematicians (like Enrico Bombieri, who wrote of "*an arcane music and a secret harmony composed by the prime numbers*" [4]) often resort to musical metaphors when discussing these matters.

In the illustration which began this chapter, we could imagine the lengths of the angels' trumpets relating to the "frequencies" of the notes which they produce. Even though spiral waves don't have fixed frequencies in the sense that sine waves do (the distance between peaks is continually increasing), they have something directly related, as I'll explain in the second volume. In the illustration below, the trumpets have been drawn in proportion to the actual "frequencies" needed for the waves which occur in the reconstruction of the "primeness count staircase" illustrated on page 290. These numbers are, in a very real sense, buried in the distribution of prime numbers, in our counting number system and hence in the very structure of reality.

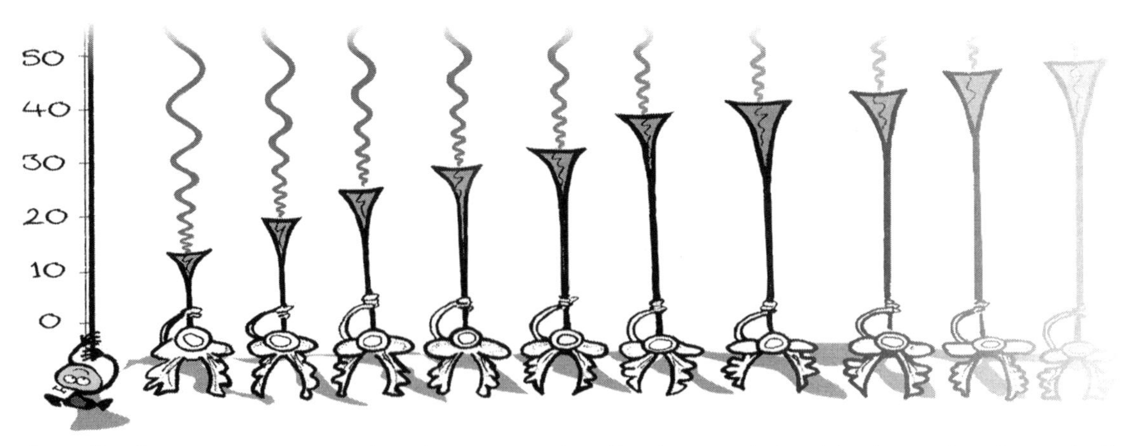

In reality, longer trumpets produce lower notes (*smaller* frequencies) - this picture is only symbolic.

As far as we know, humans have only been aware of this harmonic structure lurking behind the primes since 1859. Since then, a relatively tiny crew of interested people have been using the most powerful methods available to calculate the "frequencies" of the mysterious spiral waves. As of 2010, they've calculated billions of them. Riemann proved that like the primes, there are infinitely many of these numbers, but that *unlike* the primes, which tend to thin out, these numbers tend to *cluster together* as they get larger. And a very important difference is that the primes are all counting numbers, whereas these "frequencies" are weirdly random-looking numbers like...

We're left confronted with the hugely significant question: *where did these particular numbers come from?*

To reiterate, these numbers are the frequencies (using that word somewhat loosely) of the pairs of spiral waves which make up the harmonic structure hidden in the distribution of prime numbers. If the primes looked "random" to you before, then these numbers should look *really* random. Mathematicians, of course, once they became aware of this sequence of numbers, immediately went to work looking for patterns in it. The extraordinary truth of what they have discovered awaits you in Volume 2!

The discovery of these spiral waves was a turning point in mathematical history, opening up an entirely new, wonderfully mysterious branch of number theory. Bernhard Riemann wouldn't usually be described as a number theorist, as he wrote only one paper on the subject in his entire, illustrious, career. Riemann worked in quite a few different areas of mathematics. His work on geometry has been particularly influential – the notion of "curved space-time" which Albert Einstein used in his theory of general relativity relies on what's known as *Riemannian geometry*. But Riemann's 1859 paper "On the number of prime numbers less than a given quantity"[5] is now regarded as an ultra-classic contribution to mathematics, and its consequences – still being worked out – will, I believe, almost certainly end up eclipsing his achievements in paving the way for Einstein's relativistic physics.

The question of prime numbers, then, has been more or less cleared up. We have a precise recipe for constructing the entire distribution. But our naive probing into the issue of whether or not they are "ordered" or contain any kind of "pattern" has now opened up a much bigger question. This concerns the spiral waves and, in particular, the origin of their "frequencies", as well as the other numbers which describe them. That will be our starting point in the next volume, *The Enigma of the Spiral Waves*.

Until then, a final thought:

> "*As archetypes of our representation of the world, numbers form, in the strongest sense, part of ourselves, to such an extent that it can legitimately be asked whether the subject of study of arithmetic is not the human mind itself. From this a strange fascination arises: how can it be that these numbers, which lie so deeply within ourselves, also give rise to such formidable enigmas? Among all these mysteries, that of the prime numbers is undoubtedly the most ancient and most resistant.*"
>
> G. Tenenbaum and M. Mendès France[6]

end of volume one

notes

CHAPTER 1 [pages 5–33]

1. I've noticed that some children's "learn to count" books are authored anonymously!

2. John N. Crossley, *The Emergence of Number* (World Scientific, 1987)

3. Neuropsychology is a relatively new interdisciplinary branch of psychology and neuroscience, involving elements of neurology, psychiatry and the "philosophy of mind". It explores how physical brain structures and functions are related to psychological processes and behaviours.

4. In the spring of 2008, during the preparation of this book, researchers in Italy demonstrated that the mosquitofish (*Gambusia affinis*) is able to distinguish between the numbers 1, 2, 3 and 4. The story can be found here: `http://www.guardian.co.uk/science/2008/feb/26/1` [also accessible as `http://tinyurl.com/2dqvt2`].

5. See, for example, Daniel Tammet's *Born on a Blue Day: Inside the Extraordinary Mind of an Autistic Savant* (Simon & Schuster, 2007). A related piece by Tammet is available online here: `http://www.npr.org/documents/2007/jan/Blueday.pdf` [also accessible as `http://tinyurl.com/nec3pw`].

6. See, for example, A. Schimmel's *The Mystery of Numbers* (Oxford University Press, 1993).

7. See Marie-Louise von Franz, *Number and Time: Reflections Leading Toward a Unification of Depth Psychology and Physics* (Northwestern University Press, 1974). As this is one of *very* few books exploring these important issues and one which contains many fascinating insights, it's unfortunate that von Franz's lack of familiarity with higher mathematics leads to dubious assertions and conclusions in several places. Consequently, the passages dealing explicitly with modern Western mathematics and physics should be read with some caution.

8. Research on adults' feelings about numbers is reported in M. Milikowski and J.J. Elshout's "Numbers as friends and villains", a 1996 University of Amsterdam research report, available here: `http://www.rekencentrale.nl/bestanden/Andere_artikelen_MM/1995_1999/pdf_files/favorites.pdf` [also accessible as `http://tinyurl.com/5b56qb`].

Also, in X. Seron *et al.*, "Images of numbers, or 'when 98 is upper left and 6 sky blue'", *Cognition* 44 (1992) pp. 159–196, the authors follow up Francis Galton's 1880 research, confirming his extraordinary discovery that 5–10% of adults experience numbers as having colours and precise spatial locations.

9. Science seeks to minimise belief. However objective it is, though, at the very least it must assume that: (1) there exists a "shared" objective reality; and (2) there exists some sort of uniformity through time.

10. Brian Eno has pointed out that *smell* (our oldest sense, and the one most closely linked to memory and mood) cannot be quantified.

11. For a beautifully written exploration of these matters, see M. Hannis, "The last refuge of the unquantifiable: Aesthetics, experience and environmentalism", available at `http://www.lancs.ac.uk/depts/philosophy/awaymave/onlineresources/last%20refuge%20of%20the%20unquantifiable%20(m%20hannis).rtf` [also at `http://tinyurl.com/yfm8lbs`].

12. In a speech to the Royal Society, Oxford, November 2006.

13. See, for example, *Science As Salvation: A Modern Myth and Its Meaning* (Routledge, 1992). An accessible article about Midgley and her philosophy is available here: `http://books.guardian.co.uk/departments/politicsphilosophyandsociety/story/0,6000,421562,00.html` [also accessible as `http://tinyurl.com/nauxpl`].

14. See note 9.

15. The Law of Large Numbers is one of the cornerstones of probability theory. It's usually stated in one of several pure mathematical forms, but its purpose is to "guarantee" the averaging-out over time of the measurable output of a repeatable random event. For example, it tells us that if we roll a die (a classic "random event") many, many times, then the average "score" will get closer and closer to 3.5 (which is 1 + 2 + 3 + 4 + 5 + 6 divided by six, an average die-throw). Similarly, if we repeatedly toss a coin (another classic "random event"), the proportion of heads (or tails) will gradually get closer and closer to 50%. Although this tends to work very well in practice, there's a philosophical problem with the use of the mathematical concept of a *limit* in the formulation and proof of the Law of Large Numbers. Limits involve the notion of infinity (or, more precisely, something "tending to infinity"). This can be meaningfully discussed in a pure mathematical setting, but it can never correspond to anything we experience in physical reality (and that includes all of experimental science). Also, as the maverick thinker Terence McKenna liked to point out, there's a major problem with the idea of anything whatsoever being "repeatable".

For a serious philosophical examination of the LoLN, see Edgar Zilsel's *The Application Problem: a Philosophical Investigation of the Law of Large Numbers and its Induction* (if you can find it!)

16. R.H. Nelson, *Economics as Religion: From Samuelson to Chicago and Beyond* (Pennsylvania State University Press, 2001). The passage which follows is from page *xv*.

17. The passage being quoted here is from page *xv* of Nelson's book *Economics as Religion* (see note 16) and, at this point, it includes the following footnote:

"It should be recalled that theologians in earlier eras not only explored the greater mysteries of God but also issued pronouncements on many questions of daily life, including economic ones. In Judaism, for example, dietary laws controlled the very practical matter of what people could eat. Theological pronouncements on 'just price' doctrine could determine the acceptable forms of market transactions. Usury rules of the church commonly prohibited the explicit charging of interest on loans. The various schools of modern economics, addressing both transcendent and mundane questions, can be traced back to previous leading schools of belief within Judaism and Christianity. For a further discussion of these themes, see Robert H. Nelson, Reaching for Heaven on Earth: The Theological Meaning of Economics."

18. R.H. Nelson, *Reaching for Heaven on Earth: The Theological Meaning of Economics* (Rowman and Littlefield, 1991). The excerpt which follows, where the author describes an argument he presented in that book, is from pages *xx–xxi* of his later work *Economics as Religion* (see note 16).

19. John McMurtry, "The global market doctrine: A study in fundamentalist theology", *Canadian Social Studies* **31** no. 2 (1997) pp. 70–72. See `http://www.arts.uwaterloo.ca/ECON/needhdata/McMurtry-2.html` [also accessible as `http://tinyurl.com/nzt2od`].
Also, see "Understanding market theology", a chapter McMurtry contributed to *The Invisible Hand and the Common Good*, edited by B. Hodgson (Springer, 2004).

20. See page 61 of Keith Tester, *The Inhuman Condition* (Routledge, 1995); page 63 of Catherine Cowley, *The Value of Money: Ethics and the World of Finance* (Continuum International, 2006); George Ritzer, *Enchanting a Disenchanted World: Revolutionizing the Means of Consumption* (Pine Forge Press, 2004); Beth Gill, "Temples of consumption: Shopping malls as secular cathedrals" (1999), available at `http://tinyurl.com/nra665`; Erin Kenzie, "Browsing for fulfillment: Shopping malls as sacred in contemporary American society" (2001), available at `http://tinyurl.com/mkxowb`.

21. In *Economics as Religion*, Robert Nelson refers to the following article, which involves a survey of MIT and Harvard University economics graduates, among whom "*the scientific status of economics is clearly in doubt*": D. Colander and A. Klamer, "The making of an economist", *Journal of Economic Perspectives* **1** (1987) p. 102

As of 2008, numerous pieces can be found online debating this question. A quite dramatic development in recent years has been the emergence of a "post-autistic economics" movement from the economics departments of elite universities in France, the UK and the US which is challenging (among other things) the scientific status of the subject.

22. See Annemarie Schimmel, *The Mystery of Numbers* (Oxford University Press, 1994) and Marie-Louise von Franz, *Number and Time: Reflections Leading Toward a Unification of Depth Psychology and Physics* (Northwestern University Press, 1974).

Ancient China deserves a special mention. Dynasties governing a vast territory and dealing with agriculture, trade, taxation, military planning, *etc.* at that sort of scale obviously had a pretty solid understanding of number in the "secular" quantitative sense. But, at the same time, many aspects of ancient Chinese life were governed by a (to Western minds) curiously qualitative approach to number, involving traditional systems wherein certain numbers were associated with parts of the body, seasons, plants, animals, metals, *etc.* "Religion" is not a word whose meaning can easily be pinned down, but I don't think it would be unreasonable to suggest that number played a significant role in the religious life of ancient China. For more on the role of "qualitative number" in ancient Chinese culture, see Volumes 2 and 3 of Joseph Needham's *Science and Civilisation in China* (Cambridge University Press, 1956–59).

23. Abraham Seidenberg, "The ritual origin of counting", *Archive for History of Exact Sciences* 2 (1962) pp. 1–40

24. The following are all good starting points for exploring these matters:
John Michell, *The Dimensions of Paradise,* 3rd Ed. (Inner Traditions, 2008)
Keith Critchlow, *Time Stands Still: New Light on Megalithic Science,* Revised Ed. (Floris, 2007)
R.A. Schwaller de Lubicz, *The Temple of Man* (Inner Traditions, 1998)

25. Jung first used the term in 1916 in his lecture "The structure of the unconscious" (*Collected Works*, volume 7, paragraph 449). However, recorded remarks dating back to at least 1911 suggest that the idea was beginning to take shape in his mind. In 1911, Jung had talked about "dementia praecox introversion" loosening up the historical layers of the unconscious mind and claimed that people with this condition "suffer from the reminiscences of mankind".

26. See note 7.

27. The "Pythagorean Theorem" goes like this. Suppose you have a triangle with a right angle in it:

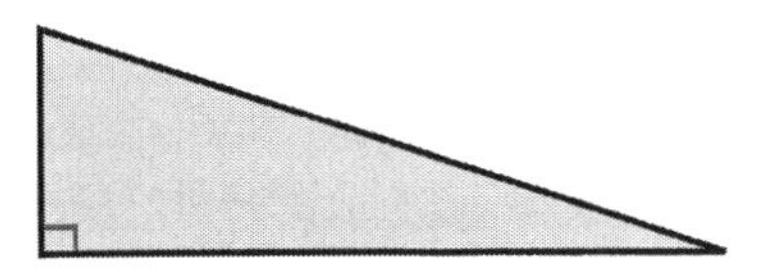

Two sides of the triangle meet at that angle, yes? Take the length of each of these two sides, and multiply it by itself, then add the two results together. This will always give the same result as multiplying the length of the third side (the sloping one in the figure above) by itself.

28. P.J. Davis and R. Hersh, *Descartes' Dream: The World According to Mathematics* (Penguin, 1988) p. 6

29. P.J. Davis and R. Hersh, *Descartes' Dream: The World According to Mathematics* (Penguin, 1988) p. *xv*

30. P.J. Davis and R. Hersh, *Descartes' Dream: The World According to Mathematics* (Penguin, 1988) p. 23

31. John N. Crossley, *The Emergence of Number* (World Scientific, 1987), p. 3

CHAPTER 2 [pages 35–49]

1. Peano's gentle, optimistic nature is described in H.C. Kennedy's *Peano: Life and Works of Giuseppe Peano* (Reidel, 1980).

2. If you look up "Peano axioms" in a book or on the Web, you might find that they're presented quite differently. You might even see more than five. The extra axioms work at a deeper level, in the sense that the five you've just seen rely on certain assumptions about what it means for things to be "equal" and other such subtleties. Closer examination of these issues will take you into some quite deep philosophical territory, but we won't have to concern ourselves with that in these books.

3. This exception involves the fifth axiom, which allows mathematicians to use an important technique called *induction*. Induction works as follows. Suppose that you're considering a particular statement about counting numbers, which, for each counting number, is either true or false. Suppose, further, that you can show:

(a) the statement is true for 1, and

(b) the statement being true for a counting number always means that it must also be true for that number's successor.

The statement is then true for ALL counting numbers. Induction is one of the more common techniques used in mathematical proofs. Its power lies in the fact that with finitely many logical steps, you can prove that something is true for infinitely many numbers.

4. The term *number system* gets used in a variety of (often extremely technical) ways. In fact, there are many different kinds of "number systems", some impossible to imagine without years of study. Here, we're going to limit its meaning to the one that's just been described.

5. The fact that addition can be more directly related to Peano's axioms than multiplication can ties in with something we'll consider later – the "uneasy" relationship between addition and multiplication.

CHAPTER 3 [pages 51–80]

1. We'll see an important exception to this claim in a few pages.

2. This was achieved by H. Rademacher in 1936, building on the pioneering work of the Indian mathematical prodigy Srinivasa Ramanujan and his Cambridge University sponsor G.H. Hardy. To give you some idea of the difficulty involved in this problem, the formula in question looks

like this:

$$p(n) = \frac{1}{\pi\sqrt{2}} \sum_{k=1}^{\infty} A_k(n)\sqrt{k}\frac{d}{dn}\frac{\sinh\left(\frac{\pi}{k}\sqrt{\frac{2}{3}(n-\frac{1}{24})}\right)}{\sqrt{n-\frac{1}{24}}}$$

where

$$A_k(n) = \sum_{0 \le m < k;(m,k)=1} \exp\{\pi i[s(m,k) - 2nm/k]\}$$

[see `http://en.wikipedia.org/wiki/Integer_partition` for more information].

Ramanujan's story is among the most interesting in the history of mathematics. He claimed that his Hindu family goddess Namagiri revealed such formulas to him in his sleep. The theory of additive partitions is a fascinating sub-branch of number theory.

3. Some would have been much less likely than others, though. At the bottom of the list would be the Pirahã people of the Amazon rainforest who have absolutely no number words or concepts whatsoever (or colour words, history or mythology). Their language even appears to lack certain basic linguistic structures which were, until recently, believed to be universal. See Daniel Everett's fascinating account of life with the Pirahã, *Don't Sleep There Are Snakes: Life and Language in the Amazonian Jungle* (Profile Books, 2008).

4. In his 1949 book *Le Temple dans l'Homme* [later translated as *The Temple In Man: Sacred Architecture and The Perfect Man* (Inner Traditions, 1981)], the controversial Egyptologist R.A. Schwaller de Lubicz argued that a certain section of the Rhind Papyrus can be interpreted as a simple prime number "sieve" (see p. 70). It has also been noted that the fraction expansions in the Rhind Papyrus have significantly different forms for primes and for composites.

5. Apostolos Doxiadis, *Uncle Petros and Goldbach's Conjecture* (Bloomsbury, 2000)

6. `http://www.fivedoves.com/revdrnatch/Does_God_think_1_is_prime.htm` [also accessible as `http://tinyurl.com/yu4kvn`].

7. D. Zagier, "The first 50 million prime numbers", *The Mathematical Intelligencer* 0 (1977) p. 8. This article is a translated transcription of a public lecture given by Zagier at Bonn University in 1975.

8. G.H. Hardy, *A Mathematician's Apology* (Cambridge University Press, 1940), p. 70

9. D. McDaniels, J. Simmons and J. Mizell, "It's Like That" (Warner/Chappell Music, 1983)

10. Every whole number, positive or negative, can be divided by -1, just as it can by 1 (one 200 Rupee debt is equal to two hundred 1 Rupee debts: $-200 = -1 \times 200$), so any meaningful

definition of "primeness" involving negative numbers would have to allow divisibility by −1. Similarly, every negative number is divisible by its positive counterpart. Divide your −200 rupee debt into 200 equal debts. They're 1 rupee each: −200 can be divided by 200 to give −1.

11. From an American Mathematical Society news article entitled "International team shows that primes can be found in surprising places" (December 1997), available at `http://www.ams.org/ams/mathnews/prime.html` [also accessible as `http://tinyurl.com/l7fqy7`].

CHAPTER 4 [pages 81–103]

1. Very little is known about Euclid of Alexandria, although historical records suggest that he was active during the reign of Ptolemy I (323–283 BCE).

2. There is some ambiguity here. Although Euclid is widely credited with being first to note the unique factorisation of counting numbers, J.J. Tattershall, in his *Elementary Number Theory in Nine Chapters* (Cambridge University Press, 1999) is more cautious:

"*It is quite possible that the... result was known to Euclid. However, since he had no notation for exponents and could not express a number with an arbitrary number of factors, it was not included in the* Elements. *Nevertheless, it is very similar to Proposition 14 in Book IX. The result was first stated explicitly by Gauss, who included a proof of the result in his doctoral thesis.*"

The 2000 year gap between Euclid's (possible) statement and Gauss' (1798) proof can be explained by the rather subtle matter of proving the "uniqueness" part. This is explored in detail in Appendix 2 of this book.

3. A *theorem* is a mathematical statement which has been *proved*, based on previously established or accepted statements, such as axioms (or other theorems). There will be more on axioms and proofs in Chapter 7.

4. Mathematicians prefer to call 0 the *additive identity* and 1 the *multiplicative identity*.

5. A. Knauf, "Number theory, dynamical systems and statistical mechanics", *Reviews in Math. Physics* 11 (1999) p.1027

6. Or possibly it was his teacher Leucippus – it's impossible to tell.

7. Just as it was believed that the atom was unsplittable for a very long time before someone split one, we can't rule out that there might be some as-yet-undiscovered mathematical context in which the primes can be "split" (I've had some interesting thoughts about this over the years). But as far as straightforward multiplication of counting numbers goes, there's no such issue – they're unsplittable.

8. I. Stewart, "Jumping champions", *Scientific American* 283, no. 6 (December 2000) p.80

9. H. Weyl, *Philosophy of Mathematics and Natural Science* (Princeton University Press, 1949) p.7

10. This was an unfortunate choice of a word on the part of the translator. In higher mathematics, a *manifold* is a very specific kind of geometric object. Weyl didn't intend his word in that sense.

CHAPTER 5 [pages 105–115]

1. G.H. Hardy, *A Mathematician's Apology* (Cambridge University Press, 1940), p.70

2. Jane Muir, *Of Men and Numbers: The Story of the Great Mathematicians* (Dover Publications, 1996), p.6

3. Just web searching with the keywords "God" and "prime numbers" will produce all sorts of weird and wonderful documents.

4. Peter Plichta, *God's Secret Formula: Deciphering the Riddle of the Universe and the Prime Number Code* (Element, 1998)

5. It could be argued that "feelings" include intuitions and "hunches". These certainly have a role to play in the process of mathematical discovery. But the real point I'm making here is that what you *feel* about a mathematical statement has no bearing whatsoever on whether on not it is *true*.

6. Unless you want to get *really* fundamental, including such things as the observation that "there's something rather than nothing".

7. One of the proofreaders of this book raised the extremely good point: "What about the teachings of the Buddha concerning emptiness? This is universally applicable and not anthropocentric (but deals with consciousness)."

8. "*In 1941, Sir James Jeans suggested that searchlights be used to signal prime numbers to the Martians during the planet's close approach to Earth.*" [from page 10 of David Lamb's *The Search for Extraterrestrial Intelligence: A Philosophical Inquiry* (Routledge, 2001)]

9. Carl Sagan, *Contact* (Simon & Schuster, 1985), Chapter 4

10. Raymond A. Chamberlin, "What sort of 'signal' would satisfactorily announce an extraterrestrial 'intelligence', as detected by modulated electromagnetic (radio or light) emission?", available at `http://home.znet.com/raych/Signal_vs_Noise.htm`
[also accessible as `http://tinyurl.com/met74y`].

11. Garrett Barden, "Bipeds and prime numbers" (address given at a degree ceremony, University

College Cork, September 2001) – see http://www.ucc.ie/opa/conferspech/gbarden.html [also accessible as http://tinyurl.com/mkjo7h]. Barden actually confuses the Arecibo message and a (related) pictorial plaque attached to NASA's Pioneer 10 space probe which was launched a couple of years earlier, in 1972.

12. David Darling, "Drake's Cryptogram", available at http://www.daviddarling.info/encyclopedia/D/Drakecrypto.html [also accessible as http://tinyurl.com/mw84wb].

13. Richard Dawkins, *River Out of Eden: A Darwinian View of Life* (Basic Books, 1995) p. 21

CHAPTER 6 [pages 117–134]

1. L. Kauffman, "Virtual logic – Formal arithmetic", *Cybernetics & Human Knowing* 7 (2000) pp. 91–95. See http://www.imprint.co.uk/C&HK/vol7/kauffman_7-4.pdf [also available as http://tinyurl.com/nnfqtp].

2. Here are some examples:

"*I hope that...I have communicated a certain impression of the immense beauty of the prime numbers and the endless surprises which they have in store for us.*"

D. Zagier from "The first 50 million prime numbers", *The Mathematical Intelligencer* 0 (1977) p. 16

"[*Primes*] *are full of surprises and very mysterious... They are like things you can touch... In mathematics most things are abstract, but I have some feeling that I can touch the primes, as if they are made of a really physical material. To me, the integers as a whole are like physical particles.*"

Y. Motohashi, quoted in K. Sabbagh, *Dr. Riemann's Zeros* (Atlantic, 2002), p. 17

"*No branch of number theory is more saturated with mystery than the study of prime numbers: those exasperating, unruly integers that refuse to be divided evenly by any integers except themselves and 1.*"

M. Gardner, "The remarkable lore of the prime numbers", *Scientific American*, March 1964

3. This is perhaps misleading. The study of additive partitions is a hugely complicated matter, as I explained in Chapter 3. But, as pointed out earlier in this chapter, partitions deal with *numbers of ways of combining numbers*. As with multiplication, this "turning of the number system on itself" represents another level of innovation beyond the Peano-style construction of a purely additive system of counting numbers.

4. Andreas Knauf, "Number theory, dynamical systems and statistical mechanics", *Reviews in Mathematical Physics* 11 (1999) p. 1027

5. A. Doxiadis, *Uncle Petros and Goldbach's Conjecture* (Bloomsbury, 2000), pp. 183–184

6. In a 1921 speech in Copenhagen [according to L.J. Mordell, "Hardy's 'A Mathematician's Apology'", *The American Mathematical Monthly* 77 no. 8 (Oct. 1970) pp. 831–836]

7. See O. Ramaré, "On Schnirelmann's constant", *Ann. Scuola Norm. Sup. Pisa Cl. Sci.* 22 no. 4 (1995) pp. 645–706. There's a widely cited claim that in the 1930s, Lev Schnirelmann proved that every even number can be expressed as no more than 300 000 primes. Ramaré says he has not seen this substantiated in "any decent literature" and suggests that the 1963 proof of Shanin and Sheptitskaya is the earliest such result, with the earliest *published* result being due to N.I. Klimov, bringing the 20 000 000 000 down to 6 000 000 000 [see `http://tinyurl.com/ygoucbg`].

CHAPTER 7 [pages 135–148]

1. The calculations are as follows. This book is set out so that 30 lines of 67 digits could fit onto a single page and 30 × 67 = 2010. Dividing 12 978 189 by that gives 6456.810... pages. This book has about 360 pages, and 6456.810... divided by 360 gives 17.935... books.

2. C.F. Gauss, *Disquisitiones Arithmeticae* (Latin original first published in 1801)

3. D. Zagier, "The first 50 million prime numbers", *The Mathematical Intelligencer* 0 (1977) p. 13

4. A. Doxiadis, *Uncle Petros and Goldbach's Conjecture* (Bloomsbury, 2000), p. 34

5. We didn't need the "uniqueness" part of the FTA, just the existence part.

6. Talking about percentages of infinite quantities is somewhat problematic as *any* percentage of the counting numbers will be infinitely large. However, the percentage being discussed here is less than 100% (for that would be *all* counting numbers), yet it is *larger than any percentage which is less than 100%*. To make this precise, we'd need to get into the formal matter of "infinitesimal" quantities as described in *nonstandard analysis,* but even without that, the point being made in the text should be clear. In the next paragraph, the percentage being discussed is larger than 0% (for that would be *no* counting numbers), but it *less than any percentage which is larger than 0%.*

CHAPTER 8 [pages 149–165]

1. U. Dudley, *Elementary Number Theory*, 2nd edition (W.H Freeman & Company, 1978), p. 163

2. Manfred Schroeder, *Number Theory in Science and Communication*, 2nd edition (Springer, 1986) p. 307

3. Paolo Ribenboim, *The New Book of Prime Number Records* (Springer, 1996) p. 213

4. There's a huge amount of literature concerning the Fibonacci sequence. Recommended places to start would be Philip Ball's *The Self-Made Tapestry* (Oxford University Press, 1999) or Ian Stewart's *Nature's Numbers* (Basic Books, 1997).

5. I. Stewart, "Jumping champions", *Scientific American* 283 no. 6 (December 2000) pp. 80–81.

6. quoted in G.F. Simmons, *Calculus Gems* (MAA, 2007) p. 276

7. A. Doxiadis, *Uncle Petros and Goldbach's Conjecture* (Bloomsbury, 2000) p. 84

8. D. Zagier, "The first 50 million prime numbers", *The Mathematical Intelligencer* 0 (1977) p. 7

9. P.J. Davis and R. Hersh, *The Mathematical Experience* (Birkhäuser, 1981) p. 213

CHAPTER 9 [pages 167–186]

1. They're also known as "logarithmic spirals" and sometimes "growth spirals".

2. This account of why moths spiral into campfires, lightbulbs and candles isn't universally accepted but certainly has the status of current "favoured explanation". The technical name for the navigation involved is *transverse orientation*. An archived radio discussion of this matter is available at `http://www.npr.org/templates/story/story.php?storyId=12903572` [also accessible as `http://tinyurl.com/lrulwa`].

3. In geometry, the formal description of a "line" always involves infinite extension in both directions. A "line" which has a starting point and thus extends infinitely in only one direction would be called a "half-line" or sometimes a "ray".

4. J. Purce, *The Mystic Spiral: Journey of the Soul* (Thames and Hudson, 1974)

5. It's only convenient because our culture uses a "base ten" system for writing down numbers. This is explained in Appendix 1, if you're interested. The base-10 logarithm of 10 is 1, of 100 is 2, of 1000 is 3, of 10 000 is 4, *etc.* So you can immediately know how many digits a number has by looking at its base-10 logarithm. However, if we happened to use, say, "base six" arithmetic, then base-6 logarithms would be the most convenient for calculating with.

6. 6.854... is equal to $1.618... \times 1.618... \times 1.618... \times 1.618...$ This means that if you travel along a quarter of a coil of this spiral, your distance from the centre is multiplied up by the golden ratio. Based on the many remarkable properties of this number, the spiral's "base", 6.854..., also turns out to equal both $(3 \times 1.618...) + 2$ and $(1 + 1.618...) \times (1 + 1.618...)$.

7. If you happen to be checking with your calculator, you'll have found that 2.7182 × 2.7182 × 2.7182 equals 20.0837..., not 20.0855... That's because 2.7182 is a rounding off the actual value of where the ladybird will be after one hour, which is, more accurately, 2.718271827459045...

8. The definition describes e as the "limit" of the sequence which begins (1+1/2)×(1+1/2), (1+1/3)×(1+1/3)×(1+1/3), (1+1/4)×(1+1/4)×(1+1/4)×(1+1/4),... (these numbers get closer and closer to 2.71828182845...). Formally, this would be written

$$e = \lim_{n\to\infty}(1+1/n)^n.$$

CHAPTER 10 [pages 187–201]

1. For example the approximation tells us that there should be about 2.85 primes less than 4. There are, in fact, two of them. Although this is out by less than a single prime, in terms of percentage error (44.27%), it's terrible. Looking at bigger numbers, we get much better results. For example, when we're dealing with ten-digit numbers, the percentage error is always less than 5%. See the table on page 197 and the comments which precede it.

2. The calculations are as follows. This book has about 360 pages and is set out so that 30 lines of 67 digits could fit onto a single page. 360 × 30 × 67 = 723600. Dividing 6514417 by that gives 9.002...

3. In the US and UK, it's certainly very possible to go through an undergraduate degree without encountering this. Some universities may offer optional number theory modules which do include it. And it may have been taught in the past when educational standards were higher. At postgraduate level, you might encounter it (depending on your area of specialisation) but I managed to complete a Ph.D. in mathematics without ever having properly learnt the Prime Number Theorem – I must have vaguely heard of it, but I couldn't have told you what it said.

4. P.J. Davis and R. Hersh, *The Mathematical Experience* (Birkhäuser, 1981) p.213

5. From a letter Gauss wrote to Franz Encke on December 24, 1849.

CHAPTER 11 [pages 203–229]

1. When I use the word "another", that doesn't mean that it has to be different from the number put in. In some cases the output can be the same as the input.

2. [*This will require familiarity with certain mathematical notation. Otherwise, you'll just have to take my word for it.*] According to the PNT, the number of primes less than a given x is $x/\ln x$.

Given a counting number n, if we want to find an approximation for the nth prime, we therefore need an approximate solution for the equation $x/\ln x = n$. And $x = n \ln n$ will work since $n \ln n / \ln(n \ln n) = n \ln n / (\ln n + \ln \ln n)$, which, due to the relatively small size of $\ln \ln n$, is approximately $n \ln n / \ln n$, or n.

3. D. Zagier, "The first 50 million prime numbers", *The Mathematical Intelligencer* 0 (1977) p. 7

4. D. Zagier, "The first 50 million prime numbers", *The Mathematical Intelligencer* 0 (1977) p. 9

5. http://en.wikipedia.org/wiki/Mathematical_Intelligencer

6. It could be argued that "feelings" include intuitions and "hunches", which certainly play a role in the process of mathematical discovery. But the real point I'm making here is that what you *feel* about a mathematical statement has no bearing whatsoever on whether on not it is *true*.

7. D. Zagier, "The first 50 million prime numbers", *The Mathematical Intelligencer* 0 (1977) p. 16

8. D. Zagier, "The first 50 million prime numbers", *The Mathematical Intelligencer* 0 (1977) p. 7 (emphasis mine)

9. M. du Sautoy, "The music of the primes", *Science Spectra* 11 (1998) (emphasis mine).

10. "Nature" is a problematic word – what *isn't* part of nature, after all? I'm using the fairly common meaning of "everything in the world minus humans and what they have created".

11. H.P. Aleff, from Section 1.3 of *Prime Passages to Paradise* (e-book available at http://www.recoveredscience.com/primes1content.htm). This work appears to relate to Egyptological speculation and would therefore tend to get categorised as "pseudo-science". I have yet to see the whole book, but, in any case, Aleff's quotation here is perfectly valid.

CHAPTER 12 [pages 231–247]

1. Actually, Legendre experimented with subtracting various numbers. He erroneously concluded that the choice giving the most accurate approximation would be 1.08366. While this initially does seem to be the case, it was later shown that in the long run, 1 turns out to be the best choice. A detailed history of this matter can be found here: F.L. Bauer, "Why Legendre made a wrong guess about $\pi(x)$, and how Laguerre's continued fraction for the logarithmic integral improved it", *The Mathematical Intelligencer* 25 (2003) pp. 7–11.

2. In a popular magazine article, the senior Princeton-based mathematician Enrico Bombieri invoked the philosophical technique known as *Occam's razor* to argue that it "makes sense" for primes to be counted logarithmically. He explains that William of Occam "*elevated to a method the idea that when one must choose between two explanations, one should always choose the simpler. Occam's razor, as the principle is called, cuts out the difficult and chooses the simple. When things get too complicated, it sometimes makes sense to stop and wonder: Have I asked the right question?*

Here the choice is between two functions that count primes... Surely William of Occam would have chosen to study [*the logarithmic one*]." ("Prime territory: Exploring the infinite landscape at the base of the number system", *The Sciences,* Sept./Oct. 1992)

3. Chebyshev was an accomplished 19th century mathematician, one of the fathers of Russian mathematics. Due to the ambiguities involved in translating from the Cyrillic alphabet to the Roman, you'll sometimes see his name spelled "Chebychev", "Tchebycheff", "Tschebyscheff", "Tchebicheff", "Čebyšev", "Tschebyscheff" or "Chebishev", among other possibilities.

4. Chebyshev introduced this approach to prime-counting in the article "Mémoire sur les nombres premiers" [*J. Math. Pure Appl.* 17 (1852)]. This is available in both the French and Russian editions of his *Collected Works*. Some sources suggest that the paper was first published in 1850.

5. Actually, it hangs around a horizontal line just below the number line, at a distance of about 1.837877. The origins of that number will be revealed in Chapter 15. At the sorts of scales we've been looking at, though, this line is barely distinguishable from the number line.

6. This has nothing to do with *chaos theory* – I'm using the word "chaos" in the looser, more familiar sense, meaning "the opposite of order". Chaos theory will come up in Volume 3, though.

CHAPTER 13 [pages 249–265]

1. The number line is measured out in generic "units", not in seconds, minutes or hours. We could choose to associate a unit of time with it, but any such choice would be arbitrary.

2. I've not heard it, but on page 279 of his popular book *The Music of the Primes* (HarperCollins, 2003), Marcus du Sautoy reports, having heard something which the mathematical physicist Michael Berry generated, that it sounded like "*low rumbling white noise*". In an earlier interview (E. Klarreich, "Prime time", *New Scientist*, 11 November 2000), Berry himself described the result of an earlier attempt at generating the sound as "*a horrible cacophony*".

3. In other ways it behaves like a stream of particles. This "dual nature" of light is one of the central issues of quantum theory, but we won't really be going into that in these books.

4. E. Bombieri, "Prime territory: Exploring the infinite landscape at the base of the number system", *The Sciences* (Sept./Oct. 1992) p. 36

CHAPTER 14 [pages 267–286]

1. For this to make sense, we have to pretend that the bicycle is infinitely small or else just refer to a fixed point on the bike, for example, the centre of the front axle.

2. A similar reversal can be applied with the sine wave construction in order to get the left half of the wave (that part to the left of zero).

3. There's a mathematical subtlety here. In the way I've explained the construction of spiral waves, there's no way to obtain a conventional sine wave: even if the Ferris wheel became a circle (a "degenerate spiral"), the bicycle is still accelerating at an exponential rate, so although we'd have a fixed amplitude, the wavelength would still be stretching. However, if we allow a wider range of possibilities for the bicycle's motion (including constant speed), we end up with a broader definition of "spiral wave" and this would include all conventional sine waves.

[*To be mathematically precise*: we can define a *spiral wave* to be anything of the form $e^{ag(x)}\sin(g(x))$ for real parameter a, where $g(x)$ is a continuous, monotonically increasing function on $x>0$. The bicycle's motion can be described in terms of the inverse function of g, so for the spiral waves we've seen, we would use nonzero a and $g(x)=k\log x$ for some k. Classical sine waves correspond to the case where $a=0$ and $g(x)=kx$.]

4. Some infinite collections of things (the points in a line, for example) *cannot* be "enumerated" in a list. This is related to the meaningful distinction which is made in mathematics between "countable infinities" and "uncountable infinities". Different types of infinities are studied as part of the "foundations of mathematics". Our infinite set of spiral waves is a countable infinity.

CHAPTER 15 [pages 287–294]

1. This "hook" is an infinite spike, but it's gradually eroded away by the addition of all the spiral waves. Infinitely many negative numbers (representing what the waves do at 1) get added to it, resulting in a final value of zero (which is what the staircase does at 1).

2. An infinite process can't really have an "after", of course. But a mathematical sum involving infinitely many pieces *can* have a finite result, in a very well-defined sense. I'm having to use words loosely here in order to communicate the essence of a precise mathematical idea.

3. Depending on how materialistic you are – at least "everything quantifiable".

4. E. Bombieri, "Prime territory: Exploring the infinite landscape at the base of the number system", *The Sciences* (Sept./Oct. 1992) p. 36

5. This was originally published in German, as "Über die Anzahl der Primzahlen unter einer gegebenen Grösse" in *Monatsberichte der Königlich Preußischen Akadademie der Wissenschaften zu Berlin* (November, 1859). An English translation is available online here:
`http://www.maths.tcd.ie/pub/HistMath/People/Riemann/Zeta/EZeta.pdf`
[also accessible as `http://tinyurl.com/3k4yau`].

6. Gérald Tenenbaum and Michel Mendès France, *The Prime Numbers and Their Distribution* (AMS, 2000) p. 1

appendix 1

base ten and other bases

(not too difficult to follow – no specialist knowledge required)

"23" or "twenty-three" means "twenty and three" or **20 + 3**, which is **2** × **10** plus **3** × **1**. If you look at a pile of 23 beans, though, nothing about it immediately suggests the "2" and "3" which appear in the representation "23" – it's just that the currently dominant human culture has got into the habit of counting things out in *tens*, presumably because we have ten fingers and toes. Our pile of **23** beans yields **two** piles of ten with **three** beans left over.

If, for some reason, we preferred to count things out in *eights* (we'd then be using a "base eight" system rather than the usual "base ten"), our pile of 23 beans would yield two piles of 8 with seven beans left over: 23 = (2 × 8) + 7 which we can think of as **2** × 8 plus **7** × 1.

This means that twenty-three would be represented as "27" in base eight notation. If that seems confusing, don't worry too much. The main things to keep in mind here are:

☆ We've got into the habit of counting things out in tens.

☆ There's nothing particularly special about the number ten[1]. Think of a pile of ten beans. The fact that ten (as "10") is the first two-digit number is simply the result of our choosing to work with the base ten system – it's in no way evident in the pile of ten beans. The system we've chosen provides only ten digit-symbols to work with: 0, 1, 2, 3, 4, 5, 6, 7, 8 and 9. Counting out beans starting with 0 (the absence of any), then 1, 2, 3,..., by the time we get to the tenth bean, we've run out of symbols, so we have to start using *pairs* of digits: 10, 11, 12, 13,...

☆ There are other, equally valid, ways of representing numbers which involve counting in other ways.

☆ *Whether or not a number is prime is entirely unaffected by the system we use to represent it.*

That "binary" representation of nineteen given in Chapter 3 is an example of "base two", the system which computers use to represent numbers:

The binary number "10111" means (reading the number right-to-left):

$$
\begin{aligned}
&\mathbf{1}\\
&+\mathbf{1}\times 2\\
&+\mathbf{1}\times 2\times 2\\
&+\mathbf{0}\times 2\times 2\times 2\\
&+\mathbf{1}\times 2\times 2\times 2\times 2\\
&=1+2+4+0+16\\
&=23
\end{aligned}
$$

This breakdown is also helpful in understanding "base ten" notation.

In the usual base ten system, "21 695" means (again, reading the number right-to-left):

$$\begin{aligned}&\mathbf{5}\\&+\mathbf{9}\times 10\\&+\mathbf{6}\times 10\times 10\\&+\mathbf{1}\times 10\times 10\times 10\\&+\mathbf{2}\times 10\times 10\times 10\times 10\\&=5+90+600+1000+20000.\end{aligned}$$

We can do this for any counting number base (other than 1 – "base one" doesn't work). Base twelve has certain advantages for ease of calculation. There's even a small (and entirely serious) "dozenalist" movement actively dedicated to the promotion of *duodecimal* or base twelve notation as an alternative to the dominant *decimal* or base ten system[2].

Notice that in base *two*, there are only *two* digits used: "0" and "1". As I've already pointed out, in base *ten*, we use *ten* digits: "0", "1", "2", "3", "4", "5", "6", "7", "8" and "9".

As well as binary, computer scientists sometimes work with "hexadecimal" or base sixteen. For this, we need *sixteen* digits, so we use "0", "1", "2", "3", "4", "5", "6", "7", "8", "9", "A", "B", "C", "D", "E" and "F". Although it might initially seem puzzling to see letters of the alphabet used as digits, "A" here represents ten, "B" eleven, "C" twelve, "D" thirteen, "E" fourteen and "F" fifteen.

In hexadecimal notation, you encounter representations of numbers like "5DF03B". Converting this to a familiar base ten representation works like this (again, think right-to-left):

$$\begin{aligned}&\mathbf{11}\\&+\ \mathbf{3}\times 16\\&+\ \mathbf{0}\times 16\times 16\\&+\mathbf{15}\times 16\times 16\times 16\\&+\mathbf{13}\times 16\times 16\times 16\times 16\\&+\ \mathbf{5}\times 16\times 16\times 16\times 16\times 16\\&=\ 11+48+0+61440+851968+5242880\\&=\ 6156347.\end{aligned}$$

Whether you call it "5DF03B" or "6 156 347" doesn't really matter. The number in question can always be represented as a pile of beans – so clearly it's either a prime or it isn't, entirely independently of any name that it's given.

"There are 10 kinds of people in the world: those who understand binary notation, and those who don't."
(origin unknown)

notes

1. To the mediaeval Qabalists whose "Tree of Life" involved ten *Sephiroth* and to the Pythagoreans with their *Tetractys*...

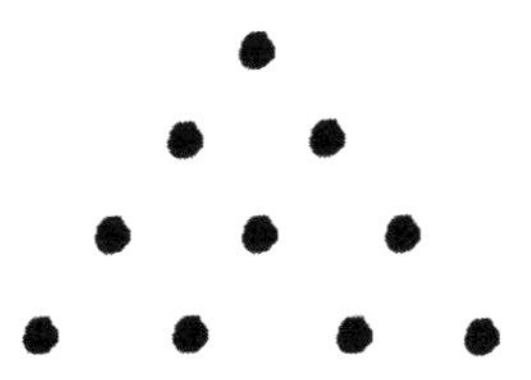

...the number ten had a profound mystical importance – see Robin Waterfield's translation of Iamblichus' *The Theology of Arithmetic* (with a foreword by Keith Critchlow, Phanes Press, 1988). These are examples of "qualitative" views of a number, about which mathematics can tell us nothing, and of which modern scientific culture is (for better or worse) entirely dismissive.

The cultural dominance of base ten is usually attributed to the fact that humans have ten fingers. The idea of percentages is based on 10×10 (a percentage is all about dividing things up into 100 equal parts – see Appendix 6 for more details). Similarly, the metric system is based entirely on tens (unlike traditional systems of measurement which involve various combinations of threes, tens, twelves, twenties, *etc.*). This is what makes it easier to work with – but it's only easier because it's based on the same number (ten) on which our system of writing numbers is based.

2. Dozenal Society of Great Britain: `http://www.dozenalsociety.org.uk`;
Dozenal Society of America: `http://www.dozenal.org`

appendix 2

proof of the Fundamental Theorem of Arithmetic

(requires a certain amount of experience with mathematical notation and reasoning)

The proof consists of two parts: an *existence* part and a *uniqueness* part. The *existence* part must show that given any counting number, it has a prime factorisation (that is, it can be built by multiplying some combination of prime numbers together). The *uniqueness* part must show that if two factorisations of the same number can be found, then they are necessarily the same (in the sense that a bowl containing two bananas, an apple and a mango is the "same" as a bowl containing an apple, a banana, a mango and another banana).

Both parts involve *proof by contradiction*. This is explained in Chapter 7. If you're not already familiar with this way of thinking, then you'll probably struggle to follow this proof.

The **existence** part is fairly easy.

We start by assuming that there exist counting numbers greater than 1 which *can't* be built by multiplying primes together. If that were the case, then there would have to be a *smallest* such counting number. It can't be a prime number since any prime acts as its own factorisation (in the sense that the factorisation of 167 is just "167"). Because it can't be prime, it must be divisible by some counting number other than 1 and itself, so we can write it as $m \times n$, where m and n are counting numbers greater than 1 but smaller than our supposed "smallest unfactorisable counting number".

Because they're smaller than it, m and n must themselves be factorisable. That is, m can be written as a product of prime numbers, and so can n. By combining these two collections of prime factors, we're then able to build our "smallest unfactorisable counting number", contradicting the supposed fact that it has no prime factorisation.

As we have arrived at a contradiction, our initial assumption must be flawed – there *cannot* exist counting numbers greater than 1 which don't have prime factorisations.

The *uniqueness* part is a bit trickier. The following argument is based on Carl Gauss' original proof from 1798[1]. Again, we start by assuming the opposite of what we seek to prove, then reason until we reach a logical contradiction. This time, we assume that there exist counting numbers with two or more genuinely distinct prime factorisations. If that were the case, then there would have to be a *smallest* such counting number. Call it *a*. We'll choose two distinct factorisations of *a*. Note that they can't have any prime factors in common, otherwise, by removing these factors from both, we'd end up with a *smaller* counting number with two distinct prime factorisations. In this way, we can be sure that one of the two factorisations will contain a prime factor smaller than all of the prime factors in the other. Write the first as $p_1 \times \cdots \times p_m$ and the second as $q_1 \times \cdots \times q_n$, where p_1 is less than all of $q_1, \ldots, q_n$. If we divide q_1 by p_1, we'll get a non-zero remainder (otherwise q_1 wouldn't be prime), so we can write $q_1 = d \times p_1 + r$, where d is a counting number (that is, at least 1) and r is a counting number less than p_1 (as it's a remainder).

We have

$$a = p_1 \times p_2 \times \cdots \times p_m = q_1 \times q_2 \times \cdots \times q_n$$

$$= (d \times p_1 + r) \times q_2 \times \cdots \times q_n.$$

Using the distributive law $(u + v) \times w = u \times w + v \times w$, we get

$$a = p_1 \times p_2 \times \cdots \times p_m = d \times p_1 \times q_2 \times \cdots \times q_n + r \times q_2 \times \cdots \times q_n.$$

Now define $b = a - d \times p_1 \times q_2 \times \cdots \times q_n$, so that we also have $b = r \times q_2 \times \cdots \times q_n$.

From the first of these representations of b, because $d \times p_1 \times q_2 \times \cdots \times q_n$ is a positive number, b must be less than a. From the second representation, b must be a counting number greater than 1.

Replacing a in the first representation of b, we can write

$$b = p_1 \times p_2 \times \cdots \times p_m - d \times p_1 \times q_2 \times \cdots \times q_n.$$

So, by the distributive law, we have

$$b = p_1 \times (p_2 \times \cdots \times p_m - d \times q_2 \times \cdots \times q_n).$$

Because b is known to be positive, we can be sure that $p_2 \times \cdots \times p_m - d \times q_2 \times \cdots \times q_n$ must be at least 1. If we replace it here by its prime factorisation (which it must have, by the "existence" part of the proof), we see that there is a prime factorisation of b involving p_1.

But we *also* know that $b = r \times q_2 \times \cdots \times q_n$, where r is less than p_1 and none of the $q_2, \ldots, q_n$ can equal p_1. Replacing r by its prime factorisation (none of the factors involved being p_1), we obtain *another* prime factorisation of b which *doesn't* involve p_1.

We've just deduced the existence of a counting number, b, which is less than a and which has two distinct prime factorisations. This contradicts our assumption that a was the smallest such counting number. Our original assumption that there exist counting numbers with two of more genuinely distinct prime factorisations must therefore be wrong. This proves uniqueness.

note

1. C.F. Gauss, *Disquisitiones Arithmeticae* (Latin original first published in 1801)

appendix 3

other proofs of the infinitude of primes

(requires a little bit of experience with mathematical notation and reasoning)

The proof we've just seen can be expressed compactly as follows. Suppose that $\{p_1, p_2, \ldots, p_k\}$ constitutes the complete set of primes, where $p_1 = 2 < p_2 = 3 < \cdots < p_k$. Then $p_1 \times p_2 \times \cdots \times p_k + 1$, a number too big to belong to the set, must be divisible by a prime – but it can't be divisible by any of $p_1, \ldots, p_k$. This contradicts the supposed completeness of our set, so no such set can exist.

Here's a nice variation on this. Again, supposing $\{p_1, \ldots, p_k\}$ to be a complete set of primes, let $n = p_1 \times p_2 \times \cdots \times p_k$. Now, $n - 1 > p_k$, so it's not in the list and must therefore be composite. This means that it must share a prime factor with n (which includes *all possible primes* in its factorisation). So the difference $n - (n - 1)$ must *also* be divisible by that prime (for if p divides a and p divides b, then p must divide $a - b$, by simple arithmetic). But $n - (n - 1) = 1$. Since no prime can divide 1, we've arrived at a contradiction of the supposed completeness of the set $\{p_1, \ldots, p_k\}$.

Another simple variant starts by assuming that there's some counting number n such that no prime is greater than n. We then consider $n! + 1$, where $n!$ is the "factorial" of n [that is, $n \times (n - 1) \times (n - 2) \times \cdots \times 3 \times 2 \times 1$, so, for example, $5! = 5 \times 4 \times 3 \times 2 \times 1 = 120$ and $7! = 7 \times 6 \times 5 \times 4 \times 3 \times 2 \times 1 = 5040$]. By the FTA, $n! + 1$ must have a prime factorisation. But the "+1" guarantees that it can't be divisible by any of $1, 2, \ldots, n$, contradicting our assumption. This means that the primes are "unbounded" and must continue indefinitely.

An inventive new proof was given by Filip Saidak of the University of North Carolina in 2005[1]. It's based on the fact that two consecutive counting numbers must be *relatively prime* – that is, they can't have a prime factor in common. For if we consider n and $n + 1$, there can be no prime number p such that p divides both n and $n + 1$. If it did, it would also divide the difference $(n + 1) - n$, but that's 1, and no prime can divide 1. So the number $n \times (n + 1)$

must have at least two different prime factors. We can use the same reasoning to deduce that because $n \times (n + 1)$ and $(n \times (n + 1)) + 1$ are relatively prime, $(n \times (n + 1))((n \times (n + 1)) + 1)$ must have at least *three* different prime factors. We can continue like this, forever producing counting numbers with at least 4, 5, 6,... different prime factors. This implies that there must be infinitely many prime numbers.

Christian Goldbach, who appeared in Chapter 6 in connection with the "Goldbach Conjecture", also discovered a proof of the infinitude of primes. Published in 1730, it involves a certain result concerning the relative primeness (lack of common prime factors) of *Fermat numbers*[2]. It's available at `http://primes.utm.edu/notes/proofs/infinite/goldbach.html` [also accessible as `tinyurl.com/6nxq9r`].

A proof which surprised everyone by making use of a most unlikely branch of mathematics – *topology* – was produced by Hillel (Harry) Furstenburg in 1955[3] (while still an undergraduate student). That can be seen at `http://www.cut-the-knot.org/proofs/Furstenberg.shtml` [also accessible as `tinyurl.com/5dqtzs`].

notes

1. F. Saidak, "A new proof of Euclid's theorem", *American Mathematical Monthly* 113 (2006) pp. 937–938

2. Fermat numbers are of the form 2 + 1 = **3**, 2 × 2 + 1 = **5**, 2 × 2 × 2 × 2 + 1 = **17**, 2 × 2 × 2 × 2 × 2 × 2 × 2 × 2 + 1 = **257**, 2 × 2 × 2 × 2 × 2 × 2 × 2 × 2 × 2 × 2 × 2 × 2 × 2 × 2 × 2 × 2 + 1 = **65537**, *etc.* Notice that the number of 2's getting multiplied is, in each case, itself a power of 2 (1, 2, 4, 8, 16, *etc.*).

3. H. Furstenberg, "On the infinitude of primes", *American Mathematical Monthly* 62 (1955) p. 353

appendix 4

formulas for primes

(requires a significant amount of experience with mathematical concepts and notation)

In 1947, W.H. Mills proved[1] that there must exist a constant A such that $[A^{3^n}]$ is prime for every counting number n. The square brackets here indicate that we round A^{3^n} down to its integer part. Unfortunately, though, to find the exact value of A, you must *already know the entire sequence of primes*. So, although interesting, this is not very helpful. A is known to begin 1.3064… Incidentally, it's possible to formulate this result entirely in terms of a pair of spirals – after reading Chapter 9, you might want to try this.

A few years later, along similar lines, E.M. Wright proved[2] the existence of a constant ω such that

$$\left[2^{2^{\cdot^{\cdot^{\cdot^{2^{\omega}}}}}}\right]$$

is a prime number for every positive number of 2's in this "tower of exponents". ω is known to be roughly 1.92878, but we have the same problem of exact determination, rendering this useless. And, in any case, the quantities involved grow so ridiculously fast that it would be hopelessly impractical to work with this.

The following formula for the nth prime was published by C.P. Willans in 1964[3]:

$$p_n = 1 + \sum_{m=1}^{2^n} \left[\left(\frac{n}{\sum_{j=1}^{m} \left[\cos^2 \left(\pi \times \frac{(j-1)!+1}{j} \right) \right]} \right)^{1/n} \right]$$

If you subtract one from the denominator (below the solitary "n"), you get the exact number of primes less than or equal to m. The formula is based on *Wilson's Theorem*, which is explained in Appendix 5. The first "Σ" indicates a sum from 1 to 2^n, the second a sum from 1 to m. The square brackets indicate that we take the integer part, "cos" is the trigonometric cosine function, $\pi = 3.14\ldots$ and the "!" indicates a *factorial* (see Appendix 3). Wilson's Theorem is hopelessly inefficient due to the presence of this factorial, so the above formula is amusing at best.

J.M. Gandhi provided an "iterative" formula for the nth prime number in 1971[4]. The idea is that if we already know the first $n-1$ primes $p_1, p_2, \ldots, p_{n-1}$, then we can use his formula to find p_n. We define the product $P_{n-1} = p_1 \times p_2 \times \cdots \times p_{n-1}$ and then

$$p_n = \left[1 - \frac{1}{\log 2} \times \log \left(-\frac{1}{2} + \sum_{d|P_{n-1}} \frac{\mu(d)}{2^d - 1} \right) \right]$$

Here, the "Σ" indicates a sum over all divisors of P_{n-1}, log is a logarithm of any base, the square brackets again indicate "integer part" and $\mu(d)$ is the *Möbius function* (nothing to do with the *Möbius strip*, which you may have heard of) which is defined as follows:

$\mu(1) = 1$,

$\mu(d) = -1$ if d is the product of an odd number of *distinct* prime factors,

$\mu(d) = 1$ if d is the product of an even number of *distinct* prime factors,

$\mu(d) = 0$ if the factorisation of d contains any repeated prime factors.

Equivalently, we have

$$p_n = \left[1 - \log_2\left(\frac{1}{2} + \sum_{r=1}^{n} \sum_{1 \le i_1 < \cdots < i_r \le n} \frac{(-1)^r}{2^{p_{i_1} \times \cdots \times p_{i_r}} - 1}\right)\right]$$

where log_2 is a base-two logarithm. In theory, either of these formulas could be exploited by a computer to generate primes endlessly. Before long, though, the computations would slow down to such an extent that the exercise would become pointless. Eratosthenes' sieve provides a much more efficient way to produce a complete list of primes less than some given number.

Also in the early 1970s, it was proved[5] that the set of prime numbers is *diophantine*. This means that there must exist a polynomial with integer coefficients such that the set of positive values taken by the polynomial, as the variables range over all possible combinations of non-negative integers, is exactly the set of prime numbers.

A few years later[6], an actual polynomial was found which fulfils this condition:

$$(k+2)\{1 - [wz + h + j - q]^2 - [(gk + 2g + k + 1)(h + j) + h - z]^2 - [2n + p + q + z - e]^2$$
$$- [16(k+1)^3(k+1)(n+1)^2 + 1 - f^2]^2 - [e^3(e+2)(a+1)^2 + 1 - o^2]^2 - [(a^2-1)y^2 + 1 - x^2]^2$$
$$- [16r^2y^4(a^2-1) + 1 - u^2]^2 - [((a + u^2(u^2-a))^2 - 1)(n + 4dy)^2 + 1 - (x + cu)^2]^2 - [n + l + v - y]^2$$
$$- [(a^2-1)l^2 + 1 - m^2]^2 - [al + k + 1 - l - i]^2 - [p + l(a - n - 1) + b(2an + 2a - n^2 - 2n - 2) - m]^2$$
$$- [q + y(a - p - 1) + s(2ap + 2a - p^2 - 2p - 2) - x]^2 - [z + pl(a - p) + t(2ap - p^2 - 1) - pm]^2\}$$

Conveniently, this polynomial has 26 variables, so the English alphabet can be used in its entirety (although, unhelpfully, "*o*" looks a bit like "0", "*l*" like "1" and "*x*" like "×"). It's of degree 25, meaning that the highest combined power involved is 25. Remarkably, any *positive* value (we must ignore negative values) produced by substituting either zero or a counting number for each of *a*, *b*,…, *z* is necessarily a prime number, and *every* prime can be produced in this way.

Polynomials with fewer variables have been found which do the same job, but as the number of variables is reduced, the degree necessarily increases. For example, a polynomial with 12 variables is known, but its degree is 13697. Similarly, we can bring the degree down, but only by introducing more variables: there's a degree 5 polynomial with 42 variables known to have this prime-generating property.

The existence of these polynomials isn't quite as interesting as it might first seem, though, since it has been proved that similar "generating polynomials" can be found for *many* sequences of counting numbers. In other words, this kind of formula is not particularly relevant to the essential "primality" of the primes.

Paolo Ribenboim dedicates an entire chapter of his *The New Book of Prime Number Records* (Springer, 1966) to the issue of formulas and functions for producing prime numbers. Many other references can be found here: `http://mathworld.wolfram.com/PrimeFormulas.html` [also accessible as `http://tinyurl.com/64agsn`].

notes

1. W.H. Mills, "A prime-representing function", *Bulletin of the American Mathematical Society* 53 (1947) p. 604
2. E.M. Wright, "A prime-representing function", *American Mathematical Monthly* 58 (1951) pp. 616–618
3. C.P. Willans, "On formulae for the *n*th prime number", *The Mathematical Gazette* 48 (1964) pp. 413–415
4. J.M. Gandhi, "Formulae for the *N*th prime", *Proceedings of the Washington State University Conference on Number Theory* (WSU, 1971) pp. 96–107. Also, see C.V. Eynden, "A Proof of Gandhi's Formula for the *n*th Prime", *American Mathematical Monthly* 79 (1972) p. 625 and S.W. Golomb, "A direct interpretation of Gandhi's formula", *American Mathematical Monthly* 81 (1974) pp. 752–757.
5. Y.V. Matiyasevich, "A Diophantine representation of the set of prime numbers" [Original Russian language article: *Doklady Akademii Nauk SSSR* 196 (1971) pp. 770–773; English translation: *Soviet Mathematics – Doklady* 12 (1971) pp. 249–254]
6. J.P. Jones, D. Sato, H. Wada and D. Wiens, "Diophantine representation of the set of prime numbers", *American Mathematical Monthly* 83 (1976) pp. 449–464

appendix 5

techniques for testing primality

(fairly easy to follow if you can deal with a little bit of mathematical notation)

There is, in fact, a very simple test to determine whether or not a counting number is prime. It's called *Wilson's Theorem*:

p is a prime number if and only if $(p - 1)! + 1$ *is divisible by p.*

Here "$(p - 1)!$" is the "factorial" of $p - 1$, that is, $(p - 1) \times (p - 2) \times \cdots \times 3 \times 2 \times 1$. So, for example, $5! = 5 \times 4 \times 3 \times 2 \times 1 = 120$ and $7! = 7 \times 6 \times 5 \times 4 \times 3 \times 2 \times 1 = 5040$. Here's how the theorem works:

$(2 - 1)! + 1 = 1! + 1 = 1 + 1 = 2$, which *is* divisible by 2

$(3 - 1)! + 1 = 2! + 1 = 2 + 1 = 3$, which *is* divisible by 3

$(4 - 1)! + 1 = 3! + 1 = 6 + 1 = 7$, which is *not* divisible by 4

$(5 - 1)! + 1 = 4! + 1 = 24 + 1 = 25$, which *is* divisible by 5

$(6 - 1)! + 1 = 5! + 1 = 120 + 1 = 121$, which is *not* divisible by 6

$(7 - 1)! + 1 = 6! + 1 = 720 + 1 = 721$, which *is* divisible by 7

$(8 - 1)! + 1 = 7! + 1 = 5040 + 1 = 5041$, which is *not* divisible by 8

$(9 - 1)! + 1 = 8! + 1 = 40320 + 1 = 40321$, which is *not* divisible by 9

$(10 - 1)! + 1 = 9! + 1 = 362880 + 1 = 362881$, which is *not* divisible by 10

$(11 - 1)! + 1 = 10! + 1 = 3628800 + 1 = 3628801$, which *is* divisible by 11

$(12 - 1)! + 1 = 11! + 1 = 39916800 + 1 = 39916801$, which is *not* divisible by 12

$(13 - 1)! + 1 = 12! + 1 = 479001600 + 1 = 479001601$, which *is* divisible by 13

$(14 - 1)! + 1 = 13! + 1 = 6227020800 + 1 = 6227020801$, which is *not* divisible by 14

…and so on.

This is all very satisfying, but, unfortunately, factorials grow so fast that before long we're having to check the divisibility of numbers far too big to handle (50! already has 65 digits, 60! has 82 – imagine what you'd get when considering the sorts of numbers we're actually interested in checking).

Note that C.P. Willans' useless formula in Appendix 4 which involved factorials was based directly on Wilson's Theorem.

Although it can't be expressed as a tidy result like Wilson's Theorem, the Sieve of Eratosthenes (see Chapter 3) provides a far more efficient "algorithm" for determining whether or not a number is prime. It's simply a matter of checking whether the number is divisible by any of the counting numbers from 2 up to the *square root* of the number in question (that is, all counting numbers bigger than 1 which when multiplied by themselves produce a result less than the number being checked). So, to find out whether 20187 is prime, we need to check its divisibility by each 2, 3, 4,..., 142 since $142 \times 142 = 20164$ but $143 \times 143 = 20449$. Of course, if we find that it's divisible by any of these numbers, we can stop checking.

There are shortcuts. To begin with, any counting number ending in 0, 2, 4, 6 or 8 will be divisible by 2, and any ending in 5 will be divisible by 5[1], so we can immediately eliminate all numbers which don't end in 1, 3, 7 or 9 (that's 60% of them). After that, we're still left with a situation where the bigger the number being checked, the longer the process will take. Once we get up into the region of 30-digit numbers, this kind of checking becomes uselessly slow. All "primality testing" efforts, then, are about developing new, more efficient algorithms. No one is seriously looking for a "magic formula" that will instantly reveal whether or not a given counting number is or isn't prime.

Algorithms for primality testing have evolved considerably since Eratosthenes' time. As of 2009, computer-testing a 100-digit number for primality now takes a matter of minutes. As computer power increases, this should reduce to seconds, then fractions of a second. The progress that has occurred is a combination of improved algorithms and faster computer processing. Primality-testing speeds would be nowhere near where they now are if we were still running old algorithms on our newest computers.

Centuries ago, the current situation was entirely unimaginable. Marin Mersenne, a French theologian, mathematician and music theorist with a highly developed interest in prime numbers, wrote in 1644 that "*... to tell if a given number of 15 or 20 digits is prime or not... all time would not suffice for the test, whatever use is made of what is already known*" [2]

A huge amount of work is going on in the development and improvement of primality testing algorithms. A good starting point for exploring this would be Chris Caldwell's webpages at `http://primes.utm.edu/prove/index.html` [also accessible as `tinyurl.com/683t8y`].

notes

1. This is assuming that we're representing our counting numbers in the usual base ten system – see Appendix 1 if you're unfamiliar with that.
2. M. Mersenne, *Cogitata Physico-Mathematica* (Latin original published in 1644)

appendix 6

percentages and percentage errors

(not too difficult to follow)

The word "percent" literally means "per 100" or "for every hundred" ("cent" comes from the Latin word for 100 – think of "century"). So 37% = 37 percent = "37 for every hundred". What's 37% of 500? We have five hundred, and so the answer should be *37 for each of these hundreds*, in other words: *five* 37's, or 5×37, which is 185: 37% of 500 = 185. What's 37% of 168? Here we don't have a whole number of 100's, but $168 = 1.68 \times 100$ – we have "one-and-a-bit" hundreds. We want 37 for each hundred, so that should be one-and-a-bit 37's, or 1.68×37, to be precise. That gives 62.16, so 37% of 168 is 62.16.

The rule we're following here is (1) divide by 100, then (2) multiply by the percentage.

$$500 \rightarrow 500 \div 100 = 5 \rightarrow 5 \times 37 = 185$$

$$168 \rightarrow 168 \div 100 = 1.68 \rightarrow 1.68 \times 37 = 62.16$$

We can think of this as dividing our number into 100 equal pieces, then taking 37 of these pieces – "37 for every hundred".

Exactly the same process can be used to find, say, 0.003% of 2.17:

$$2.17 \rightarrow 2.17 \div 100 = 0.0217 \rightarrow 0.0217 \times 0.003 = 0.0000651.$$

In this situation, we're dividing 2.17 into 100 equal pieces (0.0217 each), then taking 0.003 of these pieces (that's a lot less than a single piece – in fact it's three-thousandths of a piece, hence the tiny result).

50% and 25% are familiar from everyday use: 50% is "50 for every 100" which is the same as "1 for every 2", in other words *one-half*; 25% is "25 for every 100" which is the same as "1 for every 4", in other words *one-quarter*.

Suppose we estimate tomorrow's maximum temperature to be 38.2° and it turns out to be 41.5°. How wrong were we?

One way of describing our error is simply to take the difference (the *absolute error*)

$$41.5^\circ - 38.2^\circ = 3.3^\circ.$$

In some situations, we want to express this error as a percentage of the actual (that is, the correct) figure. So what percentage of 41.5 is 3.3? To find out, we divide the error by the correct figure, and then multiply by 100, that is...

$$(3.3 \div 41.5) \times 100 = (0.0795\ldots) \times 100 = 7.95\ldots$$

...so that's just under 8% error.

Notice that we can check this by applying the method described above – to find 7.95% of 41.5, we divide 41.5 by 100, then multiply by 7.95:

$$41.5 \rightarrow 41.5 \div 100 = 0.415 \rightarrow 0.415 \times 7.95 = 3.3^{[1]}.$$

This confirms that 7.95% of 41.5 is indeed 3.3, the number of degrees our prediction differed from the actual temperature.

note

1. If you check, you'll find that it's actually 3.29925, but that's just because 7.9518072 has been rounded down to 7.95.

appendix 7

sizes of numbers needed to achieve various levels of PNT accuracy

(requires a significant amount of experience with mathematical notation and reasoning)

In 1962, J.B. Rosser and L. Schoenfeld proved[1] that the number of primes less than or equal to x is less than

$$\frac{x}{\ln x}\left(1+\frac{3}{2\ln x}\right).$$

Here, "$\ln x$" denotes the natural ("base-e") logarithm of x, and this result comes with the (perfectly reasonable) condition that x must be greater than 1.

The usual notation for "the number of primes less than or equal to x" is "$\pi(x)$". This has nothing to do with $\pi = 3.14\ldots$, though – it was adopted because π is the Greek equivalent of "p" (for "prime"). So, for x greater than 1, we can write

$$\pi(x) < (x/\ln x)\left(1+\frac{3}{2\ln x}\right).$$

We can then apply a standard set of algebraic manipulations:

$$1 < \frac{x/\ln x}{\pi(x)}\left(1+\frac{3}{2\ln x}\right)$$ [divide both sides by $\pi(x)$],

$$\frac{x/\ln x}{\pi(x)} > \frac{1}{1+\frac{3}{2\ln x}}$$ [divide both sides by $1+\frac{3}{2\ln x}$, reverse the inequality],

$$1 - \frac{x/\ln x}{\pi(x)} < 1 - \frac{1}{1 + \dfrac{3}{2\ln x}}$$

[subtract both sides from 1, switch the “>” to “<”],

$$\frac{\pi(x)}{\pi(x)} - \frac{x/\ln x}{\pi(x)} < \frac{1 + \dfrac{3}{2\ln x}}{1 + \dfrac{3}{2\ln x}} - \frac{1}{1 + \dfrac{3}{2\ln x}}$$

[rewrite the 1 on each side],

$$\frac{\pi(x) - x/\ln x}{\pi(x)} < \frac{\dfrac{3}{2\ln x}}{1 + \dfrac{3}{2\ln x}} = \frac{\dfrac{3}{2\ln x}}{\dfrac{2\ln x}{2\ln x} + \dfrac{3}{2\ln x}}$$

$$= \frac{\dfrac{3}{2\ln x}}{\dfrac{2\ln x + 3}{2\ln x}} = \frac{3}{2\ln x + 3}$$

[combine numerators, then rewrite 1 in right-hand denominator and tidy up],

$$100\left(\frac{\pi(x) - x/\ln x}{\pi(x)}\right) < \frac{300}{2\ln x + 3}$$

[multiply both sides of the inequality by 100].

The left-hand side is the percentage error associated with the function $x/\ln x$ which approximates $\pi(x)$. Therefore, to guarantee that we have less than 10% error, we just need to be sure that

$$\frac{300}{2\ln x + 3} < 10.$$

That is,

$$300 < 20 \ln x + 30$$

$$270 < 20 \ln x$$

$$\ln x > 270/20 = 13.5$$

$x > e^{13.5}$ (recalling that $y = \ln x$ means that $x = e^y$).

$e^{13.5}$ is approximately 729416, where $e = 2.718...$

Similarly, to guarantee less than 5% error, we need

$$\frac{300}{2 \ln x + 3} < 5$$

...or, equivalently...

$$300 < 10 \ln x + 15$$

$$10 \ln x > 285$$

$$\ln x > 28.5$$

$$x > e^{28.5}.$$

$e^{28.5}$ is approximately 2.38×10^{12}, a 13-digit number – so anything with more than 13 digits will guarantee less than 5% error.

The same reasoning process leads us to the conclusion that to guarantee less than 1% error, we require $x > e^{148.5}$, which is approximately 3.11×10^{64}, a 65-digit number.

The general result here is that to achieve less than k% error, you need $x > e^{150/k - 1.5}$.

Since $e^{150/k - 1.5} = 10^{150/k \ln 10 - 1.5/\ln 10}$, anything with more than

$$\left[\frac{150}{k \ln 10} - \frac{1.5}{\ln 10} + 1\right] = \left[\frac{65.144...}{k} + 0.348...\right]$$

digits (where square brackets indicate that we round down to the integer part) will suffice.

Examining tables of $\pi(x)$ values, it's evident that 5% error is achieved somewhere between 10^9 and 10^{10}, so anything more than *ten* digits will suffice. That's a thousand times better than the result deduced above. It's certain that the 1% threshold will be crossed *well* before we get into the 65-digit range, but data isn't currently available for such large numbers[2]. This approach to error estimation is not about obtaining "best possible" results (that would require some much more sophisticated mathematics) but rather about "playing it safe" and using relatively simple tools to achieve absolute certainty.

notes

1. J.B. Rosser and L. Schoenfeld, "Approximate formulas for some functions of prime numbers", *Illinois Mathematics Journal* 6 (1962) pp. 64–94

2. As of 2009, $\pi(x)$ data is available up to 4×10^{22}.

appendix 8

improved estimates for the prime counting function

(not too difficult to follow, if you have the patience – no specialised notation involved)

We're going to construct the graphs of a couple of functions which improve on the approximation to the prime counting function which we've seen (dividing each number by its logarithm, or by its logarithm minus 1). They both involve logarithms and so can ultimately be described entirely in terms of spirals, even if that description ends up being quite long and involved.

The first one's not *too* complicated.

To construct the graph we want, we'll first need to build a graph which shows what you get when you divide 1 by the logarithm of a number:

For any number greater than 1, find the corresponding point on the number line. Consider a circle centred at 0 which passes through the point.

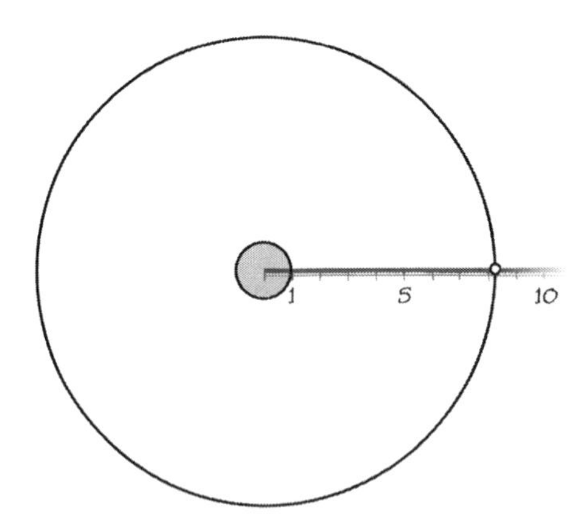

In this example the chosen number is 8.25. Count the number of coils of the "base-e" spiral between that circle and the "unit circle" shown (shaded).

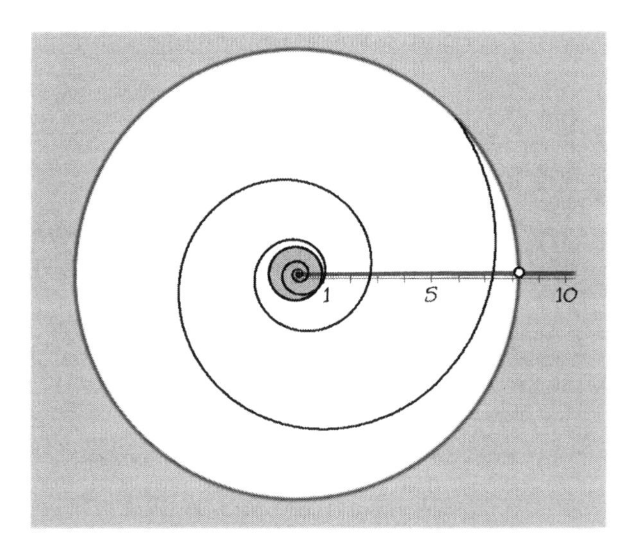

In this example it's a bit more than 2.11. We now want to divide 1 by that amount. A pocket calculator will happily oblige, but we can also carry this out graphically, as follows. The amount of coils corresponds to another point on the number line, and we draw a vertical segment of height 1 up from this point. Now join the top of that to the 0 point on the number line, making a triangle, and draw another vertical segment up from the 1 point to meet the sloping line.

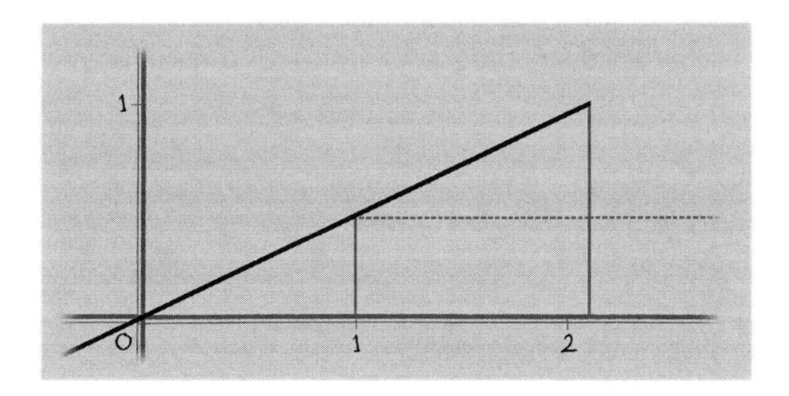

The height of the latter segment is *1 divided by the logarithm of the number we started with* (1 divided by 2.11…, or about 0.474 in this example). Shift the second (left) segment along to the position of our original number (8.25) and mark a point at its top. So, in this example, 8.25 has been transported by the function to a number a bit less than 0.5.

Doing this for all numbers bigger than 1, the set of points we get looks like this:

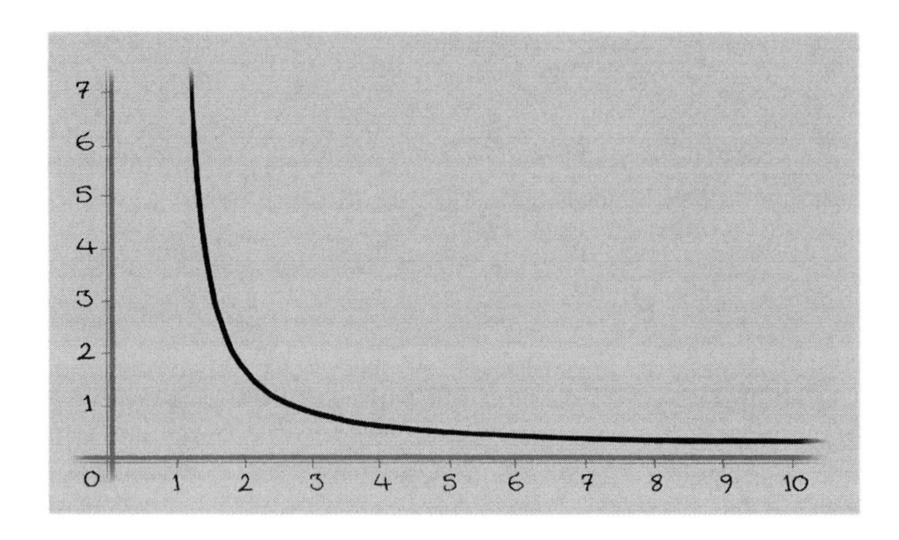

The farther out we go, the more coils there are to be counted. Dividing 1 by ever larger numbers results in ever *smaller* numbers, which explains the dwindling of this graph.

The graph has a property very directly related to equiangular spirals. That is, if we draw the "base-e" spiral centred at 0, it will cross the number line at $e = 2.718\ldots$, $e \times e = 7.389\ldots$, $e \times e \times e = 20.085\ldots$, $e \times e \times e \times e = 54.598\ldots$, *etc.* At these points, the graph's height will be, respectively, 1, 1/2, 1/3, 1/4, *etc.* The spiral also crosses the number line at 1, of course, where the graph effectively becomes infinite (this is related to the issue of dividing 1 by 0).

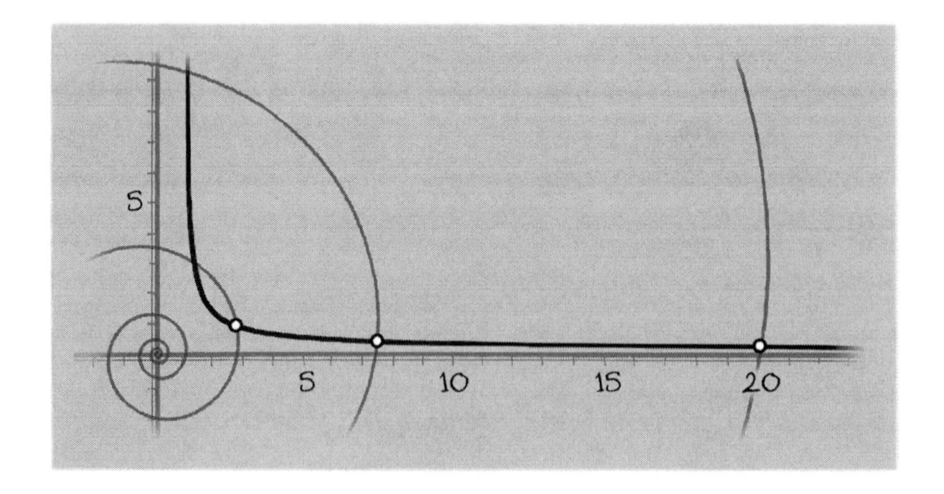

The idea now is to use the graph we've just built in order to build a second graph (which will represent the first of our improved approximations). We're going to start by working with just those points to the right of 2 on the number line. Pick any such point (corresponding to a number greater than 2) – in this example we'll again use 8.25. Now draw vertical segments up from 2 to the graph and up from the chosen point to the graph:

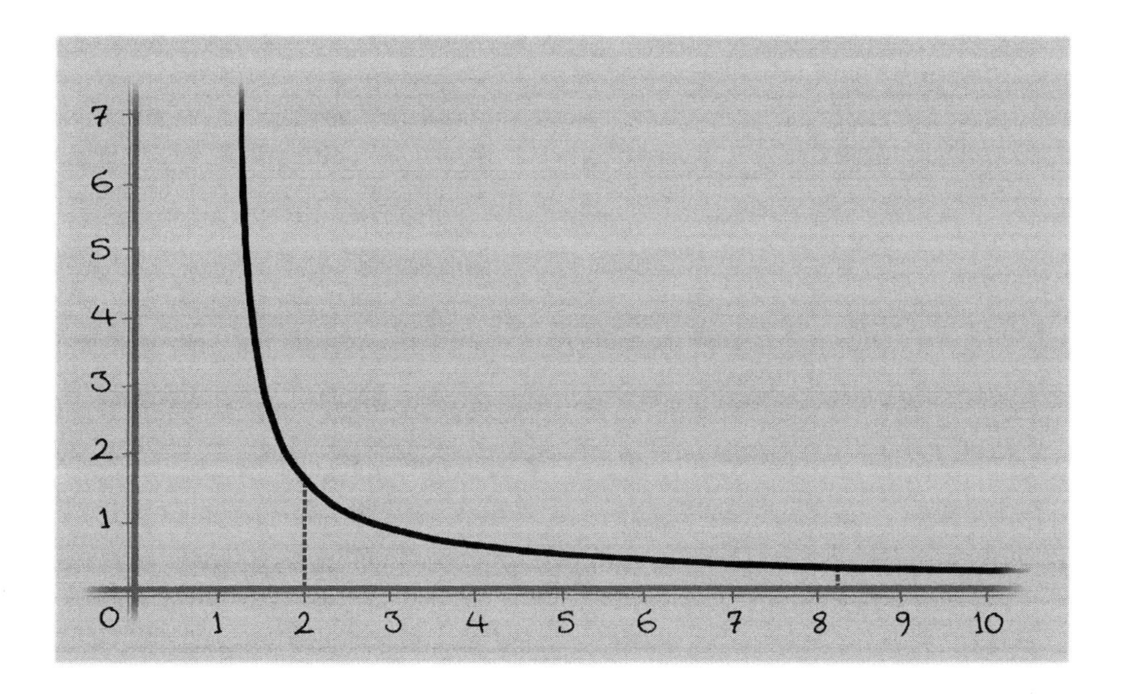

Imagine we wanted to paint the area enclosed between the two vertical segments, the curve and the number line. How much paint would we need? How would you measure that? The obvious measure would be in terms of "squares", the sides of which correspond to the distance from 0 to 1 on the number line. How many squares-worth of paint would be needed?

Using standard calculus, it's possible to work out the following:

1 square of paint will cover the area from 2 out to 2.872...

2 squares of paint will cover the area from 2 out to 4.108...

3 squares of paint will cover the area from 2 out to 5.687...

4 squares of paint will cover the area from 2 out to 7.572...

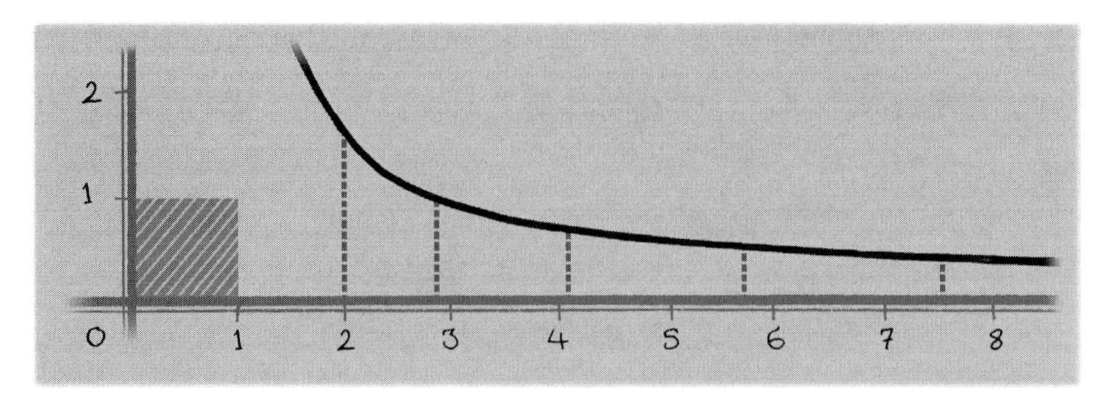

All of the subdivisions of the space under the graph between consecutive dotted vertical segments have an area of one square unit, equivalent to that of the shaded square region.

So, for every number bigger than 2, the area under the graph between 2 and that number can be expressed as an amount of squares. This won't usually be a counting number – for example, between 2 and 8.25, the number of squares of area under the graph will be just under 4.328.

Above each point on the number line to the right of 2, then, we can place a point at that height. The farther along we go, the more area there will be under the first graph, so we can expect our points to get successively higher in the new graph. But because the original graph dwindles towards zero, we can't expect them to get higher too quickly (you need to travel a bit farther each time to get another "squares-worth" of area). What we get looks like this:

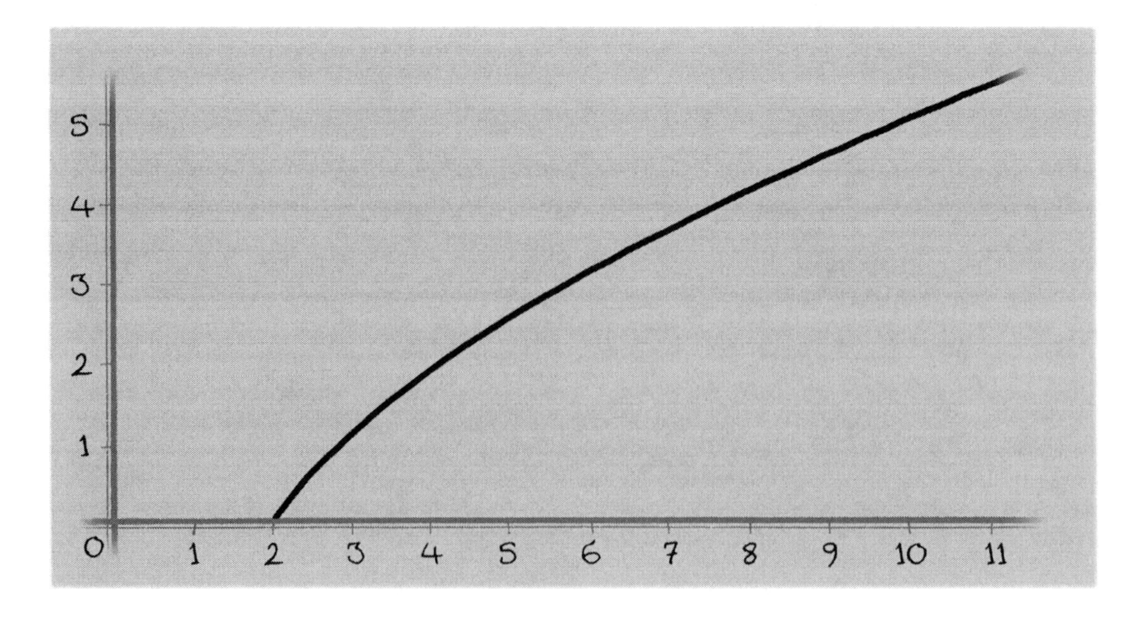

Ultimately this graph can be traced back to the geometry of equiangular spirals, although it's now several steps removed.

The function that this graph describes is called "Li" (which stands for "logarithmic integral"). It acts as a much better approximation of the prime counting function than what we get by dividing each number by its logarithm (or even by its logarithm minus 1). Carl Gauss was originally responsible for this idea, a refinement of his original guess.

There's a closely related function, just slightly different, called "li". You may have wondered why we started from 2 when measuring areas under our first graph. There's not a particularly good answer for that – the history of the "Li" and "li" functions is somewhat obscure. All you need to know is that the first graph can be extended in a mathematically coherent way to cover everything to the right of 0, like this:

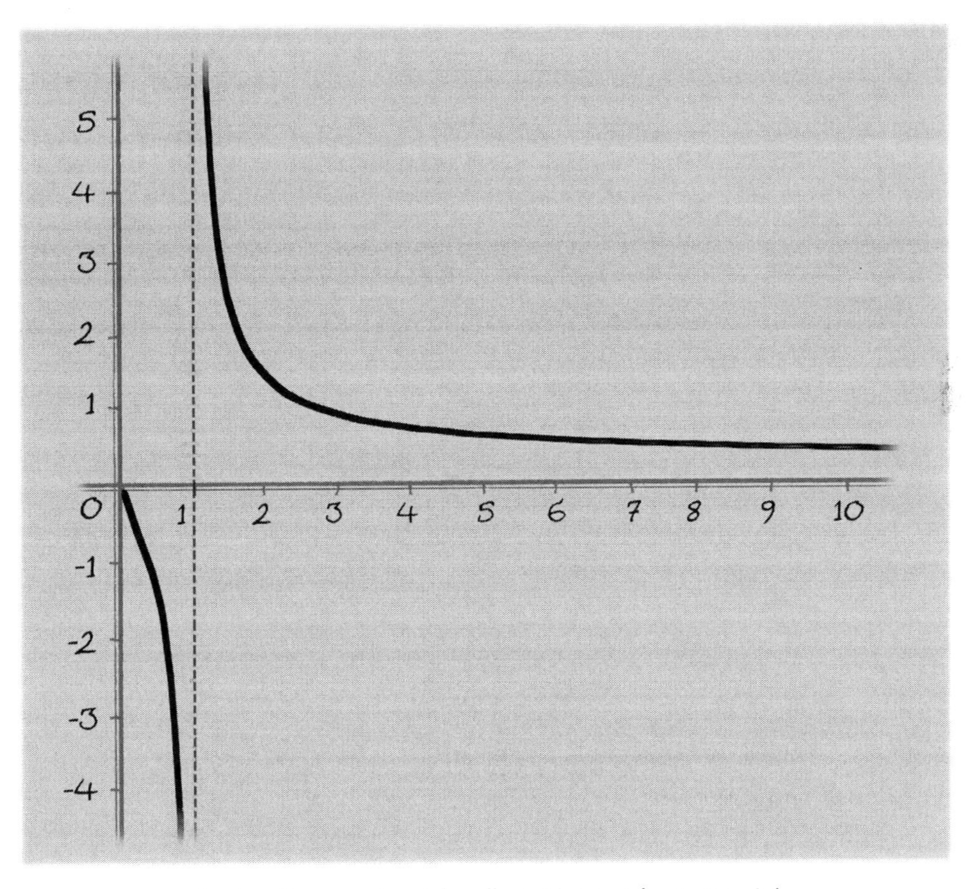

The function is defined for all positive numbers except 1.

We can then measure areas enclosed between this graph and the number line starting from 0, rather than having to start from 2. It's a bit tricky, as some of the graph is below the number line, which means we'll have *negative areas* involved. Fortunately, the positive and negative areas cancel out nicely, and the second graph we constructed can, in this way, be extended so that it also covers the 0–2 range, like this:

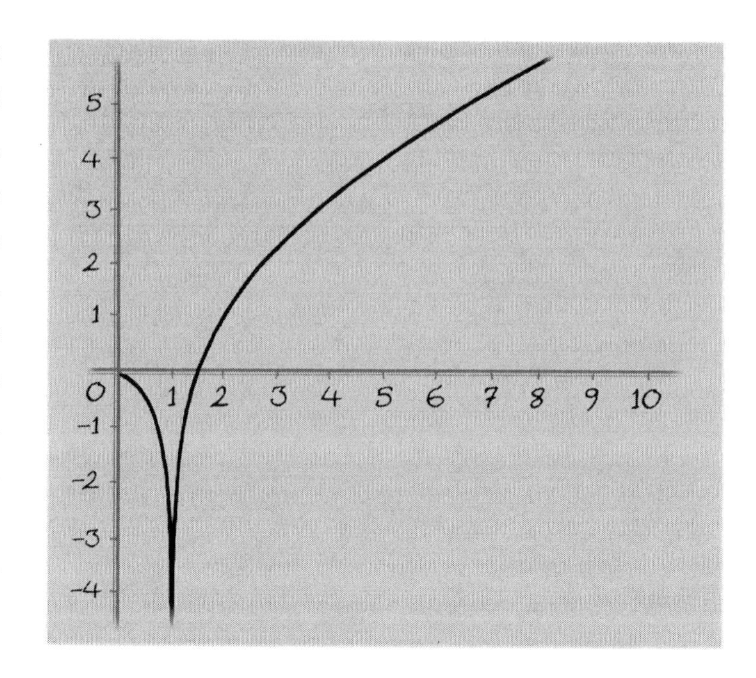

For that part of the number line to the right of 2, the "li" graph is identical to the "Li" graph apart from the fact that it's shifted up by 1.045… (which is what you get when you measure the area under the graph – positive and negative – between 0 and 2).

The following table provides a comparison between the actual prime count, the improved approximation from Chapter 10, and the "li" approximation (numbers are here rounded to the nearest counting number):

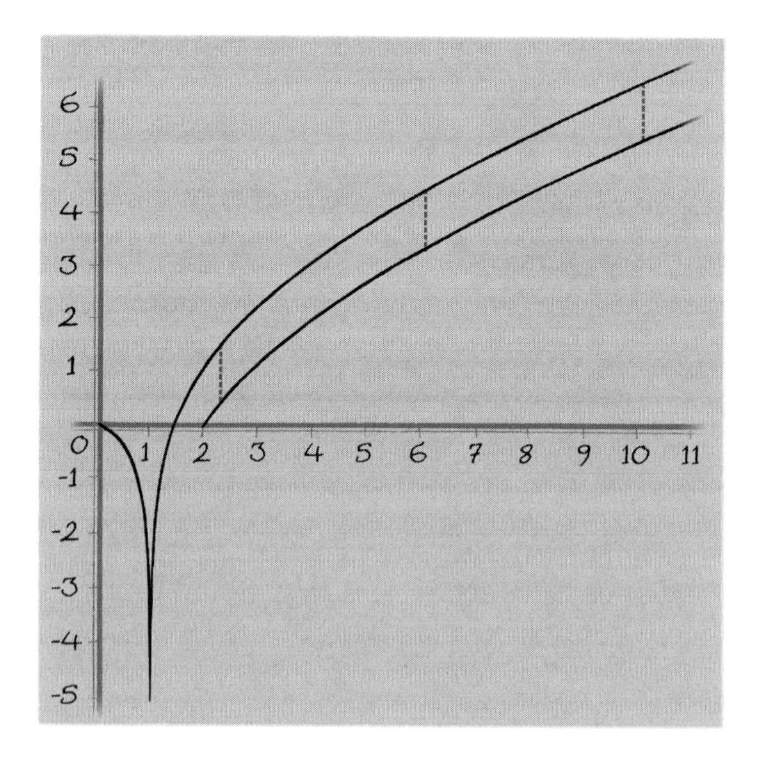

number	primes less than or equal to the number	the number divided by its logarithm	the number divided by (its logarithm − 1)	'li' function approximation for primes less than the number
10	4	4	8	6
100	25	22	28	30
1000	168	145	169	178
10000	1229	1086	1218	1246
100000	9592	8686	9512	9630
1000000	78498	72382	78030	78628
10000000	664579	620421	661459	665918
100000000	5761455	5428681	5740304	5762209
1000000000	50847534	48254942	50701542	50849235
10000000000	455052511	434294482	454011971	455055615

But we can do even better. Bernhard Riemann, a mathematician we'll meet in Chapter 14, realised that the "li" function is actually a much better approximation than it initially appears if we interpret it as counting not just prime numbers but *both primes and their powers* (where $p \times p$ counts as half a prime, $p \times p \times p$ as a third of a prime, $p \times p \times p \times p$ as a quarter of a prime, *etc.*)

Based around this kind of thinking, we can build a new graph representing a function which approximates a count of *just the primes* (and *not* their powers). The idea is to "stretch and compress" a sequence of copies of the "li" graph, and then combine all of these into a "master graph".

We start with the "li" graph itself. Left unstretched and uncompressed, it corresponds to the counting number 1.

Then, for each of the rest of the counting numbers 2, 3, 4, 5,…, we carry out a set of instructions. First, we factorise the number. If it contains any repeated prime factors, we ignore it and move on to the next number. Otherwise, we'll add a graph to our collection. We start with the "li" graph and "compress" it by dividing the height of each point by our counting number. So, for 2 we halve the height of each point on the "li" graph, for 3 we reduce each height to a third, we skip 4 (it has a repeated factor, 2), for 5 we reduce each height to one-fifth, for 6 we reduce each height to one-sixth, and so on.

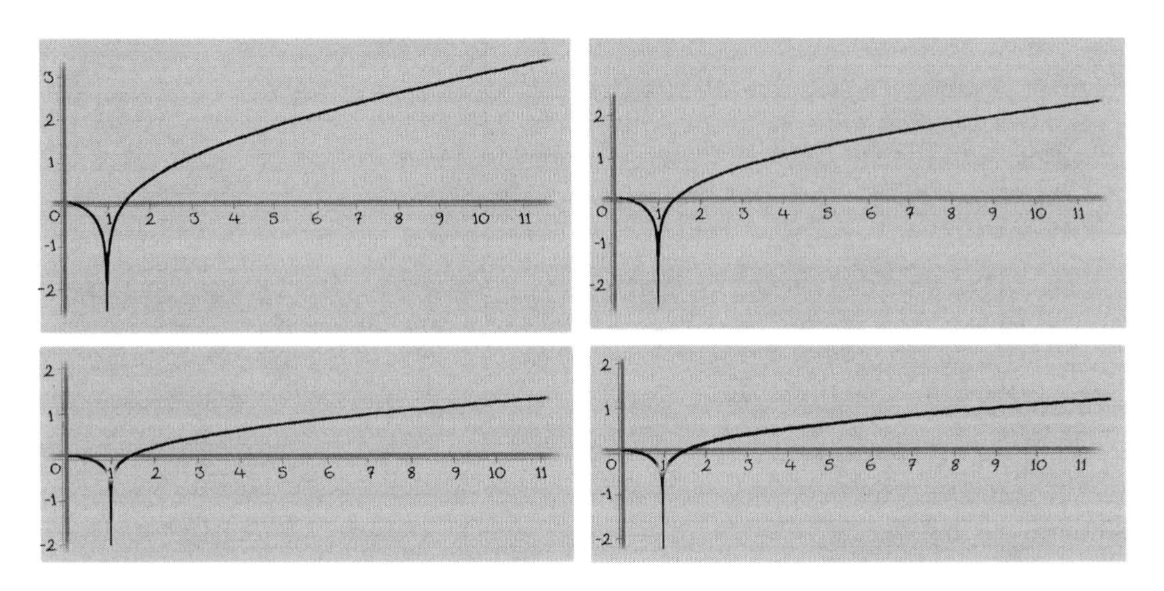

The upper-left graph shows the "li" graph compressed vertically and stretched horizontally by a factor of 2. In the upper-right graph, it's by a factor of 3, in the lower-left graph, it's by a factor of 5, and in the lower-right graph it's by a factor of 7.

That's the "compression". We're now going to stretch the number line (and the "li" graph with it) in a way where every number gets shifted along to its appropriate power (what you get by multiplying it by itself the appropriate number of times). So, if we're looking at 3, then 2 gets shifted to $2 \times 2 \times 2$, or 8; 5 gets shifted to $5 \times 5 \times 5$, or 125; a number like 1.7 similarly gets shifted along to $1.7 \times 1.7 \times 1.7 = 4.913$; the point corresponding to 1 stays put, as $1 \times 1 \times 1 = 1$ (the same will be true of 1 for any such stretch).

We can think of this "stretching" in terms of spirals. For each point on the number line, create a spiral with that number as its base (so it passes through 1, and its next crossing is that point). As we're currently considering the counting number 3, the point will get stretched to the *third* crossing point of its spiral (when we consider 5, each point will get stretched to the fifth crossing of its spiral, and so on).

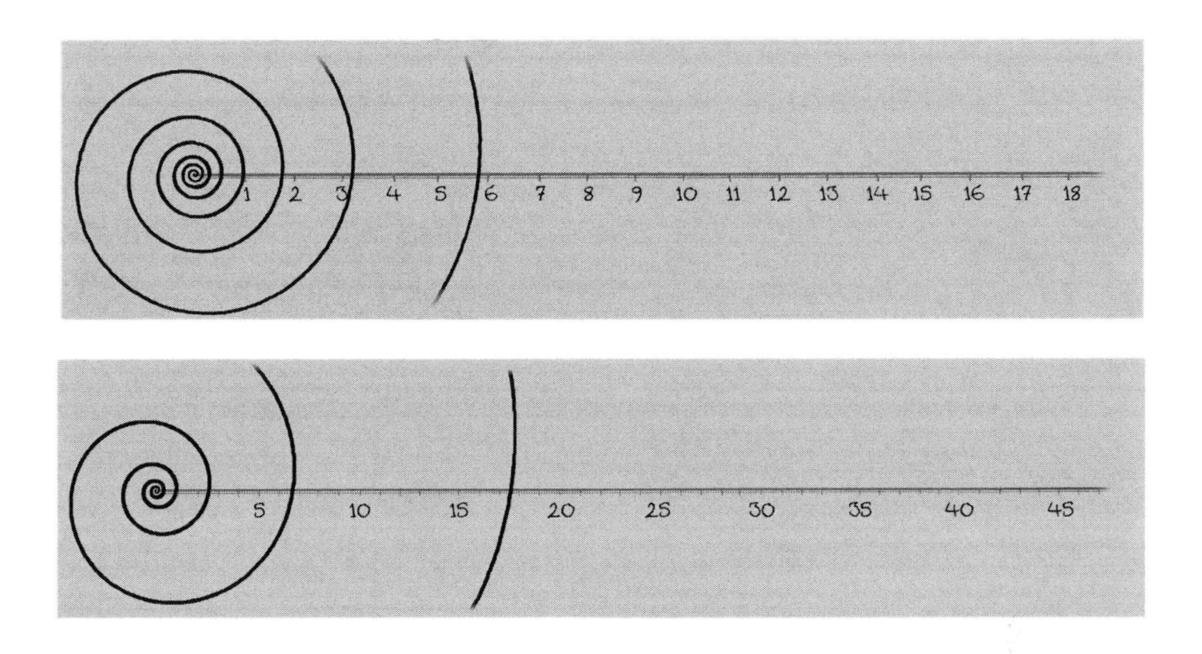

The examples here show how, when dealing with the counting number 3, the number 1.8 gets transported to $1.8 \times 1.8 \times 1.8$, or 5.832, and the number 2.6 gets transported to $2.6 \times 2.6 \times 2.6$, or 17.576.

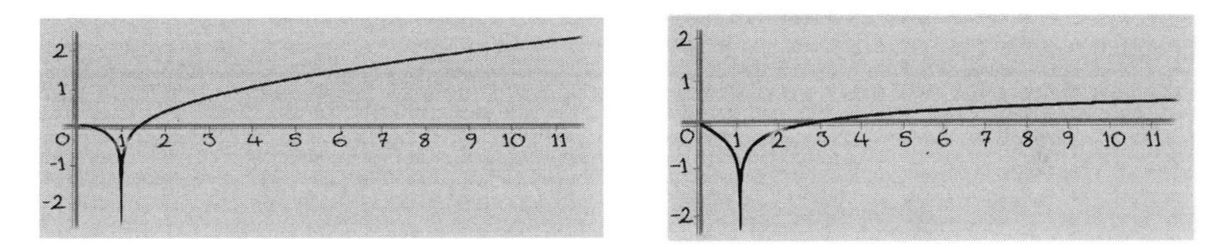

On the left, we see the "li" graph compressed vertically by a factor of 3. On the right we see the same thing, but further stretched horizontally by a power of 3 as was just described. Note that, although it looks (at this scale) like the second graph passes through the number line at 3, this isn't actually the case: at 3, the graph is 0.00823 units below the graph.

For each of 2, 3, 5, 6, 7, 10, 11, 13, 14, 15, 17, 19, 21, 22, 23, 26, 29, 30, 31, 33,... we'll get a "stretched-and-compressed" "li" graph to add to our collection (the collection began with the unstretched and uncompressed "li" graph). There's one more step: if the factorisation of the counting number has an odd number of factors – that includes all primes (which have just one factor, 1 being an odd number) but also numbers like $30 = 2 \times 3 \times 5$ and $87\,087 = 3 \times 7 \times 11 \times 13 \times 29$ – then we flip the graph across the number line[1]:

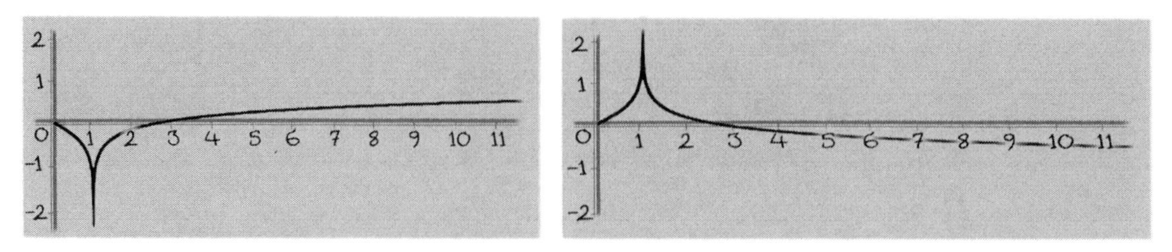

This is what happens for the number 3, which has an odd number (1) of prime factors.

We're now going to add together all of the functions which the resulting graphs represent. Remember, there will be infinitely many graphs in our collection, and I haven't really explained how you can add infinitely many functions together. Without worrying too much about the details, it works something like this:

For each point on the number line, we have a sequence of heights from the graphs corresponding to the counting numbers 1, 2, 3, 5, 6, 7, 10, 11, 13,... Some will be positive and others will be negative (sometimes because the graph got flipped, sometimes because the negative part of the graph got stretched to include the point in question). As the counting numbers grow, the associated graphs get more and more stretched and compressed, so the heights involved (positive or negative) become ever tinier. The net result of all this is that if we proceed to add the sequence of heights together, the running total gets ever closer to a fixed, well-defined, positive number[2]. Here's an example. If we choose 3.713 on the number line, the sequence of heights we get is 2.754, –0.468, –0.080, 0.096, –0.119, 0.129, –0.132, 0.130, 0.124. Adding all of these together, we get ever closer to the value 2.3088028... That will be the height above 3.713 of the "master graph" we're building here:

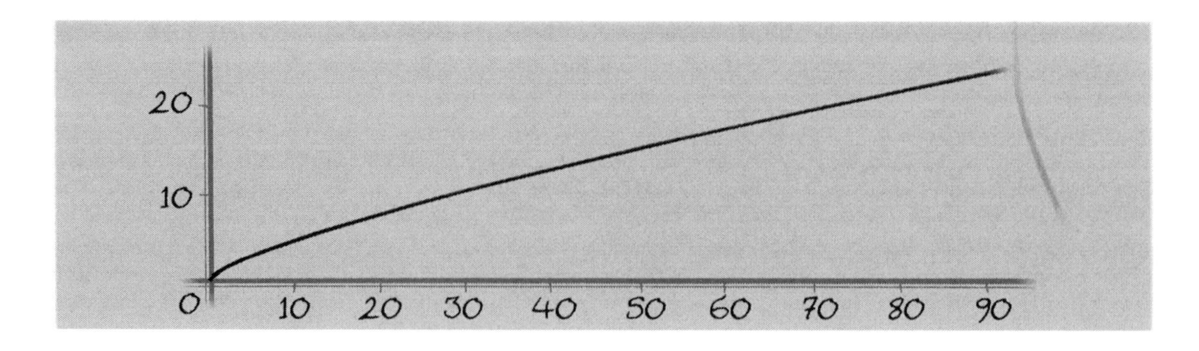

The function that this graph describes (sometimes called the *Riemann function*) provides an especially good approximation of the prime counting function.

number	primes less than or equal to the number	the number divided by its logarithm	the number divided by (its logarithm − 1)	'li' function approximation of primes less than or equal to the number	Riemann function approximation of primes less than or equal to the number
10	4	4	8	6	5
100	25	22	28	30	26
1000	168	145	169	178	168
10 000	1229	1086	1218	1246	1227
100 000	9592	8686	9512	9630	9587
1000 000	78 498	72 382	78 030	78 628	78 527
10 000 000	664 579	620 421	661 459	665 918	664 667
100 000 000	5761 455	5 428 681	5 740 304	5 762 209	5 761 552
1000 000 000	50 847 534	48 254 942	50 701 542	50 849 235	50 847 455
10 000 000 000	455 052 511	434 294 482	454 011 971	455 055 615	455 050 683

Note that table entries are rounded off to the nearest counting numbers.

notes

1. This is closely related to the *Möbius function* which makes an appearance in Appendix 4.
2. An infinite sum like this would be described by mathematicians as a *convergent series*.

appendix 9

constructing one of the "extra bits" to get added to the spiral waves

(not too difficult to follow, if you have the patience - no specialised notation involved)

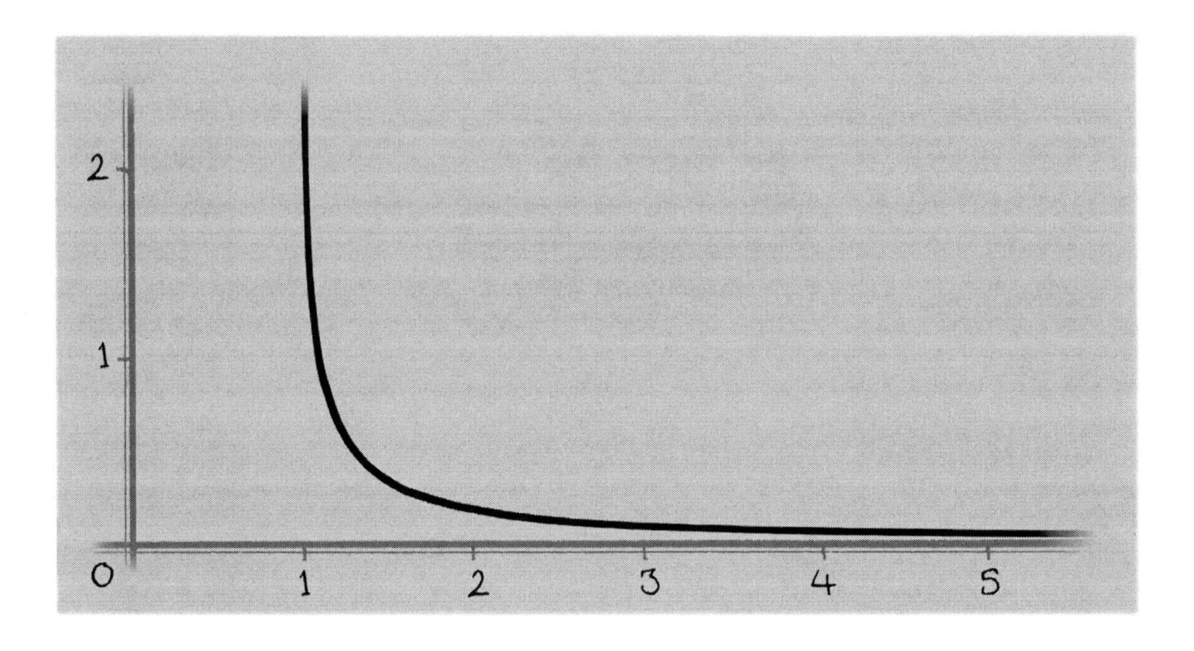

Remember that this graph represents a *function*. Given any point on the number line, it tells us where the function sends that point (the graph's height above the number line there). The function which *this* graph represents works as follows. Choose any point on the number line – let's say 3.71. Multiply it by itself (that gives 13.7641). Divide 1 by that (which gives approximately 0.07265). Now subtract this from 1 (to get about 0.9273) and take its natural logarithm. You'll find the result to be about –0.0754. Going back to the description of logarithms in terms of spirals and the counting of coils, you might just recall a brief remark I made that for numbers between 0 and 1, we get *inward* coils, counted as negative numbers.

Now take half of that −0.0754, giving −0.0377. Finally, drop the negative sign. So, the function sends 3.71 to approximately 0.0377, according to this long-winded "multiply the number by itself, divide 1 by the answer, subtract that from 1, find the logarithm, take half of that and drop the negative sign" procedure. Mathematicians would write this as

$$x \to -\frac{1}{2}\ln(1 - \frac{1}{x^2}).$$

If we make a mark 0.0377 units above the 3.71 point on the number line, then we'll find that this lies on the graph shown on the previous page. If we were to carry out this procedure for *every* point on the number line, then the set of all our marks would *be* this graph.

As it involves logarithms, this function can also be described in terms of spirals, as follows:

Given a point on the number line (to the right of 1), construct the unique equiangular spiral which passes through 1 and whose next crossing out along the number line is that point (so the number associated with our point acts as the "base" of the spiral – in this example, it's 4.85). Then, from 1, travel *inwardly* around the spiral (towards the centre), until you get to the *second* crossing of the number line.

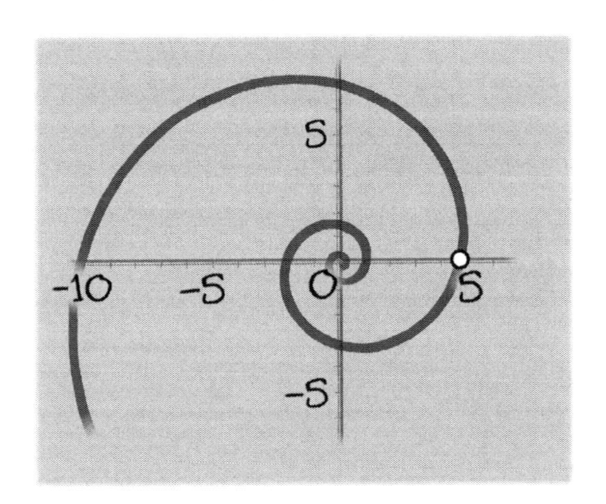

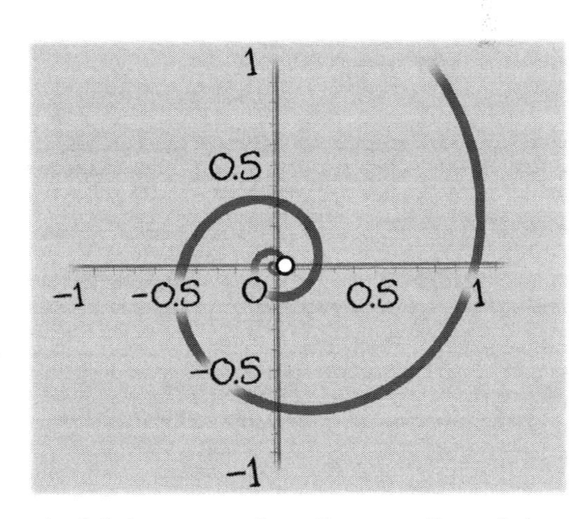

The image on the right is a close-up of the spiral on the left (compare where they pass through 1 on the horizontal axis). There's a crossing at 0.206... and another (marked) at 0.0425...

Now, flip the section of the number line between 0 and 1 around so you get a point the same distance from 1 as you just were from zero:

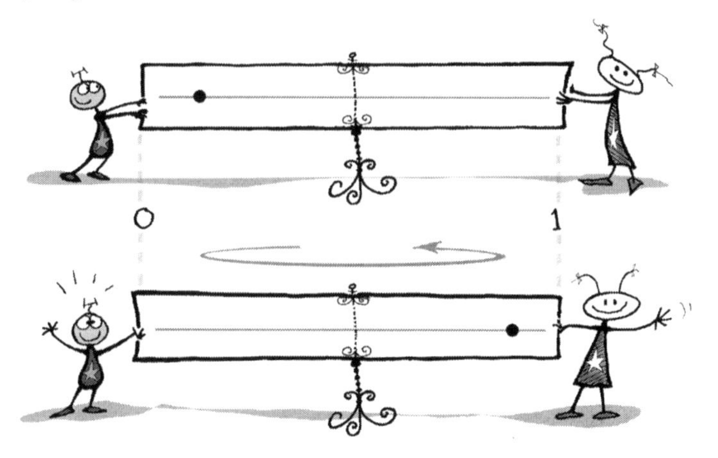

The idea here is that however far the point is to the right of 0, it should end up that far to the left of 1. In our case, 0.0425... ends up at 0.957... Basically, we're just subtracting the number from 1.

In this example, you get approximately 0.957. Next, draw a circle through that point, centred at 0. Draw another circle centred at 0, this time through 1. This produces a narrow ring:

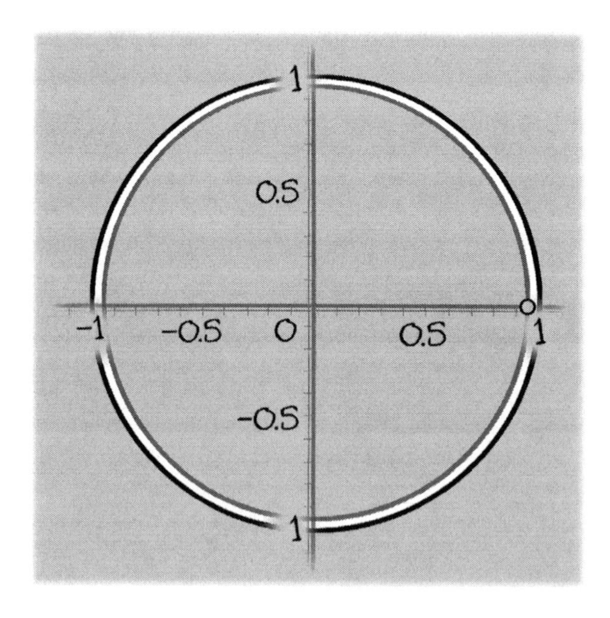

Now, forget the original spiral and draw the "base-e" spiral (the spiral associated with natural logarithms) over the top of these circles, and with the same centre.

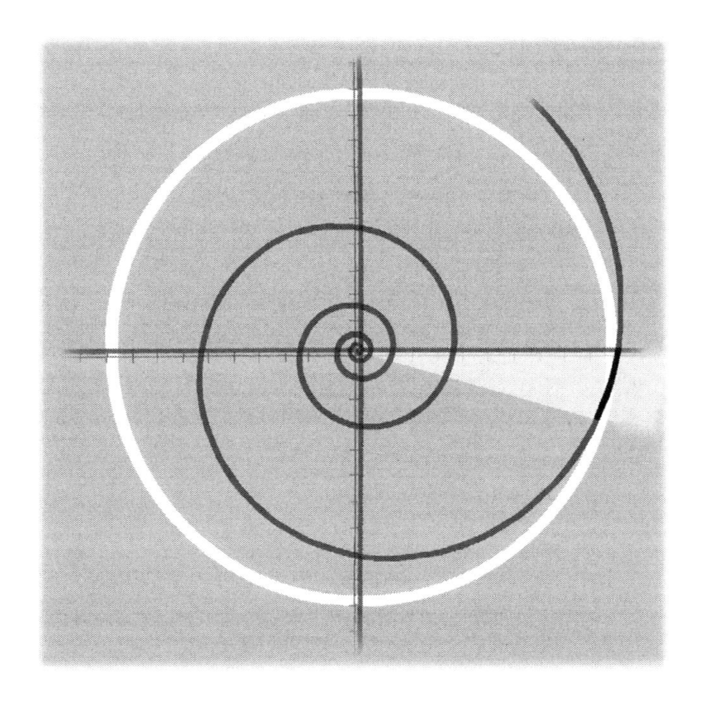

Count the coils of this spiral between the circles. Take half of that. That will be the height of the graph above the point you started with. In this example, the number of coils is about 0.0434 (that's roughly 1/23 of a coil), so the graph height at 4.85 should be about 0.0217.

What you'll find is that the farther out along the number line you choose your starting point, the more loosely wound will be the spiral that's produced, so the *closer* you'll be to 0 when you travel two coils in from 1. When you flip the section between 0 and 1, that means you'll end up very close to 1. In this way, the number of coils of the "base-e" spiral which are enclosed by the very narrow ring becomes a tinier and tinier fraction. When you take half of it, there's hardly anything left. This is why, as you proceed out along the number line, the graph of this particular function rapidly dwindles towards zero. Here it is again:

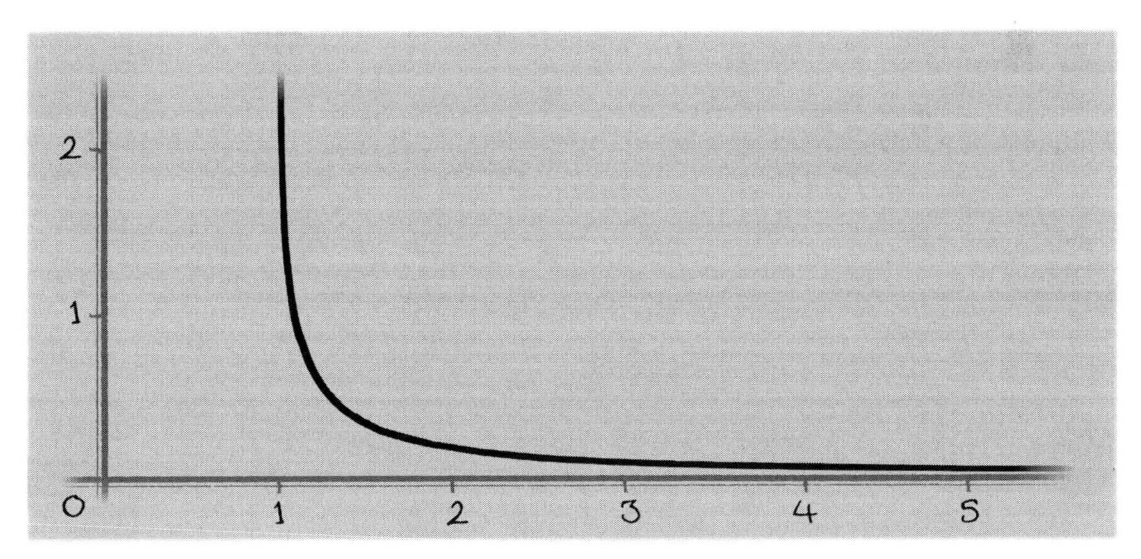

index

1, status of
analogies for 94, 95, 99, 118–9
as not composite 84
as not prime 74–5, 92–3
2, status of 73–4
absolute error
definition of 329
in prime count approximation 198, 207
acoustic science 256
additive partitions 61–3, 119
Aleff, H. Peter 228
Alexandria 70, 136
amplitude
concept applied to "spiral waves" 267, 269, 270–1, 273–4, 286
of a sine wave 254
ancient civilisations 27–9
Chinese 63, 298–9
Egyptian 63, 75
Greek 28, 33, 63–4, 88, 99
Indian 63
lack of awareness of primes 63–4, 80
numerals 75
archetypes, Jung's theory of 27–8
numbers as 28, 294
Archimedean spirals 167
area, multiplication explained in terms of 123
Arecibo message 112
"arithmetic", use of the word 89
atoms, analogy between primes and 96–9

Barden, Garrett 113
base notation, number 77, 312–15
base eight 312
base sixteen 314–5
base ten 77, 312–13, 315
base twelve 314
Bible, The 27, 75, 109
binary notation 75, 313–15
Blair, Tony 21
Bombieri, Enrico 264–5, 292, 308

Chebyshev, Pafnuty 240
China, ancient 63, 298–9
Christianity 21, 23, 24
codes
analogy with prime numbers 105–7, 156
genetic 114
prime-based, transmitted from Earth 112–14
collective unconscious, Jung's theory of the 27–8
composite numbers
analogies for 94–99
definition of 84
computers 18–19
limitations in finding primes 70, 323
used for finding primes 136, 146–7, 161, 200
used for testing "primality" 326–7
use of binary and hexadecimal numbers 313–14
consumerism 24, 25
Contact (novel and film) 111
counting numbers
definition of 38
Crossley, John 7, 33

Davis, Philip 30–1, 199
Dawkins, Richard 114
Dehaene, Stanislas 7
de la Vallèe-Poussin, Charles 201
"deviation function" 243–7
graphing 249–51
harmonic decomposition of 263–5
into "spiral waves" 274–80, 285, 288–90
the "other deviation", decomposing 282–3
diffusion theory (A. Seidenberg) 27
digital technology 18–20, 29
diophantine, proof that set of primes is 323
distribution of prime numbers 158–62
as a kind of "pattern" 162–5
illustrated via "staircase" functions 203–7
understood in terms of counting primes 190–1
divisibility 80, 85, 141
divisor 74, 141
DNA 107, 113, 114
Doxiadis, Apostolos 66, 158–9, 223
dozenalist movement 314–5
Drake, Frank 113–14

du Sautoy, Marcus 226, 309
Dudley, Underwood 149

e 178–82
base-*e* spiral 182–6
economics
as form of religion 23–5
as quantification 15–18, 30
use of logarithms in 187
education, experiences of mathematics in 32, 115, 174, 181
Egypt, ancient 63
numerals 75
Einstein, Albert 294
equiangular spirals 168–72
and distribution of primes 188–9, 192–4, 216–18, 232
and "spiral waves" 268–73
approximate, generated by primes 218–21
base-*e* 182
connection with logarithms 172–6, 188–9
"golden" 177
Eratosthenes, Sieve of 70, 157, 326
as a "formula" for primes 150–1, 323
Euclid of Alexandria
observation of unique factorisation 88
proof of infinitude of primes 136–44, 148, 201, 275
Euler, Leonhard 133, 151, 158, 159, 228
even numbers 73–4
extraterrestrial communication 110–14

factorial 319, 322, 325
factorisation 87–103
as alternative view of counting numbers 126–8
"cluster" representation of 117–19
role in building "primeness count" staircase 238–9
factors, prime 84–101
feelings about numbers 8–11, 76–7
feelings about prime numbers 107, 130, 156, 158–9
surprise 130, 225–6
Fermat number 320
Fibonacci sequence 153
financial instruments 17
formulas for primes 149–52, 321–24
foundations of mathematics 44, 123, 138, 310
Fourier analysis 261–3
Fourier decomposition 263
frequencies
concept applied to "spiral waves" 269–70, 281, 286
of sine waves 254–5
of "spiral waves" underlying number system 292–4
functions 208–9
adding 259
and graphs 209–15
Li 339–41
li 339–43, 345
Riemann 345
see also "deviation function"
Fundamental Theorem of Arithmetic (FTA) 89–92, 94
and status of 1 92–3, 119
proof of 316–18
role in proof of infinitude of primes 137, 140, 143–4
Furstenburg, Hillel (Harry) 320

galaxies, spiral 221
Gandhi, J.M. 322, 324
Gauss, Carl
proof of FTA 88, 137
and distribution of primes 200–1, 232–3
geometry 27, 123, 154–5
God 21, 24, 106, 158
Goldbach, Christian 133, 320
Goldbach Conjecture 131–4
golden mean 176–7
golden ratio 176–7
golden spiral 177
Granville, Andrew 79
graphs 208–15
adding functions in terms of 258–61, 275, 290
of "deviation function" 240–7, 249–51
of prime count staircase, 214–5, 224
of "primeness" count staircase 238–9, 243–4, 284–5, 289
Greece, ancient 28, 33, 63–4, 88, 99

Hadamard, Jacques 201
Hardy, Godfrey 76, 105, 133
harmonic decomposition 263
of distribution of primes 264–5, 274–93
harmonics 263
Hersh, Reuben 30–1, 199
hexadecimal notation 314–15
Hobbes, Thomas 25

India, ancient 63
infinitude of primes 135–48
proofs of 137–44, 319–20
infinity 37, 47, 146–8
integers, definition of 38

Ishango bone 63

Jackson, Allyn 79
Jung, Carl 27–8

Kauffman, Louis 123–4
Knauf, Andreas 94, 129–30
Kreisel, Georg 45

Law of Large Numbers 22
Legendre, Adrien-Marie 200–1, 232–3
Li (function) 339–40
li (function) 339–43, 345
logarithms 174–6, 187
  base-*e* 182, 184
  base of 175
  natural 178–184
  relating addition and multiplication 188–9
  role in Prime Number Theorem 192–4
  used in counting of primes 234–40

materialism 24–5
Mathematical Intelligencer, The 225
"mathematization" of society 30–3
measurement 13–16, 19, 20, 30
McMurtry, John 24
Mendès France, Michel 294
Mersenne, Marin 327
Mersenne primes 136
Midgley, Mary 22
Mills, W.H. 321, 324
Möbius function 322, 345
molecules, analogy with composite numbers 96–9
moth navigation 170
Muir, Jane 105
multiplication 48–9, 265
  and addition 119–25, 129–31, 148, 187
  and logarithms 188–9
  as innovation 52–3

natural logarithm, definition of 178–84
  role in approximating prime count 192–201, 231–4
  role in approximating "primeness" count 234–40
natural numbers, definition of 38
negative numbers 77–9
  in relation to functions and graphs 212–13
Nelson, Robert 23, 24
neuropsychology 7
number (as opposed to "numbers") 10
number archetypes 28
number mysticism 11, 27–9, 33
"number system", explanation of term 47–8
number theory 70, 89–90, 161
  beauty and fascination of 94, 129, 158
numerology 11, 28, 33

oscilloscope 257

partitioning counting numbers 61–3, 119
patterns 152–5
  in "frequencies" of "spiral waves" 293
  in natural world 64
  in "primeness count deviation" 240, 247, 281
  in sequence of primes
    common question regarding 149, 156
    "exact" pattern 229
    "imperfect" or "statistical" pattern 157, 162–5, 189, 192–4, 199–201
    seeming absence of 105, 131, 223, 227
  seeming absence of in sequence of factorisations 91
Peano axioms (or postulates) 35–49
  compared to "cluster" approach to number 125–9
  defining counting numbers with 53–4
  formal description 45
  unique status of 1 in 92
  use in Euclid's proof of infinitude of primes 141, 148
Peano, Giuseppe 35–7, 39, 44
percentage error, defined 328–9
phase 255–6
pi (π) 181
  relationship to *e* 185–6
pitch (of tone) 252, 256
Platonists 6
"PNT approximation function" 214
  accuracy of 330–3
  improvement on 231–4
prime counting function 190–1
  approximating 192–4, 231–4
  discontinuity of 214–5
  graphing as "staircase" 204–7
prime factorisation *see* factorisation
prime factors 84–101
"primeness", amount of 236–8, 240–1, 243–6
  and effect of "spiral waves" on 291
"primeness count deviation" *see* "deviation function"
prime numbers, definition of 64–5
Prime Number Theorem (PNT) 198–201, 205–7, 215–6
  proof of 201, 281
  psychological reactions to 223–6

proof, mathematical 88, 132–3, 137–9, 144
Pythagoras 28, 33
Pythagoreans 28, 315

Qabalists 315
qualitative approach to number 8–11, 26–9, 33, 80
quantification 13–15, 23, 26, 30–3
quantitative approach to number 9–15, 26, 29, 33, 126–9

randomness 149, 154–5
relatively prime 319–20
religion 21–9, 106–7
religious art and architecture 27, 153
Ribenboim, Paolo 152, 157, 324
Riemann, Bernhard 281, 293–4, 341
Riemann function 345
Riemannian geometry 294
Rosser, J. Barkley 330

Sagan, Carl 111
Saidak, Filip 319–20
Schnirelmann, Lev 304
Schoenfeld, Lowell 330
school, experiences of mathematics in 32, 115, 174, 181
Schroeder, Manfred 152
scientific method 22
Seidenberg, Abraham 27
SETI 111, 113–14
"shifting units" 18, 129
Sieve of Eratosthenes 70, 150–1, 157, 323, 326
sieve theory 70
signals and signal processing 257
sine waves 253–62
  compared to "spiral waves" 272–4, 286
social constructivists 6
sound engineering 256, 261
sound waves 251–2, 256–7, 261
spiral galaxies 221
spirals 165–8
  Archimedean 167
  equiangular or logarithmic 168–77
    and logarithms 172–6, 188–9
    and "spiral waves" 268–71
    approximate, created via primes 218–20
    "base-*e*" 182, 184–6
    "base" of 175
    used to approximate prime count 192–4, 217–18
"spiral waves" 267–75, 280–1, 283, 285–94
staircases
  prime count 203–7, 214–15, 223-4
    approximations of 231–4
  "primeness" count 238–40, 284–5
    and "deviation function" 243–4
    reconstructing 279–81, 288–9
Stewart, Ian 101, 130, 157
subatomic particles 14, 99

technology 18–20, 29–31
Tenenbaum, Gérald 294
tetractys 315
thirteen, fear of 11, 76–7
topology 320

Uncle Petros and Goldbach's Conjecture (novel) 66, 130–1, 133, 144, 158
unquantifiable, the 21, 29–33

von Franz, Marie-Louise 28

water molecule 96
wave addition 257–61, 274–5, 288–91
waveforms 251–2, 257
wavelength
  of sine waves 254–6
  concept applied to "spiral waves" 267, 269–70, 273–4, 286
weeds, primes compared to 163, 223, 226–9, 264
Weyl, Hermann 101–2
Willans, C.P. 322, 324
Wilson's Theorem 322, 325–6
Wright, Edward M. 321, 324

Zagier, Don 75–6, 144, 162–3, 223–6

Liberalis is a Latin word which evokes ideas of freedom, liberality, generosity of spirit, dignity, honour, books, the liberal arts education tradition and the work of the Greek grammarian and storyteller Antonius Liberalis. We seek to combine all these interlinked aspects in the books we publish.

We bring classical ways of thinking and learning in touch with traditional storytelling and the latest thinking in terms of educational research and pedagogy in an approach that combines the best of the old with the best of the new.

As classical education publishers, our books are designed to appeal to readers across the globe who are interested in expanding their minds in the quest of knowledge. We cater for primary, secondary and higher education markets, homeschoolers, parents and members of the general public who have a love of ongoing learning.

If you have a proposal that you think would be of interest to Liberalis, submit your inquiry in the first instance via the website: www.liberalisbooks.com.